CHEMICAL BONDING IN SOLIDS

CHEMICAL BONDING IN SOLIDS

□ □ □ □ □

JEREMY K. BURDETT

The University of Chicago

New York   Oxford
OXFORD UNIVERSITY PRESS
1995

Oxford University Press

Oxford New York Toronto
Delhi Bombay Calcutta Madras Karachi
Kuala Lumpur Singapore Hong Kong Tokyo
Nairobi Dar es Salaam Cape Town
Melbourne Auckland Madrid

and associated companies in
Berlin Ibadan

Library of Congress Cataloging-in-Publication Data
Burdett, Jeremy K., 1947–
Chemical bonding in solids/Jeremy K. Burdett.
p. cm. Includes bibliographical references and index.
ISBN 0-19-508991-X.—ISBN 0-19-508992-8 (pbk.)
1. Solid state chemistry. I. Title.
QD478.B47 1995
541.2'24—dc20 94-13157

9 8 7 6 5 4 3 2 1

Printed in the United States of America
on acid-free paper

To Ingrid

Preface

The question of chemical bonding in solids is one that aroused the interest of scientists even before the first days of x-ray crystallography when atomic positions could be assigned in solids. The ideas of Barlow and of his contemporaries, in terms of the packing of spheres, are well known to us all. Today, how the atoms in solids are bound together, and how this determines the structure and properties of materials, are important considerations in many areas of chemistry, physics, and materials science. Over the years diverse concepts have come from equally diverse communities, but often these ideas have remained largely within the area where they originated. One of the goals of this book is to try to bring some of these together and to show how a broader picture exists once the prejudices (or some of them) that isolate one area from another are removed. There is thus much to be gained by viewing with the same "electronic spectacles" (terminology borrowed from Hans-Georg von Schnering) structural problems from fields traditionally regarded as very different. Just as in the molecular area the orbital model will prove to be a very useful way to broaden our vision of the solid state.

The way a structure is regarded geometrically often strongly influences the way that the solid is described electronically. The concept of close-packed solids, for example, is one that has been heavily used, especially in the mineral world. Solids described in this way are also frequently described in terms of "ionic" bonding. Recent studies, however, initiated by G. V. Gibbs, have brought orbital models, via calculations on fragments of the solid silicate, to the fore in this area. The structures of metals and their alloys are often approached via modifications of the nearly-free-electron theory of the "metallic" bond. However, as for many other structural questions, very useful insights into the structures of metals are accessible via the solid-state analog of molecular orbitals, namely, tight-binding theory. Recognition of a very general utility of the orbital picture in the solid state comes after analogous progress made in studies of molecules. Encouraged by Roald Hoffmann, the molecular chemist has now come to grips with the idea of delocalized bonding as a valid way to understand chemical bonding in areas other than the π systems of organic molecules. Cluster compounds such as the boranes were for years described as being held together by electron-deficient bonding, although, as Robert Mulliken aptly stated, it was really the theory that was deficient, not the electrons. Today such delocalized ideas in inorganic, organic, and organometallic chemistry enable quite a global electronic viewpoint of molecular chemistry. The physicists' picture over the years has used very generic models of solid-state behavior, much of it phrased in terms of the free-electron model and variants of it. The addition of the influence of local orbital structure, the "chemistry," brings a wealth of additional insights into the field. Once the blinders of parochiality are removed, the broad orbital picture of chemical bonding in both molecules and solids is a very enlightening one.

The phrase *bonding model* is often used to describe the type of theoretical ideas used in this field. Even in these days when advances in computation are breathtaking, the chemist in the laboratory is quite aware that the "Quantum Mechanical Truth" is rarely achieved by calculation, and that even when it is (or appears to be) the results are not usually understandable in a readily usable way. In fact it is the simpler ideas, frequently rather drastic simplifications of this "Truth," that have been the most valuable in terms of generating new ideas for experiments and understanding the complexities of the field. Chapter 1 of Ref. 209 provides an interesting set of examples that highlight this point. This book provides a set of bonding models that have the potential to be even more useful in the future than they have in the past.

Although this work is aimed at chemists who are familiar with the traditional division of the molecular fields into organic, inorganic, and organometallic areas, each with their established sociologies and ideas of chemical bonding, it is important to realize that similar divisions split other camps. Solid-state physicists come in two flavors (at least), those with backgrounds in metals and those raised in the field of semiconductors. The outlook of solid-state chemists is often different, too, depending, for example, on whether their field is oxides or chalcogenides.

The problem of language is one that will frequently be encountered in the discussions in this book. The meaning of terminology that has been in the literature for many years has often changed in the intervening period. The idea of "close-packed solids" will be reexamined following O'Keeffe's interesting remarks on the real meaning of the phrase. Certainly the creators of the first models of bonding in solids were heavily influenced by the implications of the literal view. Localization of electrons is another piece of terminology with a variety of different meanings that will be frequently discussed here. It is a concept that links properties such as metallic conduction with electronic descriptions. The metallic bond itself and the special place reserved for it by Ketelaar and van Arkel in their classification of bonding in solids, and also prevalent in modern day texts, is replaced by definition of a metal that is just a solid with a partially filled energy band. Metallic bonding *per se* has disappeared and we suggest that the term be removed from the vernacular.

A word is in order concerning the drawings of solid-state structures in this book. The solid state frequently provides challenges in terms of the three-dimensional visualization of the structure. It is often difficult to provide a view that shows basic structure features clearly, especially since there are lines added to the picture that frame the unit cell and connections drawn between atoms to indicate close contacts or "chemical bonds." It is easy to produce a view that provides little feeling for the structure. This is a problem for chemists and physicists alike since the structures that are studied are three-dimensional ones that extend indefinitely in the direction perpendicular to the page. Here many of them have come from the *locus classicus*, namely W. B. Pearson's famous book (Ref. 280). I am particularly grateful to Dr. Pearson for his permission to use his pictures here.

Chemical bonding and molecular, or solid-state, structure are two historically interwoven fields. The observed geometrical arrangements of atoms in both molecules and solids have been used for a long time now as testbeds of theories of chemical bonding. Conversely a part of the driving force for the synthesis of new compounds has been to probe existing theoretical ideas. In the solid state there are extra sets of measurements that may be used to probe the validity of chemical bonding models.

These include measurements of electrical conductivity and other transport phenomena. Although structure and bonding are indeed intertwined in the pages that follow, this book is not a comprehensive structural chemistry. Although the goal of generating a global theory to view the structures of all systems has yet to be achieved, there are many examples of structural problems scattered throughout, which have very satisfying explanations in terms of the bonding models used.

The impetus for writing a book of this type came from a series of lectures given at The Universität Mainz in 1991. A sizeable part of the material in the first two chapters comes from an article written for *Progress in Solid State Chemistry* (Ref. 43), and I acknowledge Pergamon Press for their kind permission to reproduce it here. Much of the manuscript was typed by Nita Yack, and many of the drawings were drafted by Ed Poole. The manuscript was read by Enric Canadell, Patrick Batail, Timothy Hughbanks, Stephen Lee, and John Mitchell. I am grateful to all of them for their insights and help both before and during this project. Finally, the book is dedicated to Ingrid with love.

Chicago   J.B.

Contents

Symbols and Abbreviations

e_s, e_p, e_d, atomic energies of s, p, and d orbitals

H_{ii}, diagonal entry in the Hamiltonian matrix, usually set equal to e_s, e_p, or e_d

H_{ij}, off-diagonal entry in the Hamiltonian matrix, often related to the overlap integral S and to H_{ii} and H_{jj} by the Wolfsberg–Helmholz relationship

α, Hückel label for H_{ii}

β, Hückel label for H_{ij}

M^n, Schäfli symbol for describing nets; each node of the net is a part of n M-gons

E, electronic energy

E_T, total electronic energy

E_F, Fermi level

$\mathbf{k}$, wave vector

$\mathbf{k_F}$, wave vector corresponding to E_F

$\rho(E)$, electronic density of states

W, bandwidth

Γ, coordination number

μ_n, nth moment of electronic density of states

COOP, crystal orbital overlap population

MOOP, molecular orbital overlap population

DOS, electronic density of states

Φ_{ij}, pair potential between atoms i,j

$\square$, vacancy

e_λ, second-order energy parameter in the angular overlap model for $\lambda = \sigma$, π, or δ interactions

f_λ, fourth-order energy parameter in the angular overlap model for $\lambda = \sigma$, π, or δ interactions

Ψ, wave functions for electronic state

CHEMICAL BONDING IN SOLIDS

1 □ Molecules

The discussion starts off in this book by looking at the electronic structure of molecules of various types. It provides a very useful basis from which to move to the solid state in that many of the theoretical tools needed for later chapters have their origin here. In addition, many solid-state systems, including the Zintl phases and van der Waals solids, are at the zeroth level of approximation "molecular" systems, too, although diamond would be described as a "giant" molecule. Some of the examples discussed in this chapter might appear to be rather idiosyncratic, but their utility will become apparent in the next chapter when specific electronic concepts are transferred from the finite to the infinite régime. The orbital model that is used (Ref. 5) has had wide success in studies of molecular problems and is ideally poised to be able to provide a broadly applicable model for the solid state. Symmetry (Ref. 100, 166, 209, 330, 382) will play a major rôle in the development of molecular and solid-state energy levels. This chapter begins with the simplest molecule of all, namely H_2, and progresses to larger systems.

1.1   The H_2 molecule: molecular orbital approach

Within the molecular orbital framework (Refs. 5, 165, 209, 236, 259, 337) an expression for the molecular orbitals of this diatomic molecule may be written as

$$\psi = a\phi_1 + b\phi_2 \tag{1.1}$$

where $\phi_{1,2}$ are the two atomic hydrogen $1s$ orbitals. The electron density function associated with such an orbital is then

$$\psi^2 = a^2\phi_1^2 + b^2\phi_2^2 + 2ab\phi_1\phi_2 \tag{1.2}$$

By symmetry the electron density must be equal on both hydrogen atoms, a result that implies $a = \pm b$. If S is the overlap integral between the two hydrogen $1s$ orbitals, then the bonding, ψ_b (σ_g^+) and antibonding, ψ_a, (σ_u^+) orbitals may be written

$$\left.\begin{array}{l} \psi_b = (1/\sqrt{2(1 + S)})(\phi_1 + \phi_2) \\ \psi_a = (1/\sqrt{2(1 - S)})(\phi_1 - \phi_2) \end{array}\right\} \tag{1.3}$$

Within the framework of Hückel theory (Refs. 165, 183, 335, 393), if the Coulomb integral is defined as $\alpha = \langle\phi_1|\mathscr{H}^{\text{eff}}|\phi_1\rangle = \langle\phi_2|\mathscr{H}^{\text{eff}}|\phi_2\rangle$ and β the interaction

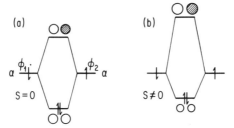

Figure 1.1   Molecular orbital diagram for the H_2 molecule (a) without and (b) with overlap included.

integral is $\langle \phi_1 | \mathcal{H}^{\text{eff}} | \phi_2 \rangle$, where $\mathcal{H}^{\text{eff}}$ is some effective Hamiltonian for the problem, then via $E_{a,b} = \langle \psi_{a,b} | \mathcal{H}^{\text{eff}} | \psi_{a,b} \rangle / \langle \psi_{a,b} | \psi_{a,b} \rangle$

$$\left. \begin{array}{l} E_a = (\alpha - \beta)/(1 - S) \\ E_b = (\alpha + \beta)/(1 + S) \end{array} \right\} \tag{1.4}$$

Ignoring S, then the simple Hückel result is

$$\left. \begin{array}{l} E_b = \alpha + \beta \\ E_a = \alpha - \beta \end{array} \right\} \tag{1.5}$$

This well-known result is shown in Figure 1.1 and for H_2 leads to the ground electronic configuration $(\psi_b)^2$.

An alternative way to generate these energies without using symmetry per se is to solve either the Hückel determinant $|H_{ij} - E| = 0$ or the extended Hückel (Ref. 174) determinant $|H_{ij} - S_{ij}E| = 0$. For the former case the result is very simple

$$\begin{vmatrix} H_{11} - E & H_{12} \\ H_{21} & H_{22} - E \end{vmatrix} = 0 \tag{1.6}$$

or

$$\begin{vmatrix} \alpha - E & \beta \\ \beta & \alpha - E \end{vmatrix} = 0 \tag{1.7}$$

leading immediately to $E = \alpha \pm \beta$ as before. Including overlap gives

$$\begin{vmatrix} H_{11} - E & H_{12} - S_{12}E \\ H_{21} - S_{21}E & H_{22} - E \end{vmatrix} = 0 \tag{1.8}$$

or

$$\begin{vmatrix} \alpha - E & \beta - SE \\ \beta - SE & \alpha - E \end{vmatrix} = 0 \tag{1.9}$$

which gives

$$E = (\alpha + \beta)/(1 + S) \quad \text{and} \quad (\alpha - \beta)/(1 - S) \tag{1.10}$$

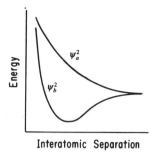

Figure 1.2 Energetic behavior of the configurations $(\psi_b)^2$ and $(\psi_a)^2$ with interatomic separation.

This state of affairs is shown in Figure 1.1(b) and leads to an understanding of why the He_2 molecule does not exist. Of the four electrons, two are stabilized on formation of the molecular orbitals, but two are destabilized and by a larger amount.

The overlap integral, and hence β, describing such orbital interactions generally increases in magnitude as the interatomic separation decreases. Thus the bonding orbital goes down in energy and the antibonding orbital goes up. However, on bringing the nuclei closer together a strong repulsion sets in. Indeed the equilibrium interatomic separation is set by the balance of this repulsion and the stabilization afforded the two electrons in the bonding orbital. A minimum is shown in the lower curve of Figure 1.2, which represents the balance of these two effects. It may be regarded as describing the energetics of the ground electronic configuration $(\psi_b)^2$. The upper curve with no minimum describes the energetics of the excited electronic configuration $(\psi_a)^2$ where both the electronic term and the internuclear interaction are repulsive.

There are two ways that these results are used in the simplest, one-electron, models of chemical bonding. The Hückel approach itself leaves α and β as parameters and the total energy is expressed in terms of them. In the extended Hückel model the parameter $H_{ii} = \alpha_i$ is most usually associated with some measure of the atomic orbital energy, e_i. (A table of some useful values is given in Ref. 41.) The parameter $H_{ij} = \beta$ may be estimated in two ways. It depends on the overlap between the two orbitals concerned and thus usually increases with a decrease in internuclear separation. The Wolfsberg–Helmholz approximation is generally used with β set proportional to either the arithmetic or geometric means of α_1 and α_2, that is, $\beta = KS(\alpha_1 + \alpha_2)/2$ or $KS(\alpha_1\alpha_2)^{1/2}$. The overlap integral S, a vital ingredient in the problem, is evaluated numerically by locating Slater-type wave functions on each atomic center. For all but the simplest systems the determination of the energy levels and molecular wave functions obtained by solution of the extended Hückel (Ref. 174) determinant $|H_{ij} - S_{ij}E| = 0$ needs to be done numerically.

The simplest many-electron wave function that could be written for the arrangement with two electrons in the bonding orbital (ignoring antisymmetrization and overlap in the normalization) for this ground electronic configuration is the one due to Mulliken and Hund and written

$$\begin{aligned}
\Psi_{MH} &= \psi_b(1)\psi_b(2) \\
&= \tfrac{1}{2}(\phi_1 + \phi_2)(1)(\phi_1 + \phi_2)(2) \\
&= \tfrac{1}{2}[\phi_1(1)\phi_1(2) + \phi_2(1)\phi_2(2) + \phi_1(1)\phi_2(2) + \phi_2(1)\phi_1(2)] \qquad (1.11)
\end{aligned}$$

with a total electronic energy of

$$E_T = 2(\alpha + \beta) + U/2 \qquad (1.12)$$

where $U = \langle \phi_1(1)\phi_1(2)|r_{12}^{-1}|\phi_1(1)\phi_1(2)\rangle$, the Coulombic repulsion between two electrons located in the same orbital on the same atom. (There is of course another term, V, which involves the intersite Coulombic repulsion, but this is smaller and has been ignored.) Notice that the energy of the electronic state [Eq. (1.12)] on this simple model is just the sum of the two types of terms, those that depend on just one electron (α, β terms) and one that depends on two electrons (U term). Often the properties of the state may be inferred from the properties of the one-electron terms (as, for example, in the strong-field limit for transition metal complexes). When U is large, then the electronic properties of the state have to be viewed in terms of the electron–electron interaction terms, too (as, for example, in the weak-field limit for transition metal complexes).

There are other electronic states that may be generated from this set of orbitals and electrons. For example, the state ψ_a^2 may be written

$$\Psi'_{MH} = \psi_a(1)\psi_a(2) \qquad (1.13)$$

with an energy $E'_T = 2(\alpha - \beta) + U/2$.

Equation (1.2) describes the electron density distribution in the molecule. Of interest is how this should be divided so as to give useful information about the molecule. Integration of Eq. (1.2) and multiplication by N, the number of electrons in the orbital, gives

$$N \int \psi^2 \, d\tau = Na^2 \int \phi_1^2 \, d\tau + Nb^2 \int \phi_2^2 \, d\tau + 2Nab \int \phi_1 \phi_2 \, d\tau$$
$$= Na^2 + Nb^2 + 2NabS_{12} \qquad (1.14)$$

This may usefully be interpreted in the following way (Refs. 177, 260). There are Na^2 electrons located on the atom holding ϕ_1, Nb^2 electrons located on the atom holding ϕ_2 and $2NabS_{12}$ electrons located in the bond between the two. The term $2NabS_{12}$ is the Mulliken bond overlap population and is a very useful device with which to describe the bonding propensity of a given energy level. The integrated or summed orbital overlap populations are a useful way to describe the total electronic interaction between a given pair of atoms. This total bond overlap population between the atoms n,m is written as $P_{mn} = 2\sum_i N_i c_{ij} c_{ik} S_{jk}$, where c_{ij} is the coefficient of orbital j located on atom m, c_{ik} the coefficient of orbital k located on atom n in molecular orbital i, and S_{jk} is their overlap integral. N_i is the number of electrons in the ith molecular orbital. Thus from Eq. (1.3) the bond overlap population for the configuration ψ_b^2 is

$$P_{12} = (2)(2)(1/\sqrt{2(1 + S)})(1/\sqrt{2(1 + S)})S$$
$$= 2S/(1 + S) \qquad (1.15)$$

and for the configuration ψ_a^2 it is analogously

$$-2S/(1 - S) \tag{1.16}$$

So in accord with the diagram of Figure 1.1 the antibonding orbital is more antibonding ($P_{12} < 0$) than the bonding orbital is bonding ($P_{12} > 0$). In terms of deciding how much electron density is located on atom m and how much on atom n, then the overlap population of Eq. (1.14) is divided equally between the two atoms. Thus in this simple case the atomic populations of the two atomic orbitals, ϕ_1 and ϕ_2, are just $Na^2 + NabS_{12}$ and $Nb^2 + NabS_{12}$. The extension to more atoms and more orbitals per center (e.g., s, p, and d) is straightforward.

1.2 The H_2 molecule: localized approach

Consider now the case when the two hydrogen atoms are far apart. The wave function of Eq. (1.11) is now inappropriate. The chance that both electrons can reside on the same atom together is small. Thus the use of a wave function that includes the first two terms of the function of Eq. (1.11) is incorrect. In its place a Heitler–London wave function is written (Ref. 236) as

$$\Psi_{HL} = (1/\sqrt{2})[\phi_1(1)\phi_2(2) + \phi_1(2)\phi_2(1)] \tag{1.17}$$

with an energy of $E_T = 2\alpha$. There are of course higher-energy states where both electrons are forced to lie on the same atom

$$\Psi'_{HL} = \phi_1(1)\phi_1(2) \quad \text{and} \quad \Psi''_{HL} = \phi_2(1)\phi_2(2) \tag{1.18}$$

These have an energy of $E_T = 2\alpha + U$.

This expression may be compared with that from the delocalized model [Eq. (1.12)] or $2(\alpha + \beta) + U/2$. Just as in many other areas of chemistry, the energy difference between the two $(2\beta - U/2)$ is a balance between the one-electron (β) and two-electron (U) terms in the energy. If $|\beta| \gg |U/4|$, then the delocalized picture is an appropriate one, but if $|\beta| \ll |U/4|$, the localized picture is the one to use. There are several ways to move theoretically from the localized to delocalized description. The first is the Hubbard approach (Refs. 16, 182), which takes the electronic states appropriate for the localized picture [Eqs. (1.17) and (1.18)] and switches on the effect of the overlap interaction between the atomic basis orbitals. Ψ'_{HL} and Ψ''_{HL} are orthogonal but both have an interaction (of $\sqrt{2}\beta$) with Ψ_{HL} when overlap is included. This leads to a secular determinant for the singlet states of the form

$$\begin{vmatrix} 2\alpha - E_T & \sqrt{2}\beta & \sqrt{2}\beta \\ \sqrt{2}\beta & 2\alpha + U - E_T & 0 \\ \sqrt{2}\beta & 0 & 2\alpha + U - E_T \end{vmatrix} = 0 \tag{1.19}$$

The ground-state energy then becomes

$$E_T \sim 2(\alpha + \beta) + U/2 - \beta\kappa^2 \tag{1.20}$$

with $\kappa = U/4\beta$. When κ is small, the MH energy is recovered [Eq. (1.12)]. When β is small, then the HL energy is found ($E_T = 2\alpha$). The new wave function becomes

$$\Psi \sim (\tfrac{1}{2})^{1/2}\Psi_{HL} + [(1 + \kappa^2)^{1/2} - \kappa](\tfrac{1}{2})(\Psi'_{HL} + \Psi''_{HL}) \tag{1.21}$$

This is a mixture of Ψ_{HL} and Ψ_{MH}. For $\kappa = 0$, $\Psi = \Psi_{MH}$, and for $\kappa \to \infty$, $\Psi = \Psi_{HL}$. Thus as β becomes larger the wave function becomes more like that of Ψ_{MH}. Note that κ, the ratio of two- to one-electron terms in the energy, is the critical parameter here.

Another approach, due to Coulson and Fischer (Ref. 102) writes a Heitler–London wave function

$$\Psi_{CF} = (1/\sqrt{2})[\xi_a(1)\xi_b(1) + \xi_a(2)\xi_b(2)] \tag{1.22}$$

but with

$$\left.\begin{array}{l} \xi_a = \phi_1 \cos \chi - \phi_2 \sin \chi \\ \xi_b = \phi_1 \sin \chi + \phi_2 \cos \chi \end{array}\right\} \tag{1.23}$$

Such a picture takes the system from a localized description ($\chi = 0$) to a delocalized one ($\chi = \pi/4$) just by changing χ, which can be determined variationally.

Yet a third way (Ref. 236) starts off from the delocalized states and allows configuration interaction between them. This leads to a lowering of the energy of the electronic ground state Ψ_{ML} [Eq. (1.11)] by mixing in a contribution from the excited-state configuration given by Eq. (1.13).

$$\begin{aligned} \Psi_{CI} &= \Psi_{MH} + \lambda\Psi'_{MH} \\ &= \tfrac{1}{2}[(1 + \lambda)[\phi_1(1)\phi_1(2) + \phi_2(1)\phi_2(2) \\ &\quad + (1 - \lambda)[\phi_1(1)\phi_2(2) + \phi_2(1)\phi_1(2)] \end{aligned} \tag{1.24}$$

If $\lambda < 0$, then such mixing of electronic states increases the contribution from terms such as $\phi_1(1)\phi_2(2)$ and reduces the contributions from terms such as $\phi_1(1)\phi_1(2)$. The latter contribute to Coulombic repulsion (U) terms in the energy. In general, therefore, the Mulliken–Hund function overestimates the contribution from terms such as $\phi_1(1)\phi_1(2)$ since the electrons are uncorrelated and move independently of each other. Mixing in of the excited-state contribution takes into account some of this electron correlation by recognizing that the contribution from $\phi_1(1)\phi_1(2)$ will in practice be somewhat smaller since the two electrons suffer an energetic penalty if they reside on the same atom.

The nature of the transition from the localized to delocalized situation is of extreme importance in solids, as will be seen later. Localized bonding where the electrons are forced to remain on one side leads to an electrically insulating state. Delocalized bonding can lead to a metal. Energetically the energy of the localized state

($E_T = 2\alpha$) should be compared with that of a delocalized one ($E_T = 2\alpha + 2\beta + U/2$). Manipulation of the critical ratio $|U/4\beta|$ in the solid state leads to very interesting aspects of metal–insulator transitions.

1.3 Energy levels of HHe

Helium is more electronegative than hydrogen, which implies that the α values of the $1s$ orbitals are not the same. The symmetry argument used for H_2 is now inapplicable, and so a secular determinant needs to be solved.

$$\begin{vmatrix} \alpha_1 - E & \beta \\ \beta & \alpha_2 - E \end{vmatrix} = 0 \qquad (1.25)$$

with roots $E \sim \{(\alpha_1 + \alpha_2) \pm [(\alpha_1 - \alpha_2)^2 + 4\beta^2]^{1/2}\}/2$. Expansion under the square root leads to $E \sim \alpha_1 + \beta^2/(\alpha_1 - \alpha_2) - \beta^4/(\alpha_1 - \alpha_2)^3$ and $\alpha_2 - \beta^2/(\alpha_1 - \alpha_2) + \beta^4/(\alpha_1 - \alpha_2)^3$. With $\alpha_1, \alpha_2 < 0$ and $\alpha_1 < \alpha_2$ this gives rise to the MO diagram of Figure 1.3. Both orbitals are now of the same symmetry (σ), a fact reflected by the necessity of using the secular determinant to evaluate their energies. Determination of the orbital coefficients shows they are unequal but have the relative sizes shown in the figure. The first term in the stabilization energy, $\beta^2/(\alpha_1 - \alpha_2)$, is simply what would be obtained using second-order perturbation theory for this problem (Refs. 5, 176). The second term extends this to fourth order. A useful result for this heteroatomic case is that the energy difference Δ between bonding and antibonding partners is given by $\Delta^2 = (\Delta\alpha)^2 + 4\beta^2$. This leads to $\Delta = 2\beta$ for the homoatomic case. This result will be useful in Chapter 7. For the present, it is interesting to label the two contributions to the energy gap as "covalent" and "ionic" ones, E_c and E_i, such that $\Delta^2 = E_i^2 + E_c^2$.

An interesting extension of the perturbation description of the interaction forms the basis for the angular overlap model (AOM) (Refs. 41, 138). The interaction integral β depends on the overlap integral between the pair of orbitals concerned. (For a table of these see Refs. 41, 329.) This is easiest to see in Figure 1.4(a), where the overlap integral varies as $\cos\theta$. When $\theta = 0$ then there is maximal overlap between the p_z orbital and the hydrogen $1s$ orbitals, but when $\theta = 90°$ the overlap integral and thus the orbital interaction is zero. If $S = S_0(r) \cos\theta$ in general (r is the interatomic separation), then $\beta = \beta_0 \cos\theta$. For the orbital situation

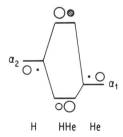

H   HHe   He

Figure 1.3 Molecular orbital diagram for the HHe molecule.

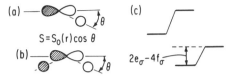

Figure 1.4 The basics of the angular overlap model (AOM). (a) Dependence of the overlap on angle. Shown is the angular dependence of the overlap between a p orbital and a ligand s or σ orbital. (b) Situation for the AB_2 molecule showing interaction between two ligand orbitals and a central atom p orbital. (c) The orbital stabilization that results from the AOM.

shown in Figure 1.4(b) the interaction energy associated with this antisymmetric orbital combination shown is in second order $\beta_0^2 \cos^2 \theta/(\alpha_1 - \alpha_2)$ or $e_\sigma \cos^2 \theta$, where the parameter $e_\sigma = \beta_0^2/(\alpha_1 - \alpha_2)$. If the fourth-order term is included by writing $f_\sigma = \beta_0^4/(\alpha_1 - \alpha_2)^3$, then the interaction energy becomes $e_\sigma \cos^2 \theta - f_\sigma \cos^4 \theta$. Figure 1.4(c) shows the orbital diagram for the linear geometry. This parametrization is particularly useful in transition metal complexes and provides a much more powerful method than the Crystal Field Model with which to look at a variety of problems (Ref. 41). For the case where there are both σ and π interactions, then two parameters, e_σ and e_π, may be used.

1.4   Energy levels of linear conjugated molecules

The energy levels of conjugated open-chain molecules (Refs. 165, 335, 393) may be derived in exactly the same way as for the diatomic H_2 molecule discussed previously. The secular determinant of Eq. (1.6) is directly applicable (Fig. 1.5) to the π system of ethylene, the basis orbitals now being the $2p\pi$ orbitals of carbon rather than the $1s$ orbitals of hydrogen. The π energy levels of allyl are obtained analogously within the Hückel framework by solution of the secular determinant

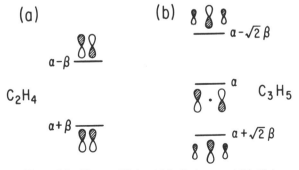

Figure 1.5   The π orbitals of (a) ethylene and (b) allyl.

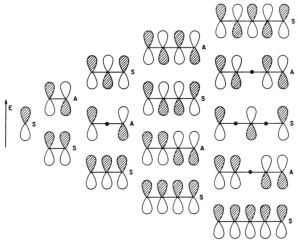

Figure 1.6 The evolution of the π-orbital picture for conjugated linear polyenes.

$$\begin{vmatrix} \alpha - E & \beta & 0 \\ \beta & \alpha - E & \beta \\ 0 & \beta & \alpha - E \end{vmatrix} = 0 \tag{1.26}$$

Notice that no entry appears in the 1,3 position since orbital interactions are included between bonded atoms only. The roots are $E = \alpha$, $\alpha \pm \sqrt{2}\beta$ and the orbital wave functions are drawn in Figure 1.5. The figure shows that the number of nodes increases as the stack of orbitals is climbed.

There is a simple expression (Refs. 165, 335) for the energy levels [Eq. (1.27)] and orbital coefficients [Eq. (1.28)] of a one-dimensional chain containing n π orbitals. For the jth level

$$E_j = \alpha + 2\beta \cos j\pi/(n + 1), \qquad j = 1, 2, 3, \ldots, n \tag{1.27}$$

and for the coefficient on the rth center

$$c_{jr} = [2/(n + 1)] \sin rj\pi/(n + 1) \tag{1.28}$$

The number of nodes in the wavefunction increases by one as the energy increases. This is shown pictorially in Figure 1.6, where the level structure of the first few numbers in the series is shown. Note that the orbital energies are symmetrically located about $E = \alpha$. This means that for odd-membered chains there is a central, exactly nonbonding, orbital with this energy.

It is interesting at this stage to anticipate the result for the infinite one-dimensional chain. From Eq. (1.27) an infinite collection of orbitals results, bounded on the bottom by one at $\alpha + 2\beta$, where the orbital coefficients are all in phase, and at the top by an orbital with an energy $\alpha - 2\beta$, with orbital coefficients that are out of phase. By analogy with Figure 1.6 the number of nodes in this infinite set of orbitals increases as the stack of levels is climbed.

1.5   Energy levels of cyclic polyenes

Group theory allows ready access (Refs. 100, 166, 209, 382) to the molecular orbital energy levels of cyclic π systems such as those in benzene and cyclobutadiene (Refs. 165, 335, 393). In general for a point group G with operations $g \in G$ a symmetry-adapted linear combination of atomic orbitals may be generated such that for the kth irreducible representation

$$\psi_k = \sum_{g \in G} \chi_k(g) g \phi_1 \tag{1.29}$$

where ϕ_1 is some member of the orbital basis and $\chi_k(g)$ is the character of the kth irreducible representation for the gth operation. The energy levels of cyclobutadiene (**1.1**) may be readily generated in this way.

1.1   1.2

The point group of the molecule is D_{4h}, but it is sufficient to use the group C_4, the cyclic group of order 4. Its character table is shown in Table 1.1. Using the four p_π orbitals as a basis for a representation, it is easy to show that they transform as $a + b + e$; that is, each irreducible representation of C_4 is contained just once in the reducible representation. This is a very general result. For a cyclic $(CH)_n$ molecule there will be $n\pi$ molecular orbitals, one belonging to each irreducible representation of the group C_n. Using the character table for C_4 and the projection operation of Eq. (1.29), it is easy to write down a normalized wave function of a symmetry

$$\psi(a) = \tfrac{1}{2}(\phi_1 + \phi_2 + \phi_3 + \phi_4) \tag{1.30}$$

Its energy from the simple Hückel approach is simply

$$\begin{aligned}
E &= \tfrac{1}{4}\langle \phi_1 + \phi_2 + \phi_3 + \phi | \mathcal{H}^{\text{eff}} | \phi_1 + \phi_2 + \phi_3 + \phi_4 \rangle \\
&= \tfrac{1}{4}(4\alpha + 8\beta) = \alpha + 2\beta \tag{1.31}
\end{aligned}$$

Similarly for the wave functions of e symmetry

$$\left. \begin{aligned}
\psi(e) &= \tfrac{1}{2}(\phi_1 + i\phi_2 - \phi_3 - i\phi_4) \\
\psi'(e) &= \tfrac{1}{2}(\phi_1 - i\phi_2 - \phi_3 + i\phi_4)
\end{aligned} \right\} \tag{1.32}$$

where in the normalization procedure it is important to use the complex conjugate in $\int \psi^* \psi \, d\tau = 1$. These two functions may be converted into two new (real) ones using

Table 1.1 The character table for the cyclic group of order 4, C_4

C_4	E	C_4	C_2	C_4^3
A	1	1	1	1
B	1	-1	1	-1
E	$\begin{cases} 1 \\ 1 \end{cases}$	$\begin{matrix} i \\ -i \end{matrix}$	$\begin{matrix} -1 \\ -1 \end{matrix}$	$\left.\begin{matrix} -i \\ i \end{matrix}\right\}$

the license allowed for degenerate wave functions

$$\begin{aligned} \psi''(e) &= 2^{-1/2}[\psi(e) + \psi'(e)] = 2^{-1/2}(\phi_1 - \phi_3) \\ \psi'''(e) &= 2^{-1/2}(-i)[\psi(e) - \psi'(e)] = 2^{-1/2}(\phi_2 - \phi_4) \end{aligned}\right\} \tag{1.33}$$

These are clearly nonbonding orbitals, and evaluation of their energy gives $E = \alpha$. Use of the characters for the b representation of Table 1.1 leads to

$$\psi(b) = \tfrac{1}{2}(\phi_1 - \phi_2 + \phi_3 - \phi_4) \tag{1.34}$$

with an energy of $E = \alpha - 2\beta$. The MO diagram for the π levels of cyclobutadiene is shown in Figure 1.7. Shown are two equally good ways of representing the degenerate pair of e orbitals. Also illustrated are the symmetry labels in the D_{4h} point group.

An identical procedure may be used for benzene (**1.2**). Here the character table is given in Table 1.2 and the same recipe gives

Table 1.2 The character table for the D_{6h} group

D_{6h}	E	$2C_6$	$2C_3$	C_2	$3C_2'$	$3C_2''$	i	$2S_3$	$2S_6$	σ_h	$3\sigma_d$	$3\sigma_v$
A_{1g}	1	1	1	1	1	1	1	1	1	1	1	1
A_{2g}	1	1	1	1	-1	-1	1	1	1	1	-1	-1
B_{1g}	1	-1	1	-1	1	-1	1	-1	1	-1	1	-1
B_{2g}	1	-1	1	-1	-1	1	1	-1	1	-1	-1	1
E_{1g}	2	1	-1	-2	0	0	2	1	-1	-2	0	0
E_{2g}	2	-1	-1	2	0	0	2	-1	-1	2	0	0
A_{1u}	1	1	1	1	1	1	-1	-1	-1	-1	-1	-1
A_{2u}	1	1	1	1	-1	-1	-1	-1	-1	-1	1	1
B_{1u}	1	-1	1	-1	1	-1	-1	1	-1	1	-1	1
B_{2u}	1	-1	1	-1	-1	1	-1	1	-1	1	1	-1
E_{1u}	2	1	-1	-2	0	0	-2	-1	1	2	0	0
E_{2u}	2	-1	-1	2	0	0	-2	1	1	-2	0	0

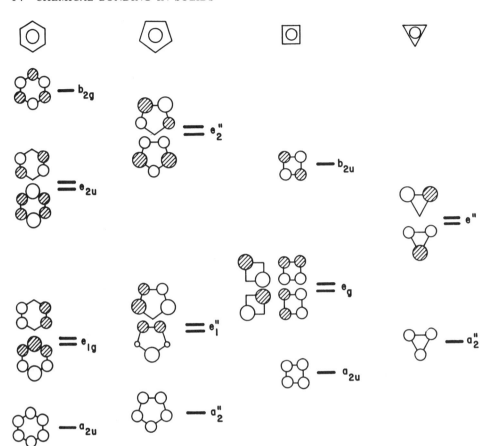

Figure 1.7   Molecular orbital diagrams for the π-orbital manifold of the first few cyclic polyenes

$$\left.\begin{aligned}
\psi(a_{2u}) &= (1/\sqrt{6})(\phi_1 + \phi_2 + \phi_3 + \phi_4 + \phi_5 + \phi_6)\\
\psi(e_{1g}) &= (1/\sqrt{12})(2\phi_1 + \phi_2 - \phi_3 - 2\phi_4 - \phi_5 + \phi_6)\\
\psi(e_{2u}) &= (1/\sqrt{12})(2\phi_1 - \phi_2 - \phi_3 + 2\phi_4 - \phi_5 - \phi_6)\\
\psi(b_{2g}) &= (1/\sqrt{6})(\phi_1 - \phi_2 + \phi_3 - \phi_4 + \phi_5 - \phi_6)
\end{aligned}\right\} \quad (1.35)$$

with energies

$$\left.\begin{aligned}
E(a_{2u}) &= \alpha + 2\beta\\
E(e_{1g}) &= \alpha + \beta\\
E(e_{2u}) &= \alpha - \beta\\
E(b_{2g}) &= \alpha - 2\beta
\end{aligned}\right\} \quad (1.36)$$

Figure 1.7 shows these two results pictorially using the group-theoretical labels for the molecular (rather than cyclic) point group. Both partners of the degenerate orbitals are shown.

The general result (Refs. 165, 335) for a cyclic system with n atoms is a very simple one. The energy of the jth orbital is given by

$$E_j = \alpha + 2\beta \cos 2j\pi/n \qquad (1.37)$$

where j runs from $0, \pm 1, \pm 2, \ldots [\pm n/2$ for n even or $(n-1)/2$ for n odd]. The simple form of Eq. (1.37) leads to a useful mnemonic (a Frost circle, Ref. 131) for remembering the energy levels of these molecules. Inscribe in a circle of radius 2β an n-vertex polygon such that one vertex lies at the bottom. The points at which the two figures touch define the Hückel energy levels of the molecule as in **1.3**.

1.3

The coefficient of the pth atomic orbital determines the form of the wave function as

$$\psi_j = \sum_{p=1}^{n} c_{jp}\phi_p = n^{-1/2} \sum_{p=1}^{n} \exp[2\pi ij(p-1)/n]\phi_p \qquad (1.38)$$

It is an expression that is very similar to that of Eq. (1.29). Indeed rewriting it as in Eq. (1.39) highlights the similarity.

$$\psi_j = n^{-1/2} \sum_{p=1}^{n} \exp[2\pi ij(p-1)/n]C_n^{p-1}\phi_1 \qquad (1.39)$$

The exponential is simply the character $\chi_j(C_n^{p-1})$ of the cyclic group of order n, as the reader may readily check for the case of $n = 4$ shown in Table 1.1. The prefactor of $n^{-1/2}$ is a (Hückel) normalization constant.

This complex form of the wave function is very useful and will be especially so in later discussions on solids. A linear combination of the wave functions of a pair of degenerate orbitals produces two new orbitals that are equivalent in every respect. Using this fact, the functions of Eq. (1.39) for the case of degenerate species may be rewritten in a somewhat nicer form by making use of the trigonometric identity $e^{ix} = \cos x + i \sin x$ [Eq. (1.40)].

$$\left.\begin{array}{l} \psi'_j = \tfrac{1}{2}(\psi_j + \psi_{-j}) = \dfrac{1}{2n^{-1/2}} \displaystyle\sum_{p=1}^{n} \cos[2\pi j(p-1)/n]\phi_p \\[3mm] \psi''_j = \tfrac{1}{2}(\psi_j - \psi_{-j}) = \dfrac{1}{2n^{-1/2}} \displaystyle\sum_{p=1}^{n} \sin[2\pi j(p-1)/n]\phi_p \end{array}\right\} \qquad (1.40)$$

It is interesting to see how the wave function of Eq. (1.38) leads to the energies of

Eq. (1.37). An expression [Eq. (1.41)] for the energy comes via $E_j = \langle \psi_j | \mathscr{H}^{\text{eff}} | \psi_j \rangle / \langle \psi_j | \psi_j \rangle$.

$$E_j = \alpha + \beta \sum_p (c_{jp} c_{j(p-1)} + c_{jp} c_{j(p+1)}) \tag{1.41}$$

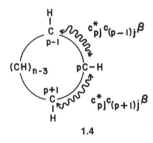

1.4

This represents the sum of all neighbor interactions (**1.4**) and

$$E_j = \alpha + \beta \sum_p \{n^{-1/2} \exp[-2\pi i j(p-1)/n]\}\{n^{-1/2} \exp[2\pi i j(p-1)/n]\}$$

$$\times (e^{2\pi i j/n} + e^{-2\pi i j/n})$$

$$= \alpha + \beta \sum_p \frac{1}{n} 2 \cos(2\pi j/n)$$

$$= \alpha + 2\beta \cos 2\pi j/n \tag{1.42}$$

Figure 1.7 shows the energy levels of the first few cyclic molecules. Note how the number of nodes increases as the level stack is climbed.

An interesting result coming from graph theory (Ref. 173) is that Eq. (1.43) holds for systems where the coordination number (Γ) is the same for all n atoms. The function μ_2 is just the second moment of the electronic density of states.

$$\mu_2 = \frac{1}{n} \sum_j (E_j - \alpha)^2 = \Gamma \beta^2 \tag{1.43}$$

This can be proven for the present case using Eq. (1.42).

$$\mu_2 = \frac{1}{n} \sum_j 4\beta^2 \cos^2(2\pi j/n) = 2\beta^2 \tag{1.44}$$

The wave functions calculated for benzene may be used to determine the overlap population contributions from the various levels. For the deepest-lying orbital (a_{2u}) the C–C overlap population (if there are two electrons in the orbital) is just

$$P(a_{2u}) = (2)(2)\frac{1}{\sqrt{6}}\frac{1}{\sqrt{6}} S_\pi = \tfrac{2}{3}S_\pi \tag{1.45}$$

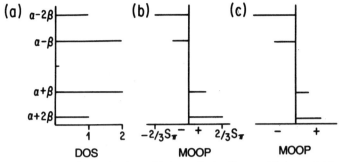

Figure 1.8   Presentation of the molecular orbital results for the π-orbital manifold of the benzene molecule. (a) Molecular density of states plot. DOS or $\rho(E)$. The length of the line is proportional to the number of energy levels at that point. (b) Molecular orbital overlap population (MOOP) plot. Here a line is drawn proportional to the contribution to the C–C overlap population. The results come from the Hückel description of the orbitals. (c) The same as (b) but with overlap included.

where S_π is the $2p\pi$–$2p\pi$ overlap integral. For the e_{1g} orbitals the overlap population from both components of the pair needs to be calculated to get the correct result. That between atoms 1 and 6 is just

$$P(e) = 0 + (2)(2)\left(\frac{1}{\sqrt{12}}\right)\left(\frac{1}{\sqrt{12}}\right)(2)S_\pi = \tfrac{2}{3}S_\pi \qquad (1.46)$$

and thus for each orbital the effective overlap population is $\tfrac{1}{3}S_\pi$. For the e_{2u} and b_{2g} orbitals the overlap population is $-\tfrac{1}{3}S_\pi$ and $-\tfrac{2}{3}S_\pi$, respectively. These results may be shown graphically as in Figure 1.8. Figure 1.8(a) shows a density of states (DOS) plot taken from the traditional energy-level diagram. A line is drawn whose length is proportional to the number of energy levels at that point. Thus the doubly degenerate e levels are represented by a line twice as long as those for the a and b levels. Figure 1.8(b) shows a molecular orbital overlap population (MOOP) diagram, where lines are drawn proportional to the overlap population contribution of each orbital. The reader can show that, in general, inclusion of overlap into the wave function leads to less bonding and increasingly antibonding overlap populations as the stack of levels is climbed just as shown for H_2. The result is shown in Figure 1.8(c).

Very similar arguments lead to the generation of the energy levels of substituted systems such as $B_3N_3H_6$ (borazine), $B_2N_2H_4$, and their derivatives. For $B_3N_3H_6$ the three B $2p\pi$ and three N $2p\pi$ orbitals transform as $a_2'' + e''$ and the form of these symmetry-adapted wave functions may simply be written down (the labeling scheme is shown in **1.5**).

1.5

For B

$$\left.\begin{array}{l} \psi_B(a_2'') = \dfrac{1}{\sqrt{3}}(\phi_1 + \phi_2 + \phi_3) \\[3mm] \psi_B(e'') = \dfrac{1}{\sqrt{6}}(2\phi_1 - \phi_2 - \phi_3) \end{array}\right\} \quad (1.47)$$

and for N

$$\left.\begin{array}{l} \psi_N(a_2'') = \dfrac{1}{\sqrt{3}}(\phi_4 + \phi_5 + \phi_6) \\[3mm] \psi_N(e'') = \dfrac{1}{\sqrt{6}}(2\phi_4 - \phi_5 - \phi_6) \end{array}\right\} \quad (1.48)$$

The energies of the molecular orbitals in $B_3N_3H_6$ may then be obtained by solving a secular determinant for the pairs of orbitals of the same symmetry.

There are two secular determinants for this case, one for each of the symmetry types. That for the a_2'' orbitals is just

$$\begin{vmatrix} \langle \psi_N(a_2'')|\mathcal{H}^{\mathrm{eff}}|\psi_N(a_2'')\rangle - E & \langle \psi_N(a_2'')|\mathcal{H}^{\mathrm{eff}}|\psi_B(a_2'')\rangle \\[2mm] \langle \psi_N(a_2'')|\mathcal{H}^{\mathrm{eff}}|\psi_B(a_2'')\rangle & \langle \psi_B(a_2'')|\mathcal{H}^{\mathrm{eff}}|\psi_B(a_2'')\rangle - E \end{vmatrix} = 0 \quad (1.49)$$

which leads to

$$\begin{vmatrix} \alpha_N - E & \tfrac{1}{3}6\beta_{BN} \\[2mm] \tfrac{1}{3}6\beta_{BN} & \alpha_B - E \end{vmatrix} = 0 \quad (1.50)$$

The resulting orbital energies then come from either the approximate algebraical solution of the determinant or by the equivalent procedure using perturbation theory.

For the two orbitals of a_2'' symmetry the second-order perturbation expression is just $\Delta E = |\langle \psi_N(a_2'')|\mathcal{H}^{\mathrm{eff}}|\psi_B(a_2'')\rangle|^2/(\alpha_N - \alpha_B) = (\tfrac{1}{3}6\beta_{BN})^2/(\alpha_N - \alpha_B) = 4\beta_{BN}^2/(\alpha_N - \alpha_B)$. The same procedure may be used for the orbitals of e'' symmetry, the result being $\Delta E = |\langle \psi_N(e'')|\mathcal{H}^{\mathrm{eff}}|\psi_B(e'')\rangle|^2/\alpha_N - \alpha_B = (\tfrac{1}{6}6\beta_{BN})^2/\alpha_N - \alpha_B = \beta_{BN}^2/(\alpha_N - \alpha_B)$, shown pictorially in Figure 1.9(a). Recalling the result for HHe of Figure 1.3, the orbitals of the substituted molecule are readily drawn in a qualitative fashion as in Figure 1.9(b). The deeper-lying (bonding) orbitals have the largest contribution from the deeper-lying starting orbital, in this case, nitrogen. Thus another way of understanding the electronic picture here is to start off with the orbitals of benzene and visualize an electronegativity perturbation, where the α values of three of the atoms go up in energy (C → B) and those for the other three go down in energy (C → N). This is shown at the left-hand side of Figure 1.9(a). In Figure 1.9(c) the orbital character of the $B_3N_3H_6$ orbitals is shown but in a different way. This is a partial density of states diagram where lines are drawn that are proportional to the local rather than total density of states. Two pictures are shown, one showing the weight of the nitrogen orbitals in the total wave function, and the other showing the weight of the boron orbitals. These are generated using the ideas described in Eq. (1.14).

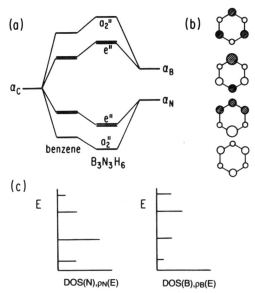

Figure 1.9 (a) Generation of the π molecular orbital picture for borazine, $B_3N_3H_6$. Two routes are shown, one (at left) starting off from the orbitals of benzene, the other assembling the diagram from the B and N basis orbitals. (b) The form of the π orbitals. (c) Partial densities of states, projected out from the calculation DOS(N), DOS(B) or $\rho_N(E)$, $\rho_B(E)$.

The orbitals of cyclobutadiene were derived previously. Notice the result that, although the e_u pair of orbitals may be drawn in several different ways, the one shown gives a description in which there are only orbital contributions from one pair of trans atoms in each component. This is especially useful when it comes to the generation of the level structure of the $B_2N_2H_2$ molecule. Using the idea of the electronegativity perturbation used for $B_3N_3H_6$, the energy of one component of e_u simply becomes equal to α_N and the other α_B, as in Figure 1.10. The energy shifts of the other two orbitals may be obtained in a manner analogous to that for borazine. A very interesting result here is that the zero HOMO–LUMO gap found for cyclobutadiene is opened up both in the boron/nitrogen analog and in substituted molecules. The latter include the "push-pull cyclobutadienes," where

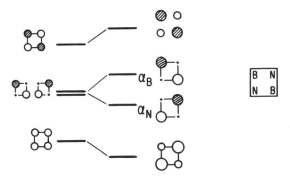

Figure 1.10 Generation of the π molecular orbital picture for square $B_2N_2H_4$ from that for square cyclobutadiene.

electron-withdrawing and -donating groups are attached alternately around the ring (Ref. 175). This result will be used later.

1.6 Energy differences and moments

In addition to an electronic description of a molecule or solid, insights into the stability of one structure relative to another are highly sought (Refs. 45, 220). Usually the inquiries fall into one of two categories. The first is the specific, and the second the general. An example of the first is the question of why the water and OF_2 molecules are nonlinear. An example of the second is the question of why the geometries of the first-row MX_2 molecules, dihydrides, and dihalides oscillate along the series, linear for BeX_2, bent for BX_2, CX_2, NX_2, OX_2 and then linear for KrF_2. Figure 1.11(a) shows molecular orbital diagrams (Ref. 54) for the linear and bent dihydrides derived using the angular overlap model. Figure 1.11(b) shows the calculated energy differences between the two structures. As another example of a general problem, the relative stability of cyclic polyenes as a function of electron count, and the origin of Hückel's $4n + 2$ rule, is a question of long-standing interest.

It should be clear that study of a specific problem demands an appreciation of the energy changes associated with the collection of occupied orbitals, whereas that of the more general problem associated with structural preferences as a function of electron count, requires a knowledge of the behavior of *all* of the orbitals of the problem. In this case a somewhat different approach is required than that usually used. It is very helpful in this context to introduce the concept of the moments (Refs. 45, 106, 220) of the electronic density of states since it will be particularly useful in the appreciation of the global structure problem. The nth moment of the collection

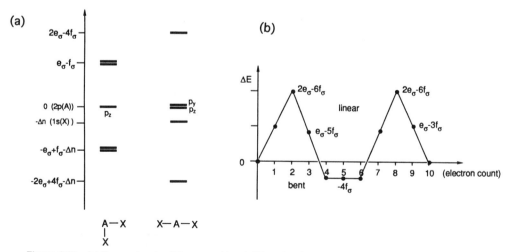

Figure 1.11 (a) Energy levels of linear and bent AX_2 molecules using the AOM. The Rundle–Pimentel approach is used, which ignores the central atom s orbital, except as a repository for two valence electrons. Δn is the energy difference between the central atom p and ligand σ orbitals. (b) The energy difference between the two as a function of electron count using the results from (a).

of energy levels $\{E_i\}$ is simply given by the expression

$$\mu_n = \sum_i E_i^n \tag{1.51}$$

The second moment occurs in Eq. (1.43) in the context of the energy levels of cyclic polyenes. Another way of writing the moments, but in terms of the H_{ij} integrals of the orbital problem, recognizes the fact that the nth moment is just the trace (sum of the diagonal elements) of $\mathbf{H}^n$ where $\mathbf{H}$ is the Hamiltonian matrix. This is easy to see. Since $\mathbf{H}$ is Hermitian, there is a unitary matrix $\mathbf{S}$ such that $\mathbf{S}^{-1}\mathbf{HS}$ is diagonal and whose diagonal elements are the eigenvalues (orbital energies of $\mathbf{H}$. As $\mathrm{Tr}(\mathbf{H}^n) = \mathrm{Tr}[(\mathbf{S}^{-1}\mathbf{HS})]$ and $\mathbf{S}^{-1}\mathbf{HS}$ is diagonal with elements E_i, then $\sum_i E_i^n = \mathrm{Tr}(\mathbf{H}^n)$. Expanding $\mathrm{Tr}(\mathbf{H}^n)$ in terms of the H_{ij} integrals leads to an interesting and general expression for μ_n. Equation (1.52) shows how the nth moment may be described in terms of a weighted sum over all the self-returning walks of the orbital problem.

$$\mu_n = \sum_{\substack{\text{walks} \\ \text{of length } n}} H_{ij} H_{jk} \cdots H_{ni} \tag{1.52}$$

The result is shown pictorially in **1.6** and allows a geometrical connection to be made

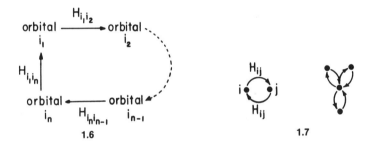

with μ_n that will prove to be very useful. Using this viewpoint, the first moment is just the sum of the "walks-in-place" and defines the values of the H_{ii} themselves. Its choice sets the "zero" of the energy scale. The second moment [Eq. (1.53)] is just the sum of the squares of the interaction integrals coupling one orbital to its neighbors, as shown schematically in **1.7**.

$$\mu_2 = \sum H_{ij} H_{ji} = \sum H_{ij}^2 \tag{1.53}$$

Equation (1.43) showed the application of this result to the cyclic polyene case.

One way, therefore, of regarding the second moment is as a measure of the coordination strength around an atom. This is a very important concept and will figure in many problems in this book. Of particular importance is the role played by the second moment in determining the equilibrium internuclear separation between the atoms, and how it changes from system to system. As a specific example consider the two structures shown in Figure 1.12 for the H_3 molecule and its ions. Shown are the Hückel energy levels of the two systems. However, there is no reason

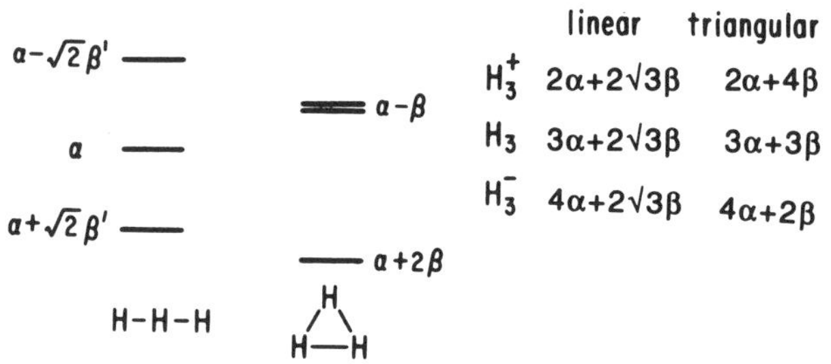

Figure 1.12 The Hückel molecular orbital diagrams for linear and triangular H_3 and the energy difference between the two as a function of electron count fixing the relationship between β and β' using the second moment scaling method.

to assume a priori that the β values of the two structures should be equal since there is no reason to believe that the bond lengths in the linear molecule (two of them) must be the same as the three in the triangle. One approach that is very useful in such energetic comparisons of the two structures is to set the second moments of the two structures equal. Although this does have a more detailed electronic justification (Refs. 179, 290, 296), discussed in Section 8.1, such second-moment scaling simply constrains the coordination strength of the sets of atoms in the two structures to be equal. Figure 1.12 shows the energy levels of the linear molecule in terms of the β values of the triangular species obtained by setting the two second moments equal. The interaction integral for the linear system (β') may be readily expressed in terms of the value for the triangle (β) since

$$3\alpha^2 + (\sqrt{2}\beta')^2 + 0^2 + (\sqrt{2}\beta')^2 = 3\alpha^2 + 2\beta^2 + (2\beta)^2$$

i.e.,

$$4\beta'^2 = 6\beta^2$$

or

$$\beta' = \sqrt{3/2}\beta \tag{1.54}$$

From the results of Figure 1.12, H_3^+ should be triangular, but H_3 and H_3^- linear. The former is true, but, although the second two species are unknown, they should indeed be unstable in the triangular geometry from Jahn–Teller considerations.

Using this constant-second-moment approach, it may also be shown how internuclear distances change with coordination number. If an atom has Γ neighbors with Γ equivalent interactions, then the second moment is just $\Gamma\beta^2 = q$, a constant. If the bond length (r) dependence of β is written as $\beta = A/r^m$ in region of chemical interest, then the equilibrium separation is simply written as (Ref. 160)

$$r_e^{2m} = \Gamma(A^2/q) \tag{1.55}$$

which shows an equilibrium bond length that increases with coordination number. This is generally true in qualitative terms and will be discussed more quantitatively in Chapter 6.

If the second moment describes the immediate coordination environment of an atom, then higher moments describe (Refs. 45, 220) how a given atom "sees" its

$\beta^2\alpha$ B A C

1.8

β^3 β^4 β^6

1.9

neighbors further away. So, as shown in **1.8** and **1.9**, the third, fourth, fifth, and sixth moments give information about the electronegativity difference between the atoms at each end of the bond, and the existence of rings of various sizes. Note that the fourth moment via its walk of length 4 is the first that allows atom B to "see" atom C via their chemical bonds to atom A. The fourth moment will also depend in general on the CAB bond angle. Since the influence of an atom on its neighbors is expected to decrease as the distance between them increases, the importance of the higher moments should drop off quickly as their order increases.

An example that shows the utility of the moments concept and some of its implications involves examination of the energy differences as a function of electron count between rings of organic π systems of different sizes (Ref. 45). The aim is to see the electronic origin of Hückel's rule. The calculations are simple. All that needs to be computed is the energy difference between 2 six-rings and p $12/p$ rings as a function of electron count. [The fractional orbital occupancy, defined as x (empty, $0 \le x \le 1$ full) is the abscissa.] They are shown (Refs. 220) in Figure 1.13. The figure shows two important features. First, the amplitude of the three curves decreases in the order $(12/p =)$ three rings > four rings. This is easy to understand in the light of the prior discussion using walks to measure moments. It may be seen that, when comparing a $12/p$-membered ring with the 6-membered ring $(p < 12)$, the first self-returning walk (moment) that will be different between the two will be the walk around the $12/p$-membered ring of length $12/p$ (**1.9**). Thus the three plots of Figure 1.13, labeled by triangle, square, and hexagon, represent third-, fourth, and sixth-moment problems, respectively, since this is the order of the first moment that is different between the two calculations. Second, the number of nodes in these curves (including those at $x = 0,1$) is equal to the order of this first disparate moment between the two structures. The structure that is more stable at the earliest orbital occupancy is the one with the largest moment. As a general result the shape and amplitude of the energy difference curve $\Delta E(x)$ between two structures is determined by the way the atoms are connected together and thus the orbital walks that are possible.

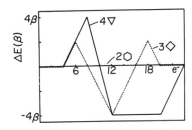

Figure 1.13 The energy difference between a 12-ring and p $12/p$ rings as a function of electron count. The fractional orbital occupancy is defined as (empty) $0 \le x \le 1$ (full). (Reproduced from Ref. 220 by permission of The American Chemical Society.)

It is very profitable to pursue the implications of Figure 1.13. The three-membered rings are most stable at early electron counts. In fact, the maximum in their stability curve comes at 8 electrons ($x = 0.33$) when there are the Hückel's rule quota of $(2n + 1)$ electron pairs per π system, with $n = 0$ in this case. There are two regions of stability for the four-membered rings, at $x = 0.25, 0.75$. These correspond to totals of 6 and 18 electrons or two and six electrons per four-membered ring, just the electron counts expected from Hückel's rule with $n = 0,1$. For the six-membered rings the three regions of stability are at $x = 0.167, 0.5$, and 0.833, or two, six, and ten electrons per ring appropriate to the Hückel values of $0,1,2$. However, comparison of all three plots leads to the following generalizations. At low electron counts (and especially at one-third filling), the three-rings are stable. At the half-filled point, six-membered rings are stable. At higher electron counts (especially around three-quarters filling) four-rings are stable.

To conclude this section, these moments ideas will be used to generate insights into the AH_2 energy difference curve of Figure 1.11(b). As noted, the fourth moment via its walk of length 4 is the first that allows atom B to "see" atom C via their chemical bonds to atom A. It will also, therefore, be the first moment that allows the AH_2 molecule to "know" about its bond angle (**1.10**). The energy difference between

the two structures as a function of electron count should then have four nodes. Indeed it does (Ref. 54) from Figure 1.11(b). How the fourth moment controls the structural problem may best be seen by study of the linear and 90° bent molecules. Figure 1.11 shows the orbitals of the two structures. In the linear geometry notice that there is a self-returning walk (**1.11**) using one of the A atom p orbitals, which is absent because of orbital orthogonality in the bent geometry (**1.12**). Thus the linear and bent structures do differ first at the fourth moment, the first walk, as noted, which allows one H atom to "see" the other. It is the linear structure, therefore, that has the larger fourth moment and therefore the one that is predicted to be more stable at early x, as indeed found from Figure 1.11(b).

1.7   The Jahn–Teller effects

Some of the most interesting results in structural chemistry are concerned with the predictions of the Jahn–Teller theorems (Refs. 41, 42, 194). Some of the mathematical details are discussed in Section 7.7. Strictly associated initially with the instability of a degenerate electronic state, it is now more generally stated in terms of the instability associated with the asymmetric electronic occupation of degenerate orbitals. Thus cyclobutadiene with its (e_g^2) configuration is "Jahn–Teller unstable" because, as shown in Figure 1.14, there is a distortion pathway that will lower the electronic energy of the molecule because of the removal of the orbital degeneracy. In this case the distortion is an alternating shortening and lengthening of the C–C distances

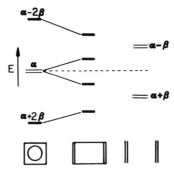

Figure 1.14 The Jahn–Teller distortion of cyclobutadiene. [Strictly speaking, this is a pseudo-Jahn–Teller distortion (Ref. 42), but the subtlties of the electronic differences here will not be of great importance.]

round the ring. The molecules S_4^{2+}, Se_4^{2+}, and Te_4^{2+}, where the e_g orbitals are full $[(e_g)^4]$, lead to no stabilization on distortion and so are Jahn–Teller stable and perfectly square. [The situation for cyclobutadiene is a little more complex in that electron correlation leads to a square singlet lying deeper in energy than the square triplet (Ref. 193).]

The basis of the Jahn–Teller approach lies in symmetry arguments (discussed in Section 7.7) that do not predict the details of the molecular distortion, but Figure 1.14 shows how the energy levels change on "dimerization". Notice the choice of the form of the degenerate cyclobutadiene levels that enables this process to be visualized easily. The dimer structure lies somewhere along the pathway to total fragmentation into two double-bonded units. As two of the linkages shorten (**1.13**), then $|\beta_1| > |\beta|$,

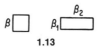

1.13

while as the other two lengthen, $|\beta_2| < |\beta|$, with the new π energies those given in Figure 1.15. By analogy with the discussion for H_3 and its ions, the second moment of the density of states of both distorted and undistorted structures should be set constant in order to generate a meaningful energy difference. This is simple to do in this case. The electronic energy, E'_T, of the distorted structure is $4\beta_1$ (i.e., $E'^2_T = 16\beta_1^2$). Since $\mu_2 = 4\beta_1^2 + 4\beta_2^2$, then $E'^2_T/4 = \mu_2 - 4\beta_2^2$. β_2 is smaller in the distorted structure $(|\beta_2| < |\beta|)$, and thus this structure is more stable. However, if the relationship

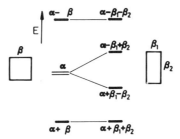

Figure 1.15 Hückel energies for the π molecular orbitals of cyclobutadiene on distortion.

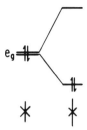

Figure 1.16 Energy level changes on distortion for the MX_6 octahedron. (The electron configuration for the low-spin d^8 system is shown.).

$\beta_1 + \beta_2 \approx 2\beta$ is used, then the π stabilization on distortion is particularly simple and equal to $\Delta E_T = 2(\beta_1 - \beta_2)$.

Notice that the $B_2N_2H_4$ molecule of Figure 1.10 is Jahn–Teller stable, even though it has the same electron count as cyclobutadiene. Here the B,N substitution of C opens up a gap and removes the asymmetric occupation of degenerate orbitals. A similar result is behind the observation of square singlets in the so-called "push–pull" cyclobutadienes (Ref. 175).

Another well-known example of a Jahn–Teller distortion comes from transition metal chemistry (Ref. 271). Octahedral transition metal complexes with low-spin d^7, d^8, and d^9 configurations are Jahn–Teller unstable, as shown in Figure 1.16. The driving force for the distortion (almost invariably towards the loss of two trans ligands) is largest for the d^8 configuration. Thus, although Cu(II) d^9 complexes are known as square planar molecules, the majority of examples show an axially elongated pair of ligands, or a square pyramid geometry with a long axial distance resulting from the loss of a single ligand from the octahedron. By way of contrast, low-spin d^8 complexes are almost always square planar. This comment applies to all complexes of Pd(II) and Pt(II). Although $Ni(CN)_4^{2-}$ is a square planar low-spin d^8 system, square pyramidal $Ni(CN)_5^{3-}$ is known but with a very long axial Ni–CN distance. There are, however, many regular octahedral complexes of Ni(II), $Ni(H_2O)_6^{2+}$ being one of them, which have no analog in Pd or Pt chemistry. This species though is a spin triplet, which is Jahn–Teller stable at the octahedral geometry (**1.14**).

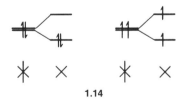

1.14

Figure 1.17 shows two possibilities for the energetic behavior on distortion of the electronic *states* that derive from the configurations of **1.14**. At the left-hand side of the diagram the energy separation between the triplet and singlet states is associated with differences in the two-electron terms in the energy. Traditionally, from the Tanabe–Sugano approach (Ref. 147) used by chemists, these are expressed in terms of the Racah interelectronic repulsion parameters B, C. At the right-hand

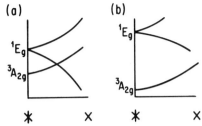

Figure 1.17 Energetic behavior on distortion away from octahedral for (a) typical first row case and (b) typical second or third row case.

side of the diagram the energy difference arises via a combination of these two-electron terms plus the one-electron energy changes of the molecular orbital levels of Figure 1.16.

It is interesting to see how these electronic parameters vary on moving down a column of the periodic table. From an analysis of the electronic spectra of transition metal complexes, the interelectron repulsion terms A,B,C are known to be larger for the first row elements than for their heavier analogs, a result usually correlated with the increase in atomic size on moving down the table and a concomitant increase in interelectronic distances. Also Δ, the e_g–t_{2g} splitting, a one-electron parameter, increases on moving down the table. (The contracted nature of the $3d$ orbitals relative to their $4d$ and $5d$ analogs is described in Section 8.5.) Thus it is to be expected that the one-electron contribution to the energy changes shown in Figure 1.17 should be larger for Pd and Pt than for Ni. As a general statement, complexes containing these elements correspond energetically to the case shown in Figure 1.17(a) with a small singlet–triplet splitting at the octahedron and large energy changes on distortion. Many nickel complexes are described by Figure 1.17(b), where the singlet–triplet splitting is large and the one-electron driving force for distortion smaller than for its heavier analogs. The result is a stable octahedral triplet. An example would be $Ni(H_2O)_6^{+2}$. $Ni(CN)_4^{2-}$ of course is described by Figure 1.17(a). Thus, again the energetics of the distortion are controlled by the interplay of one- and two-electron terms in the energy. Of course a similar effect determines whether octahedral complexes are high or low spin in the first place. If P is the pairing energy (a combination of Racah B,C parameters), then a high-spin complex is observed when $|P/\Delta| > 1$ and a low-spin complex when $|P/\Delta| < 1$.

A Jahn–Teller distortion may then be switched off in these octahedral transition metal complexes if interelectronic repulsions are high. Roughly speaking, in the triplet octahedral structure the two electrons lie in different d orbitals, but in the low-spin, singlet structure they are forced to be in the same orbital and thus experience a strong Coulombic repulsion. In the distorted, nonoctahedral structure, where the two electrons are in the same orbital, only if the one-electron stabilization energy is large enough to overcome this Coulombic penalty will the molecule distort. [Parenthetically note that Jahn–Teller distortions associated with degenerate electronic states that arise via partial occupancy of metal–ligand π (rather than σ) orbitals are generally smaller (Ref. 41) and frequently are unobservable.]

The discussion of this section has focused on the geometrical instability of degenerate electronic states, and this is how the Jahn–Teller theorem was initially

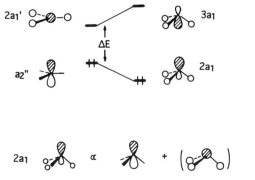

Figure 1.18 Second-order Jahn–Teller effect in planar ammonia

formulated. However, if there are two electronic states that lie close in energy, then via a second-order Jahn–Teller mechanism (Refs. 22, 41) the structure may distort along a coordinate, which allows strong mixing between them. The mathematical details are discussed in detail in Section 7.7. Figure 1.18 shows how the pyramidal geometry of the ammonia molecule is understandable in this way. At the planar geometry the HOMO and LUMO, of different symmetry, lie close in energy. Distortion to a pyramidal structure leads to strong mixing between the two, stabilization of the two electrons in the HOMO, and thus stabilization of the pyramidal structure. The generation of the lone pair at nitrogen ($2a_1$) is a natural consequence of this.

2 □ From molecules to solids

The discussions of Chapter 1 were introduced so as to be able to move as smoothly as possible from molecules, where the orbital problem is a finite one, to solids, where it is infinite in extent. Many of the same electronic features are seen in both areas (Refs. 5, 43, 298, 299), and an understanding of the basic electronic structure of molecules greatly simplifies discussion of the solid state. There are discussions of the basic electronic structure of solids in several places (Refs. 12, 16, 76, 78, 159, 160, 199, 210) and descriptions of tight-binding theory (Refs. 39, 43, 177, 205, 213, 314, 329, 364, 372). Of interest too throughout the rest of the book are compendia of solid-state structures (Refs. 2, 17, 127, 190, 256, 276, 280, 365, 366) and properties (Refs. 116, 367).

2.1 The solid as a giant molecule

As noted in Section 1.4, extension of the result for the open-chain molecule of Eq. (1.27) leads to an infinite collection of orbitals bounded by the energies of $\alpha \pm 2\beta$ shown in Figure 2.1. This energy band has a width $W = |4\beta|$. However, probably the most useful way to move from molecule to solid (Refs. 43, 139) is the use of the earlier results for the cyclic system. The first step in the process is to imagine that the infinite solid may be represented as in **2.1** by tying together the ends of the chain

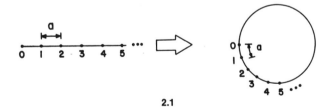

2.1

such that at no place does the "molecule" feel any curvature at all. (This comment is made in a figurative sense. Mathematically there is no problem achieving this.) Equation (1.42) shows that there will be n energy levels whose energies are given by

$$E_j = \alpha + 2\beta \cos(2j\pi/n) \tag{2.1}$$

where j takes all integral values from $0, \pm 1, \pm 2, \ldots, \pm n/2$. This is a very unwieldy

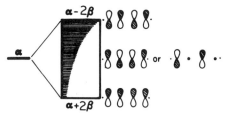

Figure 2.1 The evolution of the π energy levels of an infinite one-dimensional chain $(-CH-)_n$ by extrapolating the results of Figure 1.6.

expression, since $n/2$ is extremely large for the infinite solid, but it may be rewritten in a much neater fashion by defining a new index k such that

$$E(k) = \alpha + 2\beta \cos ka \tag{2.2}$$

Here a is the unit cell length of **2.1** and $k = 2j\pi/na$, called the wave vector, takes values from 0 continuously in the range $-\pi/a < k \le \pi/a$. Figure 2.2 shows the transition (Ref. 139) from the finite to the infinite case. Recall that for an n-membered ring j in Eq. (1.27) takes values from 0, ± 1, ± 2, through $(n-1)/2$ for an odd-membered ring. So for the 5-membered ring the extremal value of $|j|$ is $j_{max} = 2$. For the 15-membered ring the corresponding value of j_{max} is 7.

What happens if values of $|j|$ larger than j_{max} are used? The reader may readily show that the energy levels already derived with $j < j_{max}$ are generated: that is, the

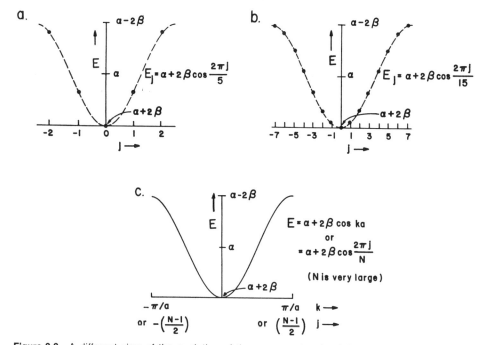

Figure 2.2 A different view of the evolution of the π energy levels of the infinite one-dimensional chain $(-CH-)_n$ by plotting E versus the quantum number j of the cyclic molecule. (a),(b) The plot for 5- and 15-membered rings and (c) the plot for the infinite material. (Adapted from Ref. 139.)

use of $j > j_{max}$ gives redundant information. Similarly in Eq. (2.2), use of values of $|k| > \pi/a$ also leads to no new information. In the crystalline state the levels between $-\pi/a < 0 < \pi/a$ lie in the (first) Brillouin zone. The point $k = 0$ is the zone center, $k = \pm\pi/a$ is the zone edge, and the variation of the energy with k is the dispersion of the band. The second, etc., zones are defined between $-2\pi/a < -\pi/a$, $\pi/a < 2\pi/a$, etc. Use of more than just the first zone is called an extended zone scheme. Figure 2.2(c) is just then a "smoothed-out." version of Figure 2.2(b), with a continuum of levels (in the limit $n = N \to \infty$) rather than a discrete set.

The wave functions of this infinite unit are also easily written down by the substitution of $k = 2j\pi/na$ in Eq. (1.39) as

$$\psi_j = n^{-1/2} \sum_{p=1}^{n} \exp[2\pi ij(p-1)/n] C_n^{p-1}\phi_1 \tag{2.3}$$

$$\psi(k) = n^{-1/2} \sum_{p=1}^{n} e^{ik(p-1)a}\phi_p \tag{2.4}$$

Group theory allows access to a similar expression for the orbitals of this infinite, periodically repeating chain. An expression analogous to Eq. (1.29) for the molecular case is given here for the translation group T.

$$\psi(k) = \sum_{t \in T} \chi_k(t) t\phi_1(\mathbf{r}) \tag{2.5}$$

$$= \sum_{t \in T} e^{ikR}\phi_t(\mathbf{r} - \mathbf{R}_t) \tag{2.6}$$

This is an infinite group made up of all translations $t \in T$. There are correspondingly an infinite number of k values, but the characters take on a simple exponential form. Here R_t is the distance along the chain which the translation operation t moves ϕ_1 and $\phi_t(\mathbf{r} - \mathbf{R}_t)$ is an orbital translationally equivalent to ϕ_1; that is, $t\phi_1(\mathbf{r}) = \phi_t(\mathbf{r} - \mathbf{R}_t)$. Writing $R_t = (p-1)a$ shows that it is obvious that Eqs. (2.4) and (2.6) are identical except for a normalization constant. This was to be expected since via the construction of **2.1** the infinite translation group and the cyclic group of infinite order are isomorphic. In other words, k, the wave vector, in Eqs. (2.4) and (2.6) are the same. In three-dimensional systems the vector nature of k becomes apparent, and, as shown later, the exponential in this equation needs to be replaced with a vector dot product $\exp(i\mathbf{k} \cdot \mathbf{R}_t)$, where $\mathbf{R}_t = \sum_i(p_i - 1)\mathbf{a}_i$, a sum over the primitive lattice vectors $\mathbf{a}_i$. Just as the vector $\mathbf{R}_t$ (with dimensions of length) maps out a direct space (x,y,z coordinates), so $\mathbf{k}$ (with dimensions of length^{-1}) is associated with a reciprocal space. The symmetry-adapted functions $\psi(k)$ are called Bloch functions $\psi(k) = \sum c_{kp}\phi_p$. Another way to view the wave function is as the simple product

$$\psi(k, \mathbf{r}) = e^{i\mathbf{k} \cdot \mathbf{R}}\zeta(\mathbf{r}) \tag{2.7}$$

The total wave function is just the product of a plane wave with $\zeta(\mathbf{r})$, a function with the periodicity of the lattice; that is, $\zeta(\mathbf{r}) = \zeta(\mathbf{r} - \mathbf{R}_t)$. As in the case of the cyclic system, all the levels of the one-dimensional chain turn up in pairs (positive and negative k values). To understand the orbital structure of the (one-dimensional) solid, only one set of values is needed; the positive set, the right-hand side of Figure 2.2(c), is chosen.

The energy levels themselves, which are called the crystal (rather than molecular) orbitals, can also be derived (Refs. 43, 364) by using an approach identical to the one in Section 1.5 and shown in **1.4**. The energy of the level of the infinite chain described by a given value of k is given within the tight-binding framework by multiplying by n, the number of orbitals in the chain, the energy contribution from the interaction of a single orbital with its neighbors.

$$E(k) = n(\langle n^{-1/2} \exp(-ikpa)\phi_p | \mathcal{H}^{\text{eff}} | n^{-1/2}$$
$$\times \{\exp[-ik(p-1)a]\phi_{p-1} + \exp(ikpa)\phi_p + \exp[ik(p+1)a]\phi_{p+1}\}\rangle)$$
$$= n\{\alpha + \beta[\exp(ika) + \exp(-ika)]\}/n$$
$$= \alpha + 2\beta \cos ka \tag{2.8}$$

The top of the band occurs at $k = \pi/a$, where $\cos ka = -1$ and $E = \alpha - 2\beta$. The bottom of the band is found at $k = 0$, where $\cos ka = 1$ and $E = \alpha + 2\beta$. The middle of the band occurs at $k = \pi/2a$, where $\cos ka = 0$ and $E = \alpha$. At $k = 0$ the phase factor linking an orbital with its neighbor is, from Eq. (2.6), equal to $+1$ and so the Bloch function looks like **2.2**. This is bonding between all adjacent atom pairs. At $k = \pi/a$ the phase factor is -1 and the Bloch function looks like **2.3**. These are, of

2.2   **2.3**

course, exactly the same functions as found from the construction of Figure 2.1.

Inclusion of overlap into the picture is straightforward. Allowing nearest-neighbor overlap (S) only, the result is similar to the molecular case and gives

$$E(\mathbf{k}) = \frac{\alpha + 2\beta \cos ka}{1 + 2S \cos ka} \tag{2.9}$$

Thus, as before, the top of the energy band $[(\alpha - 2\beta)/(1 - 2S)]$ is destabilized more than the bottom $[(\alpha + 2\beta)/(1 + 2S)]$ is destabilized.

It can be seen from the dispersion picture of Figure 2.3(a) that the largest number of states per unit increment in energy are at the bottom and top of the energy band. In fact, if the density of states is written as $[dE(k)/dk]^{-1}$, then there are singularities at these points. The density of states will thus look like that shown in Figure 2.3(b). It differs from that of Figure 1.8(b) for benzene, in that it is now continuous in the region between $\alpha \pm 2\beta$. The solid-state analog of the MOOP of Figure 1.8(c) is shown in Figure 2.3(c) as the COOP curve (crystal orbital overlap population (Refs. 185, 384). Its construction is easily visualized from Figure 2.1.

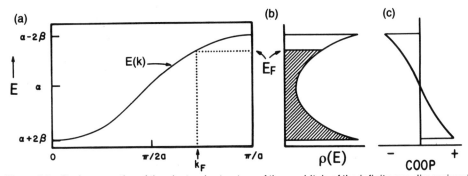

Figure 2.3   Basic properties of the electronic structure of the π orbitals of the infinite one-dimensional chain $(-CH-)_n$. Shown in (a) is the dispersion of the band, the variation in the energy with k. (b) The density of states, $\rho(E)$. (c) The crystal orbital overlap population (COOP) curve for this density of states. (b) and (c) should be compared with that for benzene in Figure 1.8.

Photoelectron spectroscopy is very useful as a probe of both molecular and solid-state energy levels. The technique may be used to show that plots such as the density of states of Figure 2.3(b) are correct. Figure 2.4 shows the observed (Ref. 304) photoelectron spectrum of the "s band" of a long-chain hydrocarbon. Of course, in heavily correlated systems (such as some transition metal oxides) where the band model begins to become inappropriate, this relationship with experiment ceases to be found.

At this stage a comparison with Figure 1.2 is a useful one. Shown in Figure 2.5 is the analogous picture for the solid. Now a continuum of levels lies between the two solid curves. Filling up the levels of the energy band with two electrons each will give a total of $2n$ electrons for the n-atom chain (n large). Normally when describing band occupancy one refers to the number of electrons per unit cell, and so this π band for the infinite chain may contain a maximum of 2 electrons.

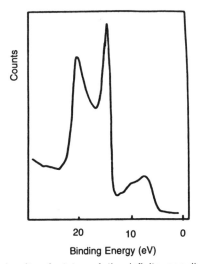

Figure 2.4   The electronic density of states of the infinite one-dimensional chain (actually a long-chain hydrocarbon) as obtained experimentally from x-ray photoelectron spectroscopy. (Adapted from Ref. 304.) Note the similarity to Figure 2.3(b).

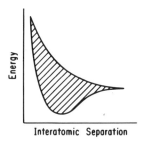

Figure 2.5 The energetic behavior of the crystal orbitals with interatomic separation. The plot is bounded at the bottom by the energy of electrons occupying the all in-phase orbital combination, and at the top by that of the electrons in the all out-of-phase orbital combination. This picture should be compared with Figure 1.2.

The collection of $p\pi$ orbitals just studied would describe the electronic π band of polyacetylene (**2.4**) in a geometry where all the C–C distances are equal. Since there

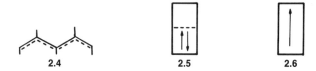

is one electron contributed per carbon $p\pi$ orbital, the π band is exactly half full. Occupied bands are shown in general by the cross-hatching of the occupied states, as in Figure 2.3(b). Throughout this book the simple representation of this situation shown in **2.5** will be used. The bandwidth depends on β, the resonance integral of Hückel theory, interaction integral in general, or hopping or transfer integral, using the language of the solid-state physicist. The hopping label is an especially useful one in a visual sense when used in conjunction with **1.6**. Note that the system of **2.5** is metallic. **2.6** shows another electronic possibility, one where all the levels are filled with unpaired electrons. A solid with this electronic arrangement would be a magnetic insulator. The interplay between the stabilities of these two electronic distributions will be discussed later. The top of the occupied stack of levels for the partially filled band is called the Fermi level, with an energy E_F. The corresponding **k** point [Fig. 2.3(a)] is labeled $\mathbf{k}_F$.

This LCAO type of approach applied to solids is called the tight-binding model by solid-state physicists (Refs. 16, 228, 314, 329). The two descriptors will be used interchangeably in this book. It should be clear from its derivation from the molecular case that it is really no different in principle from the LCAO scheme for molecules, although the traditional usage identifies tight-binding theory with the Hückel approach. Section 1.1 showed that the levels of a molecule were accessible by solution of the determinantal equation

$$|H_{ij} - S_{ij}E| = 0 \tag{2.10}$$

Here the H_{ij} and S_{ij} were the interaction and overlap integrals between a basis set of atomic orbitals $\{\phi_i\}$. In the solid state the analogous equation is (Refs. 213, 329)

$$|H_{ij}(\mathbf{k}) - S_{ij}(\mathbf{k})E| = 0 \tag{2.11}$$

where now the H_{ij} and S_{ij} integrals are those between the Bloch functions [Eq. (2.6)] constructed for each atomic orbital contained in the unit cell. Whereas for the molecule Eq. (2.10) is solved just once to generate the energy levels. Eq. (2.11) needs to be solved (in principle) at an infinite number of $\mathbf{k}$ points. Sometimes an analytic expression [as in Eq. (2.9), for example] for $E(\mathbf{k})$ may be simply written down, but more usually Eq. (2.11) needs to be solved numerically for a mesh of $\mathbf{k}$ points that will represent the behavior of the whole energy band when averaged. The method will be used later to estimate the energies of structural alternatives. This "special points" method (Refs. 18, 89) is in general a very useful way of replacing a complex integration over $\mathbf{k}$ space with a set of individual calculations.

Imagine the problem of evaluation of the value of some continuous function $f(x)$ averaged over the interval shown in **2.7**. Assuming no prior knowledge of what this

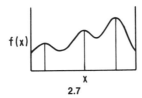

2.7

function is, the best route is to choose a small number of points that divide the interval into equal parts. Increasingly accurate values of the average energy will arise as the number of "k points" increases.

There are several different ways to estimate the needed parameters for Eq. (2.10) or (2.11). The "physicist's approach," which uses the Hückel formulation of Eq. (2.10), sometimes uses parameters taken from matches of calculated band structures and experimentally determined bandwidths, etc. There is a set of universal parameters (Ref. 160) that may be used for the distance-dependent part of the interaction, which are then combined with the known angular variation (Refs. 41, 329) of overlap integrals. [The simplest of these is shown in Fig. 1.4(b).]

The "chemist's approach" evaluates the overlap integrals between all pairs of orbitals within a large radius (to capture all significant interactions) from an atomic orbital basis usually defined by Slater orbitals at each center, just as in molecules (Ref. 5). There are also differences in the size of the orbital problem tackled by the two philosophies. Whereas the physicist's approach for the cuprate superconductors would probably just include the copper $x^2 - y^2$ orbital and the σ-type oxygen p orbitals, the chemist would include copper $3d$, $4s$, and $4p$ orbitals as well as oxygen $2s$ and $2p$. (In more detailed numerical calculations using more accurate numerical methods, the physicist's basis set is, of course, much larger.) Whatever the approach used, one of the aims of any calculation should be to extract a useful theoretical picture.

For a solid-state system the density of states may be obtained by using the technique in Figure 2.6. Solution of Eq. (2.11) at a large number of $\mathbf{k}$ points will generate an even larger number of energy levels, which may be arranged in order of increasing energy. The number of such levels contained within a small energy increment can be evaluated and used to produce a histogram for $\rho(E)$. A little cosmetic smoothing produces the desired result. Generation of a dispersion diagram

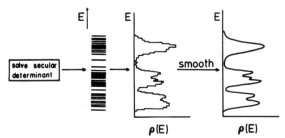

Figure 2.6 Schematic showing the method of generating the band structure of the solid. Solution of Eq. (2.10) at a large number of **k** points leads to a collection of energy levels. The number of levels contained within a small energy increment can be evaluated and used to produce a histogram for $\rho(E)$. Smoothing produces a more appealing result.

(behavior of E with respect to **k**) proceeds in the same manner. Smooth curves are drawn between the sets of points representing the energies at different k values.

In Chapter 1 the Hückel approximation was extensively used by putting $S_{ij} = \delta_{ij}$. A similar approach will be followed in much of this chapter by writing $S_{ij}(\mathbf{k}) = \delta_{ij}$, since it will allow algebraic access to several results of interest. Later, this restriction will be relaxed when more complex systems are studied and the full orbital problem examined.

This example has used a $p\pi$ orbital on a main group atom as the basis orbital for an energy band, purely out of convenience because of its ready connection with molecular chemistry. A collection of hydrogen $1s$ orbitals could equally well have been chosen as in **2.8**. An energy band will result, the orbitals being in phase at the

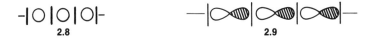

bottom and out of phase at the top. In the next section there is an example using a metal-based z^2 orbital where exactly the same criteria apply. The description of all three is topologically identical (Fig. 2.7). The quantitative differences between them lie in the different β values appropriate for the three cases.

There is an obvious difference between the energy bands derived in Figure 2.7 and that generated by a single $p\sigma$ orbital at each center **2.9**. Here, since the positive lobe of one orbital overlaps with the negative lobe of the p_z orbital carried by an adjacent atom in the chain, the interaction integral between them is positive rather than being negative. The only difference this makes to the preceding discussion is that maximum antibonding occurs at the zone center, as in Figure 2.8.

The σ band structure of a linear atomic chain containing s and p orbitals on each atom may now be qualitatively outlined. Solution of the relevant secular determinant, which will include three different values of β, one for $p\sigma$–$p\sigma$, one for s–s, and one for $p\sigma$–s interactions, is now required. This is

$$\begin{vmatrix} \alpha_s + 2\beta_{ss} \cos ka - E & 2i\beta_{sp} \sin ka \\ -2i\beta_{sp} \sin ka & \alpha_p - 2\beta_{pp} \cos ka - E \end{vmatrix} = 0 \qquad (2.12)$$

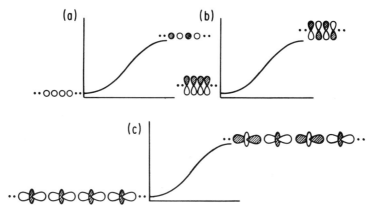

Figure 2.7   Dispersion pictures for three types of orbital. (a) $s-s$ (σ), (b) $p-p$ (π), and (c) z^2-z^2 (σ). The general form of the curve is identical in all three cases, but the width of the band will depend upon the magnitude of the interaction or hopping integral between orbitals located on adjacent centers.

The algebraic solution for the energy levels will not be written down explicitly, but note that at $k = 0$ and $= \pi/a$ the off-diagonal element is identically zero; that is, there is no $s-p$ mixing at either of these points. (This point will be discussed in terms of symmetry arguments in Section 3.4.) H_{sp} is at a maximum in fact for $k = \pi/2a$. Figure 2.9(a) shows a qualitative diagram for such a system constructed by making use of this result and the fact that $|\alpha_s| > |\alpha_p|$. The dashed lines show the dispersion in the absence of $s-p$ mixing. Here it has been assumed that the $s-p$ separation is large compared to the values of β for the s and p bands. Figure 2.9(b) shows the case where the unmixed s and p bands (dashed lines) cross in energy. As β_{sp} increases then the mixing in the middle of the zone gives rise to the energy dependence shown by the solid lines. Note that because of the different k dependence of the p and s orbital energies, the "s" band, while purely $s-s$ bonding at the zone center, is purely $p-p$ bonding at the zone edge. The two different types of behavior shown in Figure 2.9 will be of importance in Section 5.2 when discussing the structures of elemental calcium and zinc. In addition to the presence of $p\sigma$ orbitals on the chain atoms, there will be two $p\pi$ orbitals (x and y). These will be degenerate at all $\mathbf{k}$ (i.e., the π label is still a good one for the chain), and their behavior will be just like that

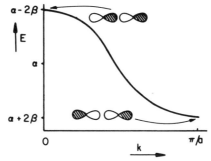

Figure 2.8   Dispersion diagram for $p-p$ (σ) interaction. The band runs "down" rather than "up" as in Figure 2.7.

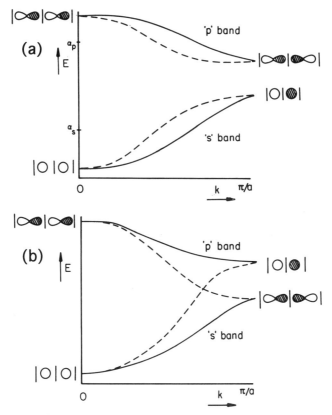

Figure 2.9   The effect of s–p mixing on the dispersion of the s and p bands of the one-dimensional chain. (a) The case where the unmixed bands do not cross with increasing k; (b) the case where they do.

found for polyacetylene. There is no overlap between these orbitals and any of the orbitals of σ type.

2.2   Some properties of solids from the band picture

There are some interesting results concerning the electrical conductivity of solids that immediately come from the band picture. Electrical conductivity arises from the presence of empty energy levels that lie close in energy to filled ones and provide a conduction pathway under an applied field. Several ways this can occur can be envisaged and are shown in Figure 2.10. A more complete discussion will be reserved for Chapter 5. Case (a) is a metal, since there are insufficient electrons to fill the energy band. Case (b) is a metal, since a band, which would be completely filled, overlaps with an empty band. Case (c) corresponds to a different type of metallic behavior. The lower band is filled, but the gap between the two bands, E_g, is small and of the order of kT. Thermal population of the upper band and the depletion of the lower band thus leads to a conducting material, a semiconductor. Case (d) shows

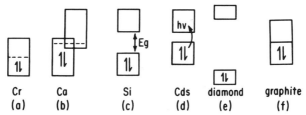

Figure 2.10 Control of the electrical properties of solids by the location and filling of their energy bands. (a) Typical metal with a partially filled band. (b) Metal generated by the overlap of filled and empty bands. (c) Small gap between filled and empty bands. As the temperature increases, then the upper band may become thermally populated, a semiconductor. (d) Similar to (c) except that population of the upper band arises through photoexcitation. (e) A large gap between the two bands, an insulator. (f) Two bands just touch (or overlap a little), a semimetal.

a system that operates in a very similar way, but here the excitation process is induced by photon absorption, leading to a photoconductor. Case (e) has a large gap such that thermal and photoexcitation is unavailable and is an insulator as a result. Clearly semiconductors are just small-gap insulators. Case (f) is a semimetal or zero-gap semiconductor. How the structure and composition of the material determines the type of behavior observed will be discussed in Chapter 5, but it is worthwhile to describe here the situation for potassium and calcium. Potassium, with its one $4s$ valence electron ($4s^1$) can be described by case (a). Calcium with its filled s band ($4s^2$) perhaps then should be described by case (e). In this case the material should be an insulator. However, s and p bands overlap (Fig. 2.11), a feature that leads to the introduction of holes in the s band and electrons in the p band. Calcium is thus a case (b) metal.

Materials of the various types are characterized by the behavior of their resistivity with temperature (Fig. 2.12). Metals (a) show a resistivity that increases with temperature, a result interpreted in terms of the scattering of electrons by phonons (Refs. 16, 210). Semiconductors (b) show the opposite behavior, as expected for a thermally activated process. Conduction increases with temperature. Importantly the behavior of a particular system may change with temperature. Figure 2.12(c) shows a transition from a metal to semiconductor (insulator) as the temperature falls. Figure 2.12(d) shows a transition from metal to superconductor.

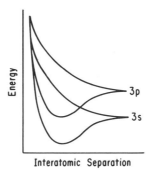

Figure 2.11 Extension of the picture of Figure 2.5, showing the energetic behavior of the s and p bands of an sp, or main group, element with distance.

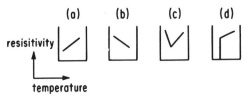

Figure 2.12 Temperature dependence of the resistivity for some of the cases of Figure 2.10. (a) A typical metal; (b) a typical semiconductor; (c) a metal–insulator transition, which occcurs on cooling; (d) a metal–superconductor transition, which occurs on cooling.

There are some fundamental observations concerning the energetics of the solid that are a natural result of the generation of an energy band that may be filled with electrons. Figure 2.13 shows (Ref. 5) the variation in the cohesive energy for solids of the main group (sp) elements and the transition (d) metals (Refs. 130, 149). This is the energy required to convert the solid to a collection of gaseous atoms. Notice the inverted parabolic dependence on electron count, the plots peaking in the middle of the series of either the half-filled $s + p$ or half-filled $s + d$ band. (The overlap of s and d bands is described in more detail in Section 4.6.) This point corresponds to 4 and 6 electrons per atom, respectively. (The plot for some of the first transition metal series may be smoothed out somewhat by noting the nature of the lowest electronic states of the gaseous atoms and thus the reference point for the plot. Some of these species have $s^2 d^n$ configurations, but others $s^1 d^{n+1}$. If the same type of atomic state is chosen, then these curves smooth out a bit.) Recall a similar dependence on bond strength or vibrational force constant [shown in Fig. 2.13(a)] for the diatomic molecules of the first row elements. A qualitative explanation for this behavior comes from the model in **2.10**.

$$E_T = \int_{-\infty}^{E_F} \rho(E) E \, dE$$

2.10

A rectangular density of states $\rho(E)$ with a bandwidth W is used for simplicity, although the general argument is not very sensitive to the actual form of the density of states (Ref. 129). Figure 2.14 compares a typical density of states for a transition metal with the rectangular one. The approximation may appear drastic, but it is a useful one. If there are M orbitals to be considered per atom, then the density of states $\rho(E)$ is just M/W.

$$\int_{-W/2}^{W/2} \rho(E) \, dE = \int_{-W/2}^{W/2} M/W \, dE = M \tag{2.13}$$

The second moment of this density of states is, from Eq. (1.43),

$$\mu_2 = \int_{-W/2}^{W/2} \rho(E) E^2 \, dE = MW^2/12 \tag{2.14}$$

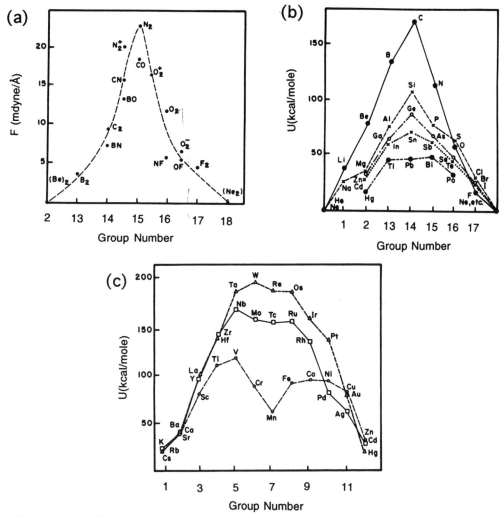

Figure 2.13 (a) The variation in AB force constant for the first row diatomic molecules. Variation in the cohesive energy of (b) the main group (*sp*) elements, (c) the transition metal (*d*) elements. (Adapted from Ref. 5.)

or just $W^2/12$ per orbital. From the definition of the second moment in Eq. (1.53), $\mu_2 = \Gamma\beta^2$, where Γ is the coordination number and β is the relevant interaction integral (in practice for these transition metals β is a composite of σ, π, and δ interactions). Thus

$$W^2 = 12\Gamma\beta^2 \tag{2.15}$$

a result showing that for the rectangular density of states the bandwidth varies as the square root of the coordination number. For the band containing five d orbitals, the d band energy may be computed (Ref. 130) as a function of the number of d

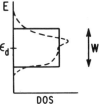

Figure 2.14 The rectangular density of states approximation (solid line) compared to a typical metallic density of states (dashed).

electrons, N

$$E_T(N) = \int_{-W/2}^{-W/2+NW/10} \rho(E)E\, dE$$

$$= -\tfrac{1}{20} WN(10\text{-}N) \tag{2.16}$$

This is a rigid band calculation, namely, one where the parameters of the model (W here) do not change with the band filling. Just as in the molecular case, this parabolic function reflects the filling first of the deepest-lying bonding orbitals followed by less bonding orbitals and then weakly antibonding orbitals followed by antibonding orbitals, which are much more strongly so. This model assumes, of course, that the bandwidth W is independent of N. In fact, the bandwidth naturally increases as the stabilization energy increases. The increasing stabilization leads to shorter inter-nuclear distances (Ref. 116) and hence larger values of W. The result is that $W(N) = WN(10\text{-}N)$, so that the cohesive energy is a more peaked function of N and should thus vary as $[N(10\text{-}N)]^2$.

Notice the important result, that the actual structure of the material has not entered into the discussion. The cohesive energies of solids are generally much larger than the energy differences between one solid-state structure and another. For example, the energy difference between the anatase and rutile forms of TiO_2 is of the order of 2 kcal/mole, but the cohesive energy is 225 kcal/mole.

The same model is useful in looking at alloys of the transition metals. Assuming that the bandwidth is the same for both partners and that their electronegativities are equal, then the heat of formation is simply

$$\Delta H = -\tfrac{1}{20} W[(N_1 + N_2)/2][10 - (N_1 + N_2)/2] + \tfrac{1}{40} WN_1(10 - N_1)$$

$$+ \tfrac{1}{40} WN_2(10 - N_2)$$

$$= -\tfrac{1}{80} W(\Delta N)^2 \tag{2.17}$$

or more generally $\Delta H = -\tfrac{1}{80} W(N_{av})(\Delta N)^2$. Figure 2.15 shows the predictions of the model with some values of ΔH taken from Miedema's empirical scheme (Refs. 239–242) (described in more detail in Section 4.6). Notice that the general trend is correct [the plot tracks the expected behavior of $W(N_{av})$] but that the simple model is an oversimplification. Some of the reasons for this are discussed in Chapter 3. The general result, however, confirms an observation known for some

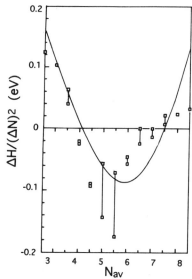

Figure 2.15 The heats of formation of transition metal–transition metal alloys. The solid curve comes from the more complete theory of Section 4.6. The squares are from Miedema's scheme for values of $\Delta N \leq 4$; points with a common N_{av} are connected together. (Adapted from Ref. 286.)

time, that the most stable alloys are those between early and late transition metals (i.e., those with large ΔN).

That the upper part of the metal d band is antibonding between the metal atoms is nicely shown by the variation in cell parameter for the series of isostructural (fcc) metals; Rh (3.80 Å), Pd (3.891 Å), Ag (4.086 Å), and Ir (3.839 Å), Pt (3.924 Å), Au (4.079 Å). The "size" of the metal atoms decreases on moving from left to right across the periodic table, but here the cell parameters increase in this direction. Figure 2.16 shows the variation in Wigner–Seitz radius across the first and second

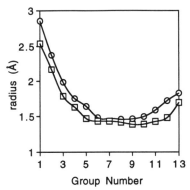

Figure 2.16 The experimental variation in Wigner–Seitz radius for the first (squares) and second (circles) row transition metal series. This radius is obtained from the atomic volume, computed by dividing the volume of the unit cell by the number of atoms in the cell. It is thus a parameter comparable between systems of different crystal structure and coordination number. It is a figure a little larger than the radius obtained by dividing the nearest-neighbor interatomic separation by two. (Adapted from Ref. 251.)

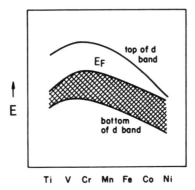

Figure 2.17 The variation in bandwidth and Fermi level for the elements of the first transition metal series. (Adapted from Ref. 14.)

transition metal series. They are similar to those from calculation (Ref. 251). The skewed behavior represents the balance between the parabolic form demanded by this form of the stabilization energy and the monotonically decreasing atomic "size" on moving from left to right across the series. The same balance controls (Refs. 13, 14) the position and width of the d band across the series. Figure 2.17 shows the variation for the first-row elements.

More evidence concerning the electronic description of the energy band comes from the structural changes on doping of materials such as $K_2Pt(CN)_4$. Here the energy band is generated by interaction of the z^2 orbitals of **2.11**. The ion $Pt(CN)_4^{2-}$ is square planar and, in the solid state, the planes stack (Refs. 110, 244, 378, 386, 390) one above the next as in **2.12**. The valence orbitals of such a molecular unit are

2.11   2.12

shown in Figure 2.18(a) along with the orbital occupancy expected for a low-spin d^8 [Pt(II)] configuration. The z^2 orbitals of each unit interact with the next in the fashion shown in **2.11**, and a band, filled with two electrons, results [Figs. 2.7(c) and 2.18(b)]. The bandwidth is 4β, just as in the polyacetylene case, but here β is the interaction integral linking adjacent metal z^2 orbitals rather than adjacent $p\pi$ orbitals. $K_2[Pt(CN)_4]$ is an insulator ($5 \times 10^{-7}\,\Omega^{-1}\,cm^{-1}$) with equal Pt–Pt distances of 3.48 Å. The salt can be cocrystallized with elemental bromine, which results in oxidation of the chain to give nonstoichiometric material $K_2(Pt(CN)_4)Br_\delta \cdot 3H_2O$. The chain ion now has to be formulated as $Pt(CN)_4^{\delta-2}$, and, as a result, electron density is removed from the z^2 band [Fig. 2.15(c)]. This material is metallic, and, since electron density has been removed from the very top of the band where [Fig. 2.7(c)] maximum antibonding interactions are found, it has a significantly shorter Pt–Pt

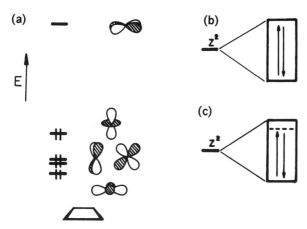

Figure 2.18 (a) The "d" orbitals of a square-planar ML_4 unit. (b) The energy band that is generated via interaction of the Pt z^2 orbitals of $K_2[Pt(CN)_4]$. (c) The effect of oxidation of the chain to give the nonstoichiometric material $K_2(Pt(CN)_4)Br_\delta \cdot 3H_2O$.

distance (2.88 Å for $\delta = 0.3$). The dependence of interatomic separation on doping level is shown (Ref. 110) in Figure 6.9.

2.3   Two-atom cells

In the previous section a repeat unit for the polyacetylene calculation was chosen that contained a single orbital. If a two-atom repeat unit is chosen as in **2.13**, how

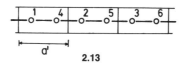

2.13

do the results change? Any observable will have the same calculated value, of course, but the $E(\mathbf{k})$ diagram will be different since there are now two $p\pi$ orbitals per cell. Equation (2.11) [with $S_{ij}(k) = \delta_{ij}$ in the Hückel approximation] needs to be solved where the Bloch basis orbitals are given by Eq. (2.4). With reference to **2.13** the Bloch functions may be written as

$$\left. \begin{array}{l} \psi_1(k) = n^{-1/2}(\cdots \phi_1 e^{-ika'} + \phi_2 + \phi_3 e^{ika'} \cdots) \\ \psi_2(k) = n^{-1/2}(\cdots \phi_4 e^{-ika'/2} + \phi_5 e^{ika'/2} + \phi_6 e^{3ika'/2}) \end{array} \right\} \qquad (2.18)$$

Now, to set up the 2×2 secular determinant the H_{ij} are calculated in the usual way

$$H_{11} = \langle \psi_1(k) | \mathscr{H}^{eff} | \psi_1(k) \rangle = n^{-1/2} n^{-1/2} (n\alpha) = \alpha \qquad (2.19)$$

and

$$H_{22} = \langle \psi_2(k) | \mathscr{H}^{eff} | \psi_2(k) \rangle = n^{-1/2} n^{-1/2} (n\alpha) = \alpha \qquad (2.20)$$

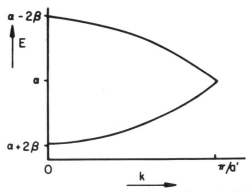

Figure 2.19   The dispersion of the π orbitals of the infinte one-dimensional chain (–CH–)$_n$ but choosing a two-carbon-atom cell.

Both of these elements are independent of $\mathbf{k}$, since there is no interaction (in the Hückel approximation) between (for example) ϕ_1 in one cell and ϕ_2 or ϕ_3 in adjacent cells. H_{12}, however, does contain β and is dependent on $\mathbf{k}$.

$$H_{12} = \langle \psi_1(k) | \mathcal{H}^{\text{eff}} | \psi_2(k) \rangle = n^{-1/2} n^{-1/2} n (e^{ika'/2} + e^{-ika'/-2}) \beta$$
$$= 2\beta \cos ka'/2 \tag{2.21}$$

The secular determinant is then

$$\begin{vmatrix} \alpha - E & 2\beta \cos ka'/2 \\ 2\beta \cos ka'/2 & \alpha - E \end{vmatrix} = 0 \tag{2.22}$$

with roots $E = \alpha \pm 2\beta \cos ka'/2$, a result shown in Figure 2.19. Remembering that a in **2.1** is half the a' of **2.13**, the relationship between Figures 2.3(a) and 2.19 is clear to see. The $E(\mathbf{k})$ diagram of the two-orbital cell is just that of the one-orbital cell with the levels folded back along $k = \pi/2a$, shown pictorially (Ref. 43) in Figure 2.20. Now there are two orbitals for each value of $\mathbf{k}$.

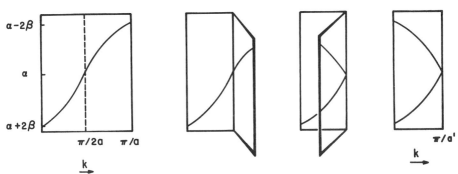

Figure 2.20   "Folding" of the orbitals of the one-atom cell to give those of the two-atom cell. (Adapted from Ref. 5.)

Another way to derive this result is to take as a basis the π and π^* levels of the diatomic unit in the unit cell (ψ_1 and ψ_2, respectively) and to set up a secular determinant as before. From Figure 2.21 it is easy to see that $H_{11} = (\alpha + \beta) + 2^{1/2} \times 2^{1/2}\beta(e^{ika'} + e^{-ika'}) = (\alpha + \beta) + \beta \cos ka'$, and that $H_{22} = (\alpha - \beta) + (2^{1/2})(-2^{-1/2}) \times \beta(e^{ika'} + e^{-ika'}) = (\alpha - \beta) - \beta \cos ka'$. H_{12} is just $2^{-1/2}(-2^{-1/2})\beta e^{ika'} + (2^{1/2})(2^{-1/2}) \times \beta e^{ika'} = -i\beta \sin ka'$ and H_{21} is $(2^{-1/2})(2^{-1/2})\beta e^{ika'} + (2^{-1/2})(-2^{-1/2})\beta e^{-ika'} = i\beta \sin ka'$. The secular determinant becomes

$$\begin{vmatrix} (\alpha + \beta) + \beta \cos ka' - E & -i\beta \sin ka' \\ i\beta \sin ka' & (\alpha - \beta) - \beta \cos ka' - E \end{vmatrix} = 0 \qquad (2.23)$$

Note that this is Hermitian (i.e., $H_{12} = H_{21}^*$). Equation (2.23) has roots

$$E = \alpha \pm \sqrt{\beta^2(2 + 2\cos ka')}$$
$$= \alpha \pm 2\beta \cos(ka'/2) \qquad (2.24)$$

which is the same result as before. Notice that both at the zone edge and the zone center $H_{12} = 0$, so the form of the wave functions is easily written down as in Figure 2.21(e). They are just simple combinations of the π or π^* functions, in phase at $k = 0$ and out of phase at $k = \pi/a'$. The one-dimensional energy band of Figure 2.16 could then be regarded as being made up of π and π^* bands as in **2.14** that touch at one

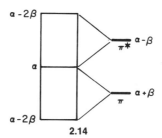

2.14

point in the Brillouin zone. There is a qualification concerning such a simple viewpoint. Because of the zero off-diagonal entry at $k = 0$, π/a' in Eq. (2.23), there is no mixing between the π and π^* functions at these points. However, the two functions can and will mix together at other k points so that a description in terms of two *separate* bands (one π, the other π^*) is not a rigorous one.

At $k = \pi/a'$ there is an orbital degeneracy. **2.15** shows one way of writing the wave

2.15   2.16

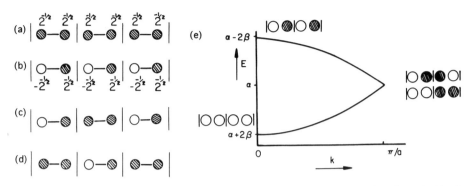

Figure 2.21 (a)–(d) The π and π^* levels of the diatomic unit in the two-atom unit cell needed for the evaluation of H_{11}, H_{22}, and H_{12}. (e) The form of the orbitals at the zone center and zone edge of the two-atom cell.

functions at this point, just employed in Figure 2.21. However, a linear combination of them as in **2.16** is just as good and emphasizes their nonbonding character. Recall, there was a similar choice for cyclobutadiene.

The orbitals of a hypothetical $(BNH_2)_n$ polymer may be readily derived using the tools established here and in Chapter 1. There are two related routes, the solution of a secular determinant and the use of perturbation theory. Assuming that the only difference between the $(CH)_n$ and $(BNH_2)_n$ polymers is the difference in the α values for the two orbitals, the secular determinant [Eq. (2.22)] becomes

$$\begin{vmatrix} \alpha_N - E & 2\beta \cos ka'/2 \\ 2\beta \cos ka'/2 & \alpha_B - E \end{vmatrix} = 0 \tag{2.25}$$

It has roots

$$E = \alpha_N + \frac{4\beta^2 \cos^2 ka'/2}{\alpha_N - \alpha_B} \quad \text{and} \quad \alpha_B - \frac{4\beta^2 \cos^2 ka'/2}{\alpha_N - \alpha_B} \tag{2.26}$$

One can also take the levels for the all carbon chain and apply a perturbation. The result is shown in Figure 2.22. Notice the gap that has opened up at the zone edge. $(2\delta\alpha = \alpha_N - \alpha_B.)$ Notice too the similarity between the orbital description at $k = \pi/a'$ with the middle pair of orbitals of square $B_2N_2H_4$ of Figure 1.10. At the zone edge the behavior on substitution of the wavefunctions is just like that of the e_g set in square C_4H_4.

2.4   The Peierls distortion

The symmetrical structure (**2.4**) was used for polyacetylene in the previous discussion, a choice that led to a metal. The polymer itself does not have this regular structure but is a semiconductor and exhibits the bond alternation shown in **2.17**. The energy bands for this arrangement are readily derived with reference to **2.18**. Now, in addition

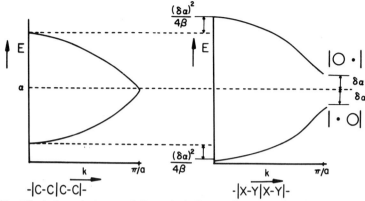

Figure 2.22 Effect of an electronegativity perturbation on the energy levels of the two-atom cell.

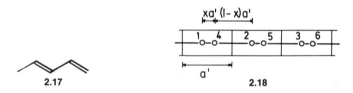

to different values of R_t in Eq. (2.6), there are two different interaction integrals, β_1 and β_2, to be included. The secular determinant is easily derived by suitable modification of H_{12} [Eq. (2.21)] and becomes

$$\begin{vmatrix} \alpha - E & \beta_1 e^{ikxa'} + \beta_2 e^{-ik(1-x)a'} \\ \beta_1 e^{-ikxa'} + \beta_2 e^{ik(1-x)a'} & \alpha - E \end{vmatrix} = 0 \qquad (2.27)$$

Solution of this determinant leads to

$$E = \alpha \pm (\beta_1^2 + \beta_2^2 + 2\beta_1\beta_2 \cos ka')^{1/2} \qquad (2.28)$$

Note that the explicit dependence on x has disappeared. (β_1 and β_2 will, of course, be x dependent.) At $k = 0$, $E = \alpha \pm (\beta_1 + \beta_2)$, and at $k = \pi/a'$, $E = \alpha \pm (\beta_1 - \beta_2)$. The $E(k)$ diagram that results now takes on the form shown in Figure 2.23(a). For the case where $|\beta_1| > |\beta_2|$ the corresponding density of states is also shown. The energy of the distorted structure of **2.17** may be estimated and compared to that of **2.4**. The function $E(k)$ should be integrated to get the best answer, but it will be sufficient here to represent the energy as being the average of that at the zone edge and that at the zone center. For the symmetric structure **2.4**

$$E_T = 2(\tfrac{1}{2})[(\alpha + 2\beta) + \alpha] = 2(\alpha + \beta) \qquad (2.29)$$

For the distorted structure **2.17**

$$E'_T = 2(\tfrac{1}{2})[(\alpha + \beta_1 + \beta_2) + (\alpha + \beta_1 - \beta_2)]$$
$$= 2(\alpha + \beta_1) \qquad (2.30)$$

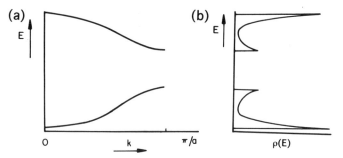

Figure 2.23 Effect of the pairing distortion of **2.18** on (a) the dispersion diagram and (b) on the energy density of states.

The energies of the two structures should be compared at constant second moment. The electronic energy E_T' of the distorted structure relative to a set of isolated atoms is $2\beta_1$, $E_T'^2 = 4\beta_1^2$. Since the second moment, just the sum of the self-returning walks around an atom in the structure, is $\mu_2 = \beta_1^2 + \beta_2^2$, then $E_T'^2/4 = \mu_2 - \beta_2^2$. β_2 is smaller in the distorted structure $(|\beta_2| < |\beta|)$, and thus this structure is more stable. Alternatively, assuming that $2\beta = \beta_1 + \beta_2$, then the stabilization energy on distortion is $\beta_1 - \beta_2$ for a unit cell containing two atoms $(|\beta_1| > |\beta_2|)$. This result is an extremely important one. With one electron per orbital the $p\pi$ band of **2.5** is half full, there is no HOMO–LUMO gap, and there is a degeneracy at the zone edge. On distortion $2.4 \rightarrow 2.17$, which results in a lowering of the orbital energy, a HOMO–LUMO gap (a band gap, in the language of the solid state) is opened up, and the degeneracy is removed. For polyacetylene (Ref. 204) the band gap in the distorted structure is about 2 eV, and the bond length difference, at 0.06 Å, is smaller than the difference between standard single- and double-bond lengths.

The situation is strongly reminiscent of that of singlet cyclobutadiene of Section 1.6. There the symmetrical structure distorted to a dimer structure. Here the atoms of the chain have dimerized in a similar way. This distortion is then the solid-state analog of the Jahn–Teller distortion. It is called a Peierls distortion (Refs. 43, 112, 119, 164, 177, 203, 205, 206, 212, 233, 244, 282, 317, 323, 327, 372). As will be shown throughout the book, there are many similarities between the two. The distortion energy in both the molecular and solid-state analogs is the same, $(\beta_1 - \beta_2)/2$ per atom using the same relationship between the β values. [In some ways this result is artificial since the zone-center and zone-edge energies have been averaged, when integration of Eq. (2.24) should have been performed. The approximation will, however, be good enough.] Figure 2.24 shows the analogy in a pictorial fashion. The structural behavior of the small polyene as it grows into the infinite polymer is described in Ref. 230. At some critical chain length the distorted structure becomes the lowest-energy arrangement. Table 2.1 shows some examples of Peierls distorted systems, some of which are described in more detail later. Notice that the system does not always distort to convert the metallic half-filled band into an insulator. Also, increasing the temperature (e.g., in the VO_2 example) is often a way to reverse the effect, by population of higher-energy levels as described in Section 5.3. Application of pressure is also effective in this regard.

How can a system such as that of **2.4** be stabilized against the Peierls distortion?

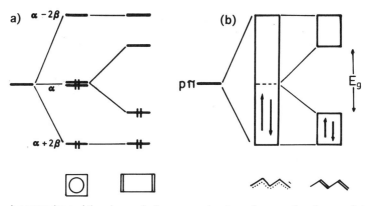

Figure 2.24 A comparison of the change in the energy levels and energy bands associated with (a) the Jahn–Teller distortion of cyclobutadiene and (b) the Peierls distortion of polyacetylene.

In the case of cyclobutadiene two extra electrons per four-atom unit remove the Jahn–Teller instability. In the present case the entire band needs to be filled with electrons. The result is the structure of fibrous sulfur and of elemental selenium and tellurium, where there are chains of atoms with equal distances between them. The chain, however, is now nonplanar, and a spiral structure is found (Fig. 5.4). Another route, suggested by the observation of the $B_2N_2R_4$ molecule, the push–pull cyclobutadienes, and the opening of the gap in $(BNH_2)_n$ of Figure 2.22, is via substitution of the chain atoms in a similar way (Ref. 175).

The dimerization result described may be generalized to other types of distortion. As shown in Figure 2.25, a one-third- and a one-quarter-filled band can lead to trimerization and tetramerization, respectively. Of course, the high-spin situation

Table 2.1 Peierls distortions in linear chains

1. Polyacetylene (bond alternation), semiconductor	Doped polyacetylene metal
2. NbI_4 chain pairing up of metal atoms	NbI_4 under pressure metal atoms equidistant
3. VO_2 chain (rutile structure) metal atoms equidistant	VO_2 at higher temperature; pairing up of metal atoms
4. Elemental hydrogen H–H dimers	High-pressure metallic behavior. Presumably···H–H–H–H···chains[a]
5. $BaVS_3$ (VS_3 chain) metallic at room temperature	Metal-insulator transition on cooling, a magnetic insulator rather than diamagnetic
6. (TTF) Br $(TTF)_2^{2+}$ dimers	(TTF) $Br_{0.7}$ metallic conduction in chain (cf. polyacetylene)

[a] In one-dimension.

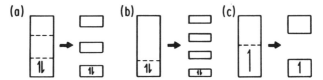

Figure 2.25 Peierls distortions resulting from partial band fillings. (a) Trimerization of a one-third-filled band, a $2k_F$ distortion; (b) tetramerization of a one-quarter-filled band, also a $2k_F$ distortion; (c) dimerization of a band half-filled with unpaired electrons (overall quarter-filled), a $4k_F$ distortion.

shown in Figure 2.25(c) leads to a dimerization even though the band is only one-quarter full. It is interesting to note that case (c) is found for $(TCNQ)_2^- (MEM)^+$ (MEM = methyl ethyl morpholinium) but case (b) where the cation is now Rb^+. (Ref. 372). The energy difference between the two cannot be very large. In general, a one-dimensional system where the band is $1/n$ or $(n-1)/n$ filled is susceptible to a distortion to an n-mer with an n-fold increase in the cell size. The driving force for the distortion drops off with the size of the fragment that has been generated. Thus a $\frac{1}{100}$-full band (obtained by doping) probably will not distort at all. This is perhaps easiest to see by returning to the molecular analogy used in Section 2.1 for the solid. Figure 2.26 shows the energy differences (ΔE_T) associated with dimerization and trimerization of cyclobutadiene and benzene[2+], respectively. These correspond to the "half-filled band" and the "one-third-filled band" respectively. The stabilization energy per atom has dropped off considerably between the two examples.

Peierls argued (Ref. 282) that distortions of this type were only likely in one dimension. These ideas are discussed further in Section 3.2 and form the basis for the electronic discussion of many important structure problems in Chapter 7.

Section 1.7 described distortions of molecules in terms of the first- and second-order Jahn–Teller effects (Ref. 22). These are characterized electronically by partially occupied degenerate energy levels and energy levels separated by small energy gaps, respectively. There is a similar situation in solids. As noted previously, there are striking similarities between the Peierls distortion of the polyacetylene chain and the Jahn–Teller distortion of square cyclobutadiene. There are, however, geometrical changes in solids that are best described as second-order Peierls distortions, since the driving force arises via the mixing of occupied and unoccupied crystal orbitals of the parent geometry, just as in the ammonia molecule of Figure 1.18. This is described in Section 2.6.

2.5   Other one-dimensional systems

In this section the electronic structures of some one-dimensional systems (Refs. 243, 244, 386–389) of increasing complexity will be derived. (Those for some other simple systems are derived in Refs. 43, 82, 177, 205, and 372.) The simplest problem, topologically identical (Fig. 2.7) to that of the polyacetylene chain, is that of a linear chain of hydrogen atoms, **2.19**. This should undergo a Peierls distortion, in an exactly analogous fashion to polyacetylene, since it has but one electron per orbital. The resulting dimers **2.20** would be in accord with traditional ideas concerning bonding

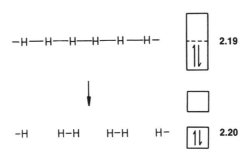

(a)

$a-2\beta$ ———

$a-\beta' =\!=\!=$

$a =\!=\!=$

$a+\beta' =\!=\!=$

$a+2\beta$ ———

$\mu_2 = 8\beta^2$ $\qquad\mu_2 = 4\beta'^2$

$E_T = 4\beta$ $\qquad E_T = 4\beta' = 4 \times \sqrt{\frac{8}{4}}\,\beta = 5.65\beta$

$\Delta E_T = 1.656\beta$ $\quad \Delta E_T/\text{atom} = 0.414\beta$

(b)

$a-2\beta$ ———

$a-\sqrt{2}\beta' =\!=\!=$

$a-\beta =\!=\!=$

$a =\!=\!=$

$a+\beta =\!=\!=$

$a+\sqrt{2}\beta' =\!=\!=$

$a+2\beta$ ———

$\mu_2 = 12\beta^2$ $\qquad \mu_2 = 8\beta^2$

$E_T = 6\beta$ $\quad E_T = 4\sqrt{2}\beta' = 4\sqrt{2} \times \sqrt{\frac{12}{8}}\,\beta = 6.928\beta$

$\Delta E_T = 0.928\beta$ $\quad \Delta E_T/\text{atom} = 0.154\beta$

Figure 2.26 The energy differences associated, respectively, with dimerization and trimerization of (a) cyclobutadiene and (b) benzene^{2+}.

—H—H—H—H—H—H—   2.19

—H   H—H   H—H   H—   2.20

in this molecule. On application of high pressure ($\gtrsim 4$ Mbar) the process can be reversed and **2.20** → **2.19**. The physical properties of **2.19** (or rather its three-dimensional analog) are interesting. Such a structure with a partially filled band should be metallic, and indeed at these high pressures it is probable that metallic conduction has been observed (Ref. 232). This particular result is of interest to geophysicists, since the "atmospheres" of several planets have been suggested to be made from such material.

There are many other materials that are built up using exactly the same principles as used for the $K_2(Pt(CN)_4)Br_\delta \cdot 3H_2O$ system above. (This is often abbreviated as TCP.) For example, the d^8 complexes of the glyoximate and dioximate ligands **2.21**

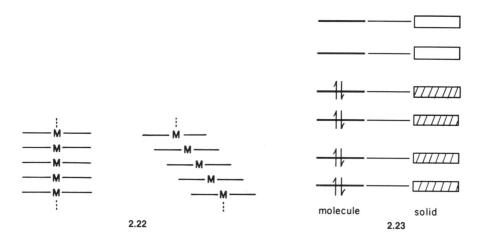

M(gly)₂ R=H

M(dpg)₂ R=Ph

Mpc

2.21

and Ni, Pd, and Pt also lead to stacks of planar units (Ref. 6). On cocrystallization with halogen, conducting materials are produced in exactly the same manner. Similar features are found in phthalocyanines (MPc) and the porphyrins. Often the crystal structure of the doped, conducting material shows regularly stacked, and that of the pure material, slipped stacks, of planar units **2.22**. Electronically, these systems are described as in **2.23**. The undoped solid is held together by van der Waals forces, but even at these large interatomic distances, where orbital overlap is small, the orbitals of the molecules broaden into bands. The bonding propensities at the top and bottom of the bands are just those of Figure 2.1. On removal of electron density, by doping with an oxidant, the system becomes metallic. There is no special "metallic

molecule   solid

2.22   **2.23**

bond" here holding the organic units together. The intermolecular forces are a combination of van der Waals interactions and the attractive interaction that arises, just as in TCP, by removal of some of the electrons occupying levels at the top of the energy band derived from the HOMO, which are antibonding between molecules.

Conceptually very similar to these TCP systems are a whole series of organic metals that contain no "metal atoms" at all (Refs 20, 320, 323, 327, 389). Such species again are typified by the stacking up of planar molecules such as those in **2.24**

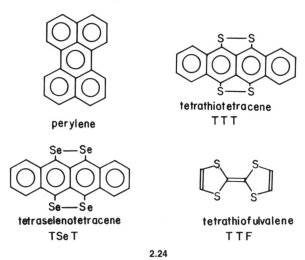

perylene

tetrathiotetracene
TTT

tetraselenotetracene
TSeT

tetrathiofulvalene
TTF

2.24

(Refs. 233, 20, 161, 164, 202). As before, in the absence of dopants the crystals are not conductors of electricity, but cocrystallization with halogen leads to conducting materials. Some examples are given in Table 2.2. A particularly interesting series is shown in Figure 2.27. The parent material is an insulator, but the nonstoichiometrically doped material $(TTF)Br_{0.7}$ is a metal. With a half-filled band as in $(TTF)_2 2Br^-$ a Peierls distortion leads to dimerization and the observation of $(TTF)_2^{2+}$ pairs in the crystal. Finally, if the band is emptied completely, individual $(TTF)^{2+}$ ions are found.

Table 2.2 Some organic metals and their resistivities

$(perylene)_{0.4}(I_3^- \cdot 2I_2)_{0.4}$	$5-50 \ (\Omega \ cm)^{-1}$
$(TTF)Br_{0.7}$	$300-500$
$(TTT)_2 I_3^-$	2700
$(TSeT)Cl_{0.5}$	2100

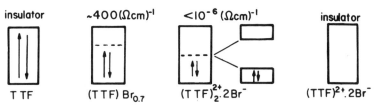

Figure 2.27 Electronic behavior of some doped TTF systems.

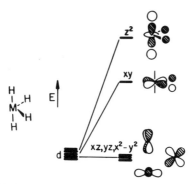

Figure 2.28 Orbitals of an octahedral cis-divacant, or butterfly, structure.

Not all of the structures of these species end up as neatly stacked planar molecules. In $(TTT)_2I_3^-$, for example, the stacks are slipped somewhat as in **2.22**. Since these materials are held together predominantly by van der Waals forces, these systems are frequently called molecular metals.

The generation of the band structure of the hypothetical transition-metal-containing MH_4 system (**2.25**) is quite illustrative (Refs. 43, 375). The chain is composed of

2.25   **2.26**

edge-sharing MH_6 octahedra. First a choice has to be made concerning the basis set to use for the problem. Perhaps the easiest route for this example is to first set up the valence orbitals of the butterfly MH_4 unit, **2.26**. With HMH angles of $90°$ and $180°$, this is easy to do and is shown in Figure 2.28. z^2 is destabilized (by $2.5e_\sigma$ using the angular overlap model, Ref. 41) more than xy (by $1.5e_\sigma$), and the three other d orbitals remain nonbonding. After all of the preceding discussion about energy bands, it is a simple matter to write down an approximate band structure. This is shown in Figure 2.29.

The energetic behavior of the levels with respect to k is easily understood. Recall that whether maximum bonding or antibonding is found at the zone center just depends on whether the overlap integral of one basis orbital is positive or negative, respectively, with its partner in the next unit cell. (The $p\pi$ orbitals of the polyacetylene chain are an example of the former; the $p\sigma$ orbitals of the chain an example of the latter.) The dispersion of the z^2 and xy bands is set by the sign and magnitude of the M–H interactions between cells. From Figure 2.28 the overlap is negative for z^2 but positive (and larger in magnitude) for xy. Notice the correspondingly larger dispersion of the xy band with the opposite behavior to that of z^2. If the energetics were dominated by metal–metal interaction, since the z^2 overlap is positive but the xy overlap negative, the dispersion of the two bands would be reversed. The xz, yz, and $x^2 - y^2$ orbitals have no hydrogen orbital character, and their dispersion is set by the positive overlap for $x^2 - y^2$ and yz and the negative overlap for xz. A tiny

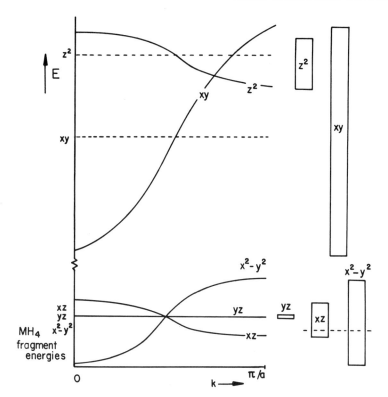

Figure 2.29 Dispersion picture for the octahedral chain of **2.25**, obtained by examining the behavior of the orbitals of the unit cell, **2.26**.

dispersion for yz is shown since it is unfavorably situated for good M–M interaction compared to the σ and π interactions of $x^2 - y^2$ and xz, respectively. The band structure ends up as a two above three pattern typical of octahedral coordination. Notice, however, that although the solid is made up of "high-symmetry" octahedral units (**2.27**) fused into a chain, the MH_4 repeat unit has a geometry of much lower symmetry.

$$\text{2.27}$$

$$\text{2.28}$$

Figure 2.29 shows the Fermi level (dashed line at right) for a d^1 metal. How might the system distort to lower its energy? The situation is not quite as simple as the one-dimensional Peierls instability of polyacetylene or elemental hydrogen, since there are three accessible energy bands to worry about. However, a pairing distortion does stabilize the system considerably, as shown in Figure 2.30, and a semiconductor is generated. This orbital problem is very similar (Ref. 375) to that of NbX_4 (X, halogen), which forms chains of the type in **2.28**. NbI_4 under pressure becomes

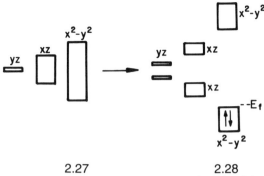

Figure 2.30   Effect on the lowest three energy bands of the distortion **2.27** → **2.28**.

metallic, and so, just as in the case of elemental hydrogen (**2.19**, **2.20**), the Peierls distortion here may be reversed (**2.28** → **2.27**) by the application of pressure.

Note that in the distortion shown in Figure 2.30, and indeed in all of the distortions of the Peierls type studied so far, the bandwidths of each of the components on distortion together are less than the undistorted width. This arises as a simple consequence of the fact that the overlap integrals decrease as the interatomic distance increases. The corresponding interaction matrix elements then become smaller on distortion.

A slightly more complex problem lies in understanding the electronic structure of the mixed-valence $Pt(L)_4 \cdot Pt(L)_4 X_2^{4+} (X = $ halide) chains (Refs. 377, 378) **2.29**, which, with $L = NH_3$ or $NEtH_2$, are present in Wolfram's red salt and Reihlen's green salt. The octahedral $Pt(L)_4 X_2$ unit may be regarded as containing Pt(IV) and the square-planar $Pt(L)_4$ unit as containing Pt(II), implying a description that is a simple distortion of the symmetric "Pt(III)" structure **2.29**. It is best to start with the unit cell of **2.31**. The energy levels of the square-planar ML_4 unit were shown in

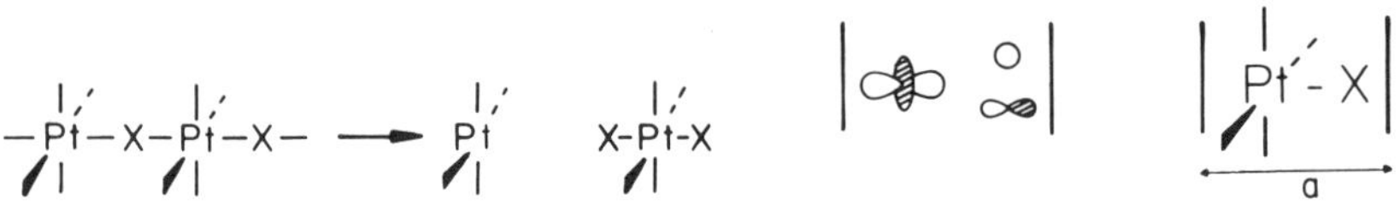

$$-\overset{|}{\underset{/|}{Pt}}-X-\overset{|}{\underset{/|}{Pt}}-X- \longrightarrow \overset{|}{\underset{/|}{Pt}} \qquad X-\overset{|}{\underset{/|}{Pt}}-X$$

2.29   2.30   2.31

Figure 2.18. Consideration of the interactions between the metal z^2 orbital and the valence s and $p\sigma$ orbitals on X will be sufficient. The secular determinant for this three-orbital problem is constructed as:

$$\begin{vmatrix} \alpha_s - E & 0 & 2\beta_{sd} \cos ka/2 \\ 0 & \alpha_p - E & -2\beta_{pd}i \sin ka/2 \\ 2\beta_{sd} \cos ka/2 & 2\beta_{pd}i \sin ka/2 & \alpha_d - E \end{vmatrix} = 0 \qquad (2.31)$$

This will not be solved in closed form as it stands. First it is interesting to see what happens if α_p and α_s are set equal as well as the two interaction integrals

$(\beta_{sd} = \beta_{pd} = \beta)$. The solutions are found to be

$$E = \alpha_p \qquad (=\alpha_s)$$

$$E = \frac{(\alpha_p + \alpha_d) \pm \sqrt{(\alpha_p - \alpha_d)^2 + 16\beta^2}}{2} \tag{2.32}$$

and show dramatically that none of the energy bands has any dispersion at all; that is, there is no k dependence. If now $\alpha_s = \alpha_p$ but setting $\beta_{sd} \neq \beta_{pd}$, the roots become

$$E = \alpha_p \qquad (=\alpha_s)$$

$$E = \frac{(\alpha_p + \alpha_d) \pm \sqrt{(\alpha_p - \alpha_d)^2 + 16[\beta_{pd}^2 - \cos^2(ka/2)(\beta_{pd} - \beta_{sd})^2]}}{2} \tag{2.33}$$

Arbitrarily assuming that $|\beta_{pd}| > |\beta_{sd}|$, the band structure may be drawn out as in Figure 2.31. By expansion (Ref. 43) of Eq. (2.33) as a power series, the roots are

$$E = \alpha_d - \frac{4[\beta_{pd}^2 - \cos^2(ka/2)(\beta_{pd}^2 - \beta_{sd}^2)]}{\alpha_p - \alpha_d} \tag{2.34}$$

$$E = \alpha_p + \frac{4[\beta_{pd}^2 - \cos^2(ka/2)(\beta_{pd}^2 - \beta_{sd}^2)]}{\alpha_p - \alpha_d} \tag{2.35}$$

The result is actually reminiscent of the behavior just noted for NbCl$_4$. Interaction of z^2 with the s orbital on X leads to a dispersion with a $\cos^2(ka/2)$ dependence with

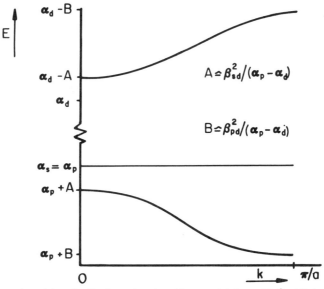

Figure 2.31 Dispersion picture for the three-band problem containing a Pt z^2 orbital and bridging X atom s and $p\sigma$ orbitals.

maximum bonding at $k = 0$, but interaction of z^2 with the p orbital on X leads to a $\sin^2(ka/2)$ dependence with maximum bonding at $k = \pi/a$. The result, if the two interactions are equal, is a dispersion-free band as found previously.

Of particular interest are the results found using a p orbital only on X and the angular overlap model to estimate the energies. Setting $\beta_{sd} = 0$ and $e_\sigma = \beta_{pd}^2/\alpha_p - \alpha_d$, the bottom of the d band lies at α_d and the top of the d band at $\alpha_d - 4e_\sigma$. The z^2 bandwidth is thus $4e_\sigma$. Recall that the energy of the z^2 orbital in a PtX_2 molecule (Ref. 41) would only be $2e_\sigma$. This turns out to be a general feature of such problems. The average band energy is indeed equal to that computed for the isolated molecule, but the top of the band lies higher in energy, and the bottom lies lower. The reader can extend this orbital problem to that of the perovskite structure (5.12) and show that the bandwidth of the e_g z^2 and $x^2 - y^2$ bands is $6e_\sigma$, and that on the p orbital only model there is no gap between e_g and t_{2g} bands.

Including both s and p orbitals on X the form of the wave functions at the top and bottom of the "z^2" band are shown in 2.32 and 2.33, respectively. The oxidation

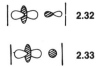

state of the platinum in 2.29 is Pt(III) and thus this "z^2" band is half full of electrons. Figure 2.32 shows the result of an extended Hückel calculation as the symmetric system 2.29 is distorted to 2.30. The initially metallic state has become insulating, and an energetic stabilization has occurred. As in the case of NbI_4 the conductivity of these salts increases markedly on the application of pressure.

A more detailed view of the form of the orbitals gives clues as to how these systems distort. Figure 2.33 shows the z^2 band for a doubled PtL_4X cell. At the zone edge the obvious choice of wave functions, intermediate in character between those at the top and bottom of the band, has been made. As in all of the degenerate orbital problems studied so far, a linear combination of these two orbitals shown in 2.34

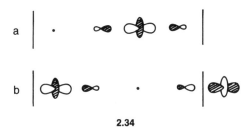

2.34

leads to another perfectly good pair. Now if the system is distorted slightly (2.29 → 2.30), then a band gap opens up at the zone edge. 2.34a goes up in energy since the Pt–X distances decrease around this metal atom, and the antibonding interactions become stronger. 2.34b goes down in energy since the Pt–X distances around this metal atom increase with a concurrent weakening of the Pt–X antibonding interactions. This is shown in Figure 2.34 in an exaggerated way. As the

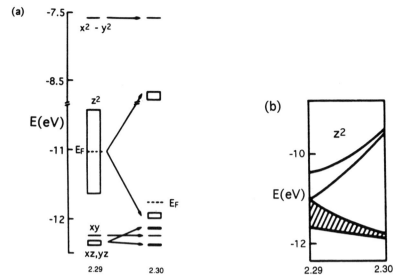

Figure 2.32 The effect of the distortion of **2.30** on the valence bands of the $Pt(L)_4 \cdot Pt(L)_4 X_2^{4+}$ (X = halide) chain. (a) The results of an extended Hückel calculation; (b) the effect of the distortion on the bandwidths.

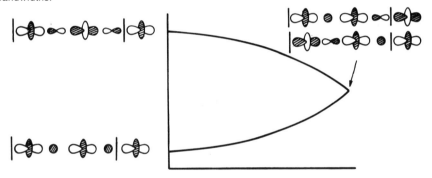

Figure 2.33 The z^2 band for a doubled PtL_4X cell.

distortion proceeds, the interaction of the z^2 orbital on the now planar Pt atom with the halogen orbitals of the bridge becomes small, and the bandwidths of both the upper and lower parts of this band become very narrow, as in Figure 2.32(b). The lower band (now full) then corresponds largely to a z^2 orbital on the square-planar Pt atom (Figure 2.34), which is now Pt(II). The upper band (now empty) corresponds to a strongly antibonding orbital on the octahedral Pt(IV) center. The result is a classical mixed-valence compound, $Pt(II)L_4 \cdot Pt(IV)L_4X_2$. The redistribution of charge that has taken place on distortion gives rise to an alternative label, that of the charge-density wave (CDW) (Refs. 249, 372, 392) for such distortions.

2.6 Second-order Peierls distortions

As noted in Section 2.4, there are geometrical changes in solids that are best described as second-order Peierls distortions (Refs. 43, 372). Here the energetic driving force

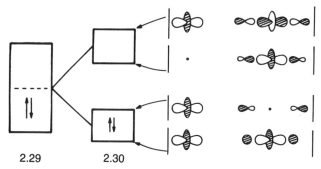

Figure 2.34 The change in the form of the wave functions as a result of the distortion **2.29** → **2.30**.

for the distortion arises via mixing of occupied and unoccupied crystal orbitals of the parent geometry as the solid is distorted. The molecular analogy that can be made is with the ammonia molecule of Figure 1.18. A nice example of a second-order Peierls distortion is that of polyacene from the symmetrical structure to the one with the bond alternation pattern shown in **2.35**. This, of course, in geometrical terms, is

2.35

an analog of the distortion of polyacetylene. First the band structure of the undistorted structure needs to be assembled (Ref. 43). This may be done in several ways, but, noting the "topological picture" of **2.36**, it will be constructed here as in **2.37** from the orbitals of "butadiene-like" units. An immediate simplification presents

2.36

2.37

itself since the solid possesses a mirror plane that bisects the chain and is shown in **2.35**. The butadiene basis functions that need to be used are divided into symmetric and antisymmetric pairs in **2.38** and **2.39**. For the symmetric orbitals the entries for

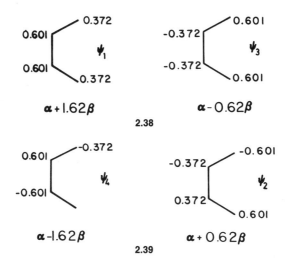

the secular determinant are

$$H_{11} = (\alpha + 1.92\beta) + (2)(2)(0.372)(0.601)\beta \cos ka$$

$$H_{33} = (\alpha - 0.62\beta) - (2)(2)(0.372)(0.601)\beta \cos ka \qquad (2.36)$$

$$H_{13} = [-2(0.372)^2]e^{ika} + [2(0.601)^2]e^{-ika}$$

This gives an expression for the bands of the symmetric orbitals as

$$E = \{(2\alpha + \beta) \pm \beta) \pm [(2.24 + 1.76 \cos ka)^2$$
$$+ 4(0.6 - 0.4 \cos 2ka)]^{1/2}\}/2 \qquad (2.37)$$

so at $k = 0$, $E = \alpha + 2.55\beta$, $\alpha - 1.55\beta$, and at $k = \pi/a$, $E = \alpha, \alpha + \beta$. The algebra is similar for the antisymmetric orbitals and at $k = 0$, $E = \alpha - 2.55\beta$, $\alpha + 1.55\beta$, and at $k = \pi/a$, $E = \alpha, \alpha - \beta$. The dispersion picture is shown in Figure 2.35. This simple picture is in accord with that from a numerical calculation (Ref. 378), but (not of importance for the discussion here) the degeneracy at the zone edge is removed if

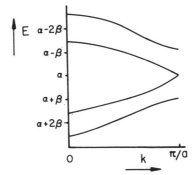

Figure 2.35 Hückel dispersion picture for the π orbitals of polyacene.

trans-annular interactions are included in the calculation. The result is that the two bands cross just before the zone edge.

The distortion of **2.35** preserves the mirror plane, and so the energetic changes are simply described via two separate secular determinants, one for the symmetric and one for the antisymmetric pair of orbitals. If one writes two values of β for the distorted structure (β_1, β_2), then the secular determinant coupling the two functions in the antisymmetric block is just

$$\begin{vmatrix} (\alpha - \beta) - E & \frac{1}{2}(\beta_1 - \beta_2) \\ \frac{1}{2}(\beta_1 - \beta_2) & \alpha - E \end{vmatrix} = 0 \qquad (2.38)$$

This has roots $E = (\alpha - \beta/2) \pm [\beta^2 + (\beta_1 - \beta_2)^2]^{1/2}/2$. Expansion leads to new values for the energies at the zone edge of

$$\left. \begin{array}{l} E = \alpha + (\beta_1 - \beta_2)^2/4\beta \\ E = a - \beta - (\beta_1 - \beta_2)^2/4\beta \end{array} \right\} \qquad (2.39)$$

Similar expressions apply to the symmetric functions, and the overall result is shown in **2.40**. The stabilization has not arisen just by the splitting apart of the degenerate

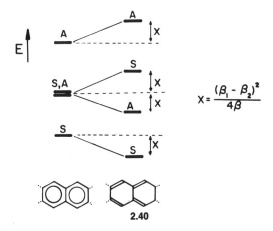

2.40

functions at the zone edge but is more complicated. This mixing of occupied and unoccupied orbitals separated in energy characterizes the distortion as a second-order Peierls distortion.

Similar considerations may be used to view the distortions of systems with more chains of atoms. If polyacene is considered as being made up of just two polyacetylene chains, then graphite may be envisaged as being composed of an infinite number. Why does graphite not distort as in **2.41**? It may be shown that the stabilization

2.41

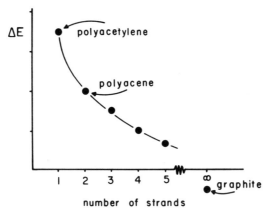

Figure 2.36 Distortion energies of coupled polyacetylene chains. ($n = 1$ corresponds to poly-acetylene itself; $n = 2$ corresponds to polyacene; $n \to \infty$ to graphite.)

energy drops off rapidly as the number, n, of strands increases and this is found too by calculation (Ref. 43) (Figure 2.36) of the variation in the distortion energy with n. Resisting the distortion described here associated with the highest occupied orbitals are the elastic forces of the underlying structure described by Eq. (5.1). Thus for polyacetylene the "electronic" forces win out, but for graphite the distortion energy from this source is not high enough.

3 □ More details concerning energy bands

This chapter contains three topics that are useful for a more detailed understanding of the electronic structure of solids.

3.1   The Brillouin zone

Probably most of the systems of interest to the chemist are not one-dimensional ones and require the extension of the ideas developed in Chapter 2 to higher dimensions (Refs. 10–12, 30, 38, 43, 76–80, 82, 177, 191, 199, 205, 210, 219, 370). In general the translation group may be written as a simple product group involving translations along the three lattice vectors, $\mathbf{a}_i$, of Section 2.1. In this case the exponential in the Bloch sum of Eq. (2.6) becomes

$$e^{i(k_1\mathbf{b}_1 \cdot l_1\mathbf{a}_1)} \, e^{i(k_2\mathbf{b}_3 \cdot l_2\mathbf{a}_2)} \, e^{i(k_3\mathbf{b}_3 \cdot l_3\mathbf{a}_3)} = e^{i(\mathbf{k} \cdot \mathbf{r})} \tag{3.1}$$

where $\mathbf{b}_i \cdot \mathbf{a}_j = 2\pi\delta_{ij}$ and the reciprocal lattice vectors $\mathbf{b}_i$ are defined by

$$
\left.
\begin{aligned}
\mathbf{b}_1 &= 2\pi \, \frac{\mathbf{a}_2 \wedge \mathbf{a}_3}{\mathbf{a}_1 \cdot (\mathbf{a}_2 \wedge \mathbf{a}_3)} \\[2mm]
\mathbf{b}_2 &= 2\pi \, \frac{\mathbf{a}_3 \wedge \mathbf{a}_1}{\mathbf{a}_1 \cdot (\mathbf{a}_2 \wedge \mathbf{a}_3)} \\[2mm]
\mathbf{b}_3 &= 2\pi \, \frac{\mathbf{a}_1 \wedge \mathbf{a}_2}{\mathbf{a}_1 \cdot (\mathbf{a}_2 \wedge \mathbf{a}_3)}
\end{aligned}
\right\}
\tag{3.2}
$$

Just as the direct lattice vectors define this lattice, so these reciprocal lattice vectors define a reciprocal lattice. (An alternative nomenclature that is often used is to label the direct lattice vectors as $\mathbf{a}, \mathbf{b}, \mathbf{c}$, and the reciprocal lattice vectors as $\mathbf{a}^*, \mathbf{b}^*, \mathbf{c}^*$.) A simple example will illustrate construction of the reciprocal lattice. Figure 3.1(a) shows a primitive orthorhombic lattice. The direct lattice vectors $\mathbf{a}_i$ are given by

$$\mathbf{a}_1 = a\mathbf{x}; \qquad \mathbf{a}_2 = b\mathbf{y}; \qquad \mathbf{a}_3 = c\mathbf{z} \tag{3.3}$$

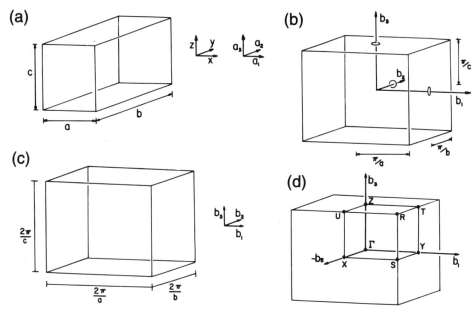

Figure 3.1 Construction of the Brillouin zone for a three-dimensional system. (a) A primitive orthorhombic lattice; (b) the reciprocal lattice; (c) the construction of the Brillouin zone for the reciprocal lattice of (b); (d) the symmetry labels given various points that lie on the faces, edges, or vertices of the Brillouin zone

where $\mathbf{x}$, $\mathbf{y}$, and $\mathbf{z}$ represent unit vectors along the x, y, and z directions. The $\mathbf{b}_i$ are then simply

$$\mathbf{b}_1 = (2\pi/a)\mathbf{x}; \qquad \mathbf{b}_2 = (2\pi/a)\mathbf{y}; \qquad \mathbf{b}_3 = (2\pi/a)\mathbf{z} \qquad (3.4)$$

and the reciprocal lattice as in Figure 3.1(b). Now the first Brillouin zone is defined as the volume enclosed by the set of planes that bisect perpendicularly all the lines drawn from one lattice point to all others in the reciprocal lattice. In practice only a small number of close points are needed. Figure 3.1(c) shows the construction for the reciprocal lattice of Figure 3.1(b). Various points that lie on the faces, edges, or vertices of the Brillouin zone are usually given symmetry labels. Figure 3.1(d) shows the conventional choice for this example. In units of $2\pi/a$ the values of k_1, k_2, and k_3 for these symmetry points are

$$\left.\begin{array}{lll} \Gamma: 0, 0, 0 & X: 0, \tfrac{1}{2}, 0 & Z: 0, 0, \tfrac{1}{2} \\ Y: -\tfrac{1}{2}, 0, 0 & T: -\tfrac{1}{2}, 0, \tfrac{1}{2} & U: 0, \tfrac{1}{2}, \tfrac{1}{2} \\ S: -\tfrac{1}{2}, \tfrac{1}{2}, 0 & R: -\tfrac{1}{2}, \tfrac{1}{2}, \tfrac{1}{2} & \end{array}\right\} \qquad (3.5)$$

Notice that the one-dimensional example, exclusively described until now, is the special case of Figure 3.1(c), with $\mathbf{b}_2 = \mathbf{b}_3 = 0$. A particularly simple zone has been chosen for the example. Other lattices give rise to zones that correspond to more complex polyhedra (Ref. 30).

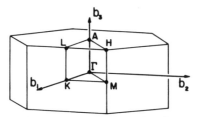

Figure 3.2   The Brillouin zone for a primitive hexagonal lattice.

3.1   3.2

For the primitive hexagonal lattice, which will be used shortly, the situation is a little more complex. If the primitive direct lattice vectors are as in **3.1**, then the reciprocal lattice vectors, by the construction of Eqs. (3.2), become those of **3.2**. The first Brillouin zone also has hexagonal symmetry and is shown in Figure 3.2. Notice that the point K is just $(\frac{1}{2}, 0, 0)2\pi/a$, but M is $(\frac{1}{3}, \frac{1}{3}, 0)2\pi/a$ by simple geometry.

First the problem of the square net, **3.3**, is tackled. This appears in several systems,

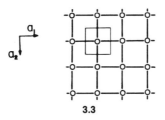

3.3

GdPS amongst them (Ref. 348). The net has a set of primitive lattice vectors $a\mathbf{x}$, $b\mathbf{y}$, and $c\mathbf{z}$, where c may be visualized as being very large and $a = b$ [of Fig. 3.1(b)]. The two-dimensional zone to be considered is therefore the one given in **3.4**

3.4

The energy band generated by an s or p_z orbital of the square net is easy to derive since it is a one-orbital problem. By analogy with the one-dimensional chain of Section 2.1, the $E(\mathbf{k})$ dependence is written as

$$\left.\begin{aligned} E(\mathbf{k}) &= \alpha + 2\beta \cos(\mathbf{k} \cdot \mathbf{a}_1) + 2\beta \cos(\mathbf{k} \cdot \mathbf{a}_2) \\ &= \alpha + 2\beta \cos k_x a + 2\beta \cos k_y a \end{aligned}\right\} \tag{3.6}$$

which therefore has a maximum energy of $\alpha - 4\beta$ and a minimum energy of $\alpha + 4\beta$.

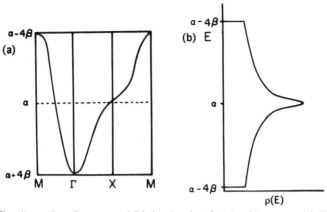

Figure 3.3   (a) The dispersion diagram and (b) the density of states for a square lattice containing a single orbital (an s or $p\pi$ orbital) at each node.

The bandwidth is thus 8β. A comparison with the bandwidth for the one-dimensional chain thus implies that for such problems the bandwidth is given by $2\Gamma\beta$, where Γ is the coordination number. There is a pictorial problem in easily showing the dispersion of the energy in more than one dimension, but the energetic behavior may be traced along lines joining symmetry points of the Brillouin zone. Using this technique the energetic dispersion is shown in Figure 3.3(a). Its density of states is shown in Figure 3.3(b) and indicates a maximum at the half-filled band. Figure 3.4 shows a three-dimensional view of the energetic behavior of $E(\mathbf{k})$. A zone four times larger than the smallest one that contains the essential energetic information (the irreducible wedge of the Brillouin zone) is used for aesthetic purposes. For situations involving more than one orbital, such pictures become so complex as to be of little use. The wave functions at some symmetry points are given in Figure 3.5. For this particular system they are easy to derive since at the zone boundaries the phase factor controlling the sign of the orbital coefficient is determined in the same way as before with **2.2** and **2.3**.

A slightly more adventurous derivation is of the π band of graphite (Ref. 43). The unit cell used is shown in **3.5** with the primitive lattice vectors that are needed in Eqs. (3.2) and (3.3). Here the vector nature of $\mathbf{k}$ needs to be taken into account and the Bloch functions evaluated using the phase factor $\exp(i\mathbf{k}\cdot\mathbf{R}_t)$ a little more carefully, since $\mathbf{a}_1$ and $\mathbf{a}_2$ are not orthogonal. Using as a basis the two $p\pi$ orbitals of the cell

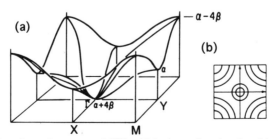

Figure 3.4   (a) A different way to represent $E(\mathbf{k})$, which shows the visual problems associated even with the behavior of a single band of a two-dimensional structure. (b) A contour diagram.

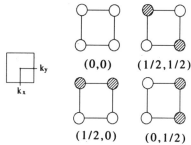

Figure 3.5 Wavefunctions at various symmetry points for the simple square lattice with an s orbital at each node (squarium). The points are in units of $2\pi/a$. An alternative notation replaces points such as $(\frac{1}{2}, \frac{1}{2})$ by (π, π).

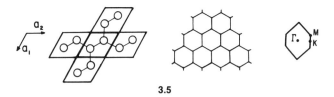

3.5

in **3.5** the secular determinant becomes

$$\begin{vmatrix} \alpha - E & \beta\{\exp[i\mathbf{k}\cdot(\frac{2}{3}\mathbf{a}_1 + \frac{1}{3}\mathbf{a}_2)] + \exp[i\mathbf{k}\cdot(\frac{1}{3}\mathbf{a}_1 - \frac{2}{3}\mathbf{a}_2)] + \exp[i\mathbf{k}\cdot(\frac{1}{3}\mathbf{a}_1 + \frac{1}{3}\mathbf{a}_2)]\} \\ H_{12}^* & \alpha - E \end{vmatrix} = 0$$

(3.7)

with roots $E = \alpha \pm A^{1/2}\beta$, where

$$A = [3 + 2\cos(\mathbf{a}_1 + \mathbf{a}_2)\cdot\mathbf{k} + 2\cos \mathbf{a}_3\cdot\mathbf{k} + 2\cos \mathbf{a}_1\cdot\mathbf{k})] \qquad (3.8)$$

Clearly again the $E(\mathbf{k})$ dependence cannot be shown in two dimensions in the way analogous to the one-dimensional case, but the energy changes can be depicted along lines in the Brillouin zone of Figure 3.6. At the three symmetry points Γ, M and K the energy is given by

$$\Gamma: E = \alpha \pm 3\beta; \qquad M: E = \alpha \pm \beta; \qquad K: E = \alpha \pm 0\beta \qquad (3.9)$$

Figure 3.6(a) shows the graphite π band structure using these results. Just as the top and bottom of the band lie at $E = \alpha \pm 2\beta$ for the polyacetylene chain with two-coordinate atoms, so here with three-coordinate atoms the top and bottom of the band lie at $E = \alpha \pm 3\beta$. (Analogously the cubium band is bounded by $E = \alpha \pm 6\beta$ and the square net by $E = \beta \pm 4\beta$.) Figure 3.6(b) shows a band structure for graphite from an extended Hückel calculation that also includes the σ bands. Notice that here the levels at $E > \alpha$ are destabilized more than the levels for which $E < \alpha$ are stabilized. This has an explanation identical to that discussed in Section 1.3. Inclusion of overlap destroys the symmetry associated with the orbitals or bands about the $E = \alpha$ level.

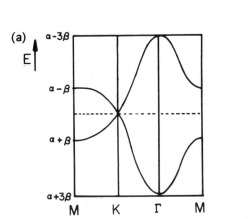
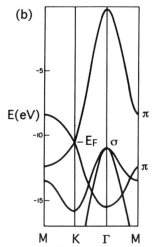

Figure 3.6 Energy bands of graphite. (a) From a Hückel calculation; (b) from an extended Hückel calculation. (Adapted from Ref. 378.)

Notice, however, that the degeneracy at $E = \alpha$ at the point K is maintained even when overlap is included. With one electron per $p\pi$ orbital in graphite, this band is half filled and $\mathbf{k}_F$ lies at the point K. Graphite is thus a zero-gap semiconductor or semimetal [Fig. 2.10(f)], one where valence and conduction bands just touch. (The term semimetal is also used in a more general way to include those systems where the bands overlap by a small amount.) That the conduction band is antibonding between nearest-neighbor carbon atoms is confirmed experimentally (Ref. 113) by the increase in C–C distance on doping (with potassium, for example). Figure 3.7 shows the wave functions at three of the symmetry points. These are obtained as before, just by inspecting the sign of the phase factors $e^{i\mathbf{k}\cdot\mathbf{r}}$. As with all pairs of degenerate functions, there is a choice available at the point K. Notice that here the bands are nearest-neighbor nonbonding. The choice made for the pair of levels will be useful for discussions in Section 5.1.

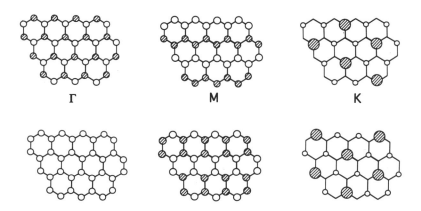

Figure 3.7 Wave functions at three symmetry points for graphite.

Exactly the same results are found (Ref. 43) if, instead of using the individual $p\pi$ orbitals as a basis, the π and π^* functions of the two-orbital cell **3.6** are used. The

3.6

3.7

secular determinant is then

$$\begin{vmatrix} (\alpha + \beta) + \beta \cos \mathbf{k} \cdot \mathbf{a}_2 + \beta \cos \mathbf{k} \cdot \mathbf{a}_1 - E & i\beta(\sin \mathbf{k} \cdot \mathbf{a}_1 + \sin \mathbf{k} \cdot \mathbf{a}_2) \\ -i\beta(\sin \mathbf{k} \cdot \mathbf{a}_1 + \sin \mathbf{k} \cdot \mathbf{a}_2) & (\alpha - \beta) - \beta \cos \mathbf{k} \cdot \mathbf{a}_2 - \beta \cos \mathbf{k} \cdot \mathbf{a}_1 - E \end{vmatrix} = 0 \tag{3.10}$$

with roots $E = \alpha \pm A^{1/2}\beta$ as before [Eq, (3.8)].

The secular determinant of Eq. (3.10) may easily be rewritten for the case of BN (**3.7**) by using two different values, α_N and β_B. The energy levels may be evaluated by expansion of the secular determinant in the usual way, as

$$E = \tfrac{1}{2}(\alpha_B + \alpha_N) \pm \tfrac{1}{2}\sqrt{(\alpha_B - \alpha_N)^2 + 4A^2} \tag{3.11}$$

The most striking result of this substitution is the removal of the degeneracy at the point K. Here the energies become $E = \alpha_N$ and α_B. Figure 3.8(a) shows the result predicted. With two π electrons per atom pair, only the lower band is filled. An extended Hückel band structure is shown in Fig. 3.8(b). BN is thus an insulator and, in contrast to the metallic sheen of graphite, BN is a plain white solid. Graphite and

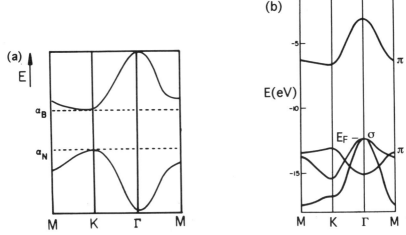

Figure 3.8   Energy bands of BN in the graphite (C) structure. (a) From a Hückel calculation; (b) from an extended Hückel calculation. (Adapted from Ref. 378.)

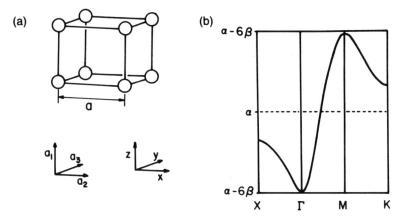

Figure 3.9 (a) The simple cubic structure and (b) the energy band that results from the location of an *s* orbital at each node (cubium).

BN have half-filled $p\pi$ bands, and it is interesting to see that the observed structure of BN is one where the boron and nitrogen atoms alternate in two dimensions. As shown in Section 8.6, such alternating structures are the ones favored for half-filled bands.

In three dimensions things can get quite complicated. Although numerical solutions are always available, there are places where some simplifications help out in the presentation of an analytical picture. First consider the simple cubic structure of Figure 3.9(a). In many ways this may be regarded as a simple sum of three one-dimensional chain problems, one lying along each Cartesian direction. For a solid composed of *s* orbitals (cubium) the energy dependence on **k** is, from Eq. (2.2) for a one-atom cell,

$$E(\mathbf{k}) = \alpha + 2\beta(\cos k_x a + \cos k_y a + \cos k_z a) \tag{3.12}$$

This leads to the dispersion curve in Figure 3.9(b), where Γ, M, K, and X represent the points $(k_x, k_y, k_z) = (0, 0, 0)2\pi/a$, $(\frac{1}{2}, \frac{1}{2}, \frac{1}{2})2\pi/a$, $(\frac{1}{2}, \frac{1}{2}, 0)2\pi/a$, and $(0, 0, \frac{1}{2})2\pi/a$, respectively. Figure 3.10 shows the wave functions at these points.

For three *p* orbitals located on a single atom another simple result applies. Neglecting $p\pi$–$p\pi$ interaction **(3.8)** between orbitals on adjacent atoms and only

considering σ overlap **(3.9)**, the energy dependence on **k** for the three orbitals is simply

$$\begin{aligned} p_x: \quad E(\mathbf{k}) &= \alpha - 2\beta \cos k_x a \\ p_y: \quad E(\mathbf{k}) &= \alpha - 2\beta \cos k_y a \\ p_z: \quad E(\mathbf{k}) &= \alpha - 2\beta \cos k_z a \end{aligned} \tag{3.13}$$

This result is shown pictorially in Figure 3.11. Just as the half-filled one-dimensional

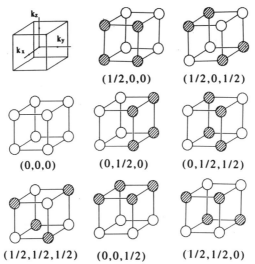

(1/2,0,0)   (1/2,0,1/2)

(0,0,0)   (0,1/2,0)   (0,1/2,1/2)

(1/2,1/2,1/2)   (0,0,1/2)   (1/2,1/2,0)

Figure 3.10 Wave functions at various symmetry points for the simple cubium structure.

chain of Section 2.3 underwent a Peierls distortion, so the simple cubic lattice with three p electrons would similarly be expected to be unstable. The structures of elemental arsenic and black phosphorus may be viewed in this way (Refs. 61, 282). Assume for simplicity that two electrons per atom reside in a deep-lying valence s orbital. For these Group 15 elements, this leaves three electrons to half-occupy the three p orbitals. In terms of the fractional orbital occupancies of the three bands, $f_1 = f_2 = f_3 = \frac{1}{2}$. Figure 3.12 shows how the energy bands change on a distortion that involves pairing up all atoms along the x, y, and z directions. There are several ways in which the pairing up may be done. It may be shown (Ref. 55) that there are a total of 36 different possibilities for a simple cubic cell containing eight atoms. Two of these correspond to the black phosphorus and arsenic (Fig. 3.13) structures. Thus these structures may be regarded as being reached via a three-dimensional Peierls distortion. On application of pressure, black phosphorus transforms to the metallic, simple cubic structure (Ref. 196). So, just as in some of the examples described earlier, the Peierls distortion is reversed by the application of pressure. Similar ideas based

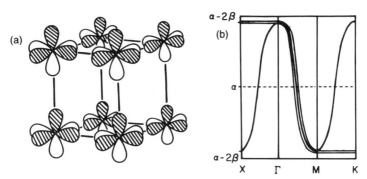

Figure 3.11 (a) The location of a set of three p orbitals at each node of the simple cubic structure. (b) The energy bands that result if interactions of π type are ignored between them.

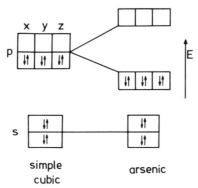

Figure 3.12 The electronic origin of the three-dimensional Peierls distortion of the simple cubic structure to give the arsenic and black phosphorus structures of Figure 3.13. (The fractional occupation of the three bands at the parent geometry are $f_1 = f_2 = f_3 = \frac{1}{2}$.)

on the Peierls instability of the simple cubic parent that lead to the diamond structure are discussed for the Group 14 elements in Chapter 6.

For some simple solids the band dispersion may be written in closed form, especially if s and p orbitals are not allowed to mix. In addition to the expressions given, those describing (Ref. 314) the s bands of some simple solids are given in the following:

Body-centered cubic:

$$E(\mathbf{k}) = \alpha + 8\beta \cos(k_x a) \cos(k_y a) \cos(k_z a) \tag{3.14}$$

Face-centered cubic structure

$$E(\mathbf{k}) = \alpha + 4\beta[\cos(k_x a) \cos(k_y a) + \cos(k_x a) \cos(k_z a) + \cos(k_y a) \cos(k_z a)] \tag{3.15}$$

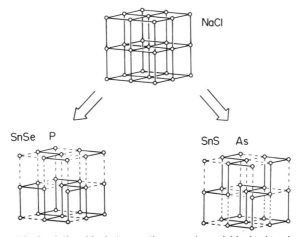

Figure 3.13 Geometrical relationship between the arsenic and black phosphorus structures and that of the simple cubic structure. The simplest AB derivative structures are also indicated, NaCl, SnS, and SnSe.

Hexagonal close-packed structure

$$E(\mathbf{k}) = \alpha + 2\beta\{\cos(k_x a) + 2\cos(\tfrac{1}{2}k_x a)\cos(\sqrt{\tfrac{3}{2}}k_y a)$$

$$\pm \cos(k_z c)[1 + 4\cos^2(\tfrac{1}{2}k_x a) + 4\cos(\tfrac{1}{2}k_x a)\cos(\sqrt{\tfrac{3}{2}}k_y a)^{1/2}]\} \quad (3.16)$$

3.2 The Fermi surface

The Fermi surface is a very interesting and useful concept. (The discussion here follows closely that of Ref. 82.) It does suffer, however, from the disadvantage that it exists in reciprocal space and so is frequently difficult to understand without recourse to the dispersion picture of the material in hand. Most chemists think in direct, not reciprocal, space. For a partially filled collection of bands, the Fermi surface is the constant-energy plot in **k** space of the highest occupied energy levels at absolute zero of temperature $(T = 0)$. Alternatively, it represents the junction between filled and empty levels at $T = 0$. For a system with filled bands there is no Fermi surface at all, and so the concept only applies to metals. (For such an insulating system the Fermi level occurs in the middle of the gap between the top of the filled band and the bottom of the empty one.) Figures 3.12 and 3.14 show how the Fermi surface varies for three different electronic situations of the two-dimensional lattice (**3.10**).

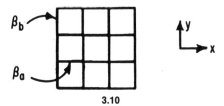

3.10

How the surface is constructed for the case of the half-filled truly square net is shown in Figure 3.15, by filling the energy levels of Figure 3.4. Even from this picture it is not clear at all why the Fermi surface is a square. This depends on the curvature of the $E(\mathbf{k})$ picture and in general needs some care in generating. Computationally it is straightforward. The Fermi surface is just the constant-energy surface at $E = E_F$. For the two-dimensional lattice (**3.10**) in general there will be two values of β in the two

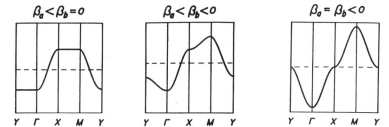

Figure 3.14 Dispersion pictures for a range of electronic situations of the square net with an *s* orbital at each node. The structure should be regarded as a set of chains along *x* linked in various ways to each other along *y* (**3.10**). (Adapted from Ref. 82.)

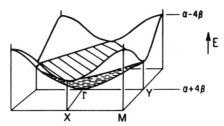

Figure 3.15 The Fermi surface for the case of half-filling the energy bands of Figure 3.14(c). The dashed region represents the occupied levels. The shaded region corresponds to the shaded area of Figure 3.16(c).

directions. If $\beta_b = 0$, then clearly the result is a series of uncoupled one-dimensional chains with no dispersion along the direction $X \rightarrow M$. For $\beta_a = \beta_b$ then the square lattice of Figure 3.3 results. The other electronic situations readily lead to the other cases shown in Figure 3.14. Figure 3.16(a)–(c) shows the Fermi surfaces for the half-filled band for these three cases. The shaded areas show the regions in **k** space of the occupied levels as in Figure 3.15. The area (volume in three dimensions) of the first Brillouin zone enclosed by the Fermi surface is proportional to the band filling. The Fermi surface varies with electron count, of course. Figures 3.14(d) and (e) show how the Fermi surface of case (c) changes for the less-than-half-full and greater-than-half-full bands. Notice that case (a) is clearly one dimensional, but that (c), (d) and (e) are different in character. In fact, they are described as being two-dimensional surfaces since they contain a closed loop. Although this is easy to see for (d), only by repeating the picture of (e) as shown in **3.11** is this clear. This electronic dimensionality of the problem dictates the metallic (or otherwise) properties of the solid. Case (a) is only metallic along x, case (e) along both x and y for the half-filled band. These two cases are experimentally distinguishable, in principle at least (Ref. 392).

The Fermi surface is most easily depicted in two dimensions, as shown in

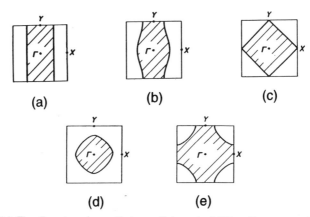

Figure 3.16 (a)–(c) The Fermi surfaces that result from half-filling the energy bands of Figure 3.14(a)–(c). The Fermi surfaces that result from (d) slightly less than half-full and (e) the slightly greater than half-full situation for the truly square net of Figure 3.14(c) are also shown. (Adapted from Ref. 82.)

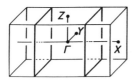 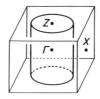

Figure 3.17 The three-dimensional Fermi surfaces of Figures 3.14(a),(d) when there is no coupling to the third dimension ($\beta_c = 0$). (Adapted from Ref. 82).

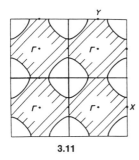

3.11

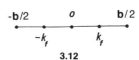

3.12

Figure 3.16. However, in one dimension it is a set of points, and for the half-filled band of [case (a)] it is given just by the two points $\pm 2\mathbf{k}_F$ as in **3.12**. In three dimensions drawing the Fermi surface is usually a complex task, but for the cases of Figure 3.14(a),(d), where there is no coupling to the third dimension ($\beta_c = 0$), then pictures such as those in Figure 3.17 result. The Fermi surface is most useful in the structural area as a tool for understanding distortions in solids. To understand this the concept of "nesting" of the Fermi surface is needed.

When a section of the Fermi surface can be moved by a vector $\mathbf{q}$ such that it is exactly superimposed on another section of the surface, then the Fermi surface is nested by this vector $\mathbf{q}$. The Fermi surface of Figure 3.16(a) is nested by an infinite set of vectors, two of which are shown in Figure 3.18(a). The surface of Figure 3.18(b) is nested by the vector shown. A useful trick in appreciating some of these nesting possibilities is to draw out more than one zone as shown for this case.

A more complex example is shown in Figure 3.19. Here in (a) is shown the

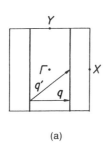

(a)

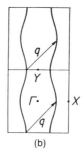

(b)

Figure 3.18 (a) Two of the infinite set of vectors that nest the Fermi surface of Figure 3.14(a). (b) The nesting vector for the Fermi surface of Figure 3.14(b). Notice here that the nesting is only clearly seen if the picture of Figure 3.14(b) is doubled in size. (Adapted from Ref. 82.)

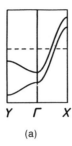
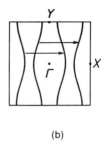

(a)   (b)

Figure 3.19 The effect on the Fermi surface (b) of the dispersion of the two bands in (a). (Adapted from Ref. 82.)

dispersion behavior of two bands of a two-dimensional structure. The Fermi surface is shown in Fig. 3.19(b), where the inner two pieces come from the lower-energy band and the outer pieces from the higher-energy band. Notice, however, that although there are two distinct pieces of the Fermi surface that are nested, the nesting vector is identical for each.

The utility of such a description lies in insights provided into electronically driven geometrical instabilities. The potential associated with the distortion may be written as

$$V = V_q \exp(i\,\mathbf{q}\cdot\mathbf{r}) + V_{-q}\exp(-i\,\mathbf{q}\cdot\mathbf{r}) \tag{3.17}$$

where $\mathbf{q}$ is a reciprocal lattice vector. The nature of $\mathbf{q}$ defines the way the structure changes. For example, in the simple one-dimensional case, if $\mathbf{q} = \mathbf{b}/2$, then a distortion that leads to a doubling of the unit cell in this direction is indicated. The wave functions at two points in $\mathbf{k}$ space, $\mathbf{k}$ and $\mathbf{k}'$, orthogonal when this distortion is absent, can now interact with one another. In general, from second-order perturbation theory the energy of interaction ΔE will be given by

$$\Delta E = \frac{|\int \psi_k^* \psi_k V\,d\tau|^2}{E(\mathbf{k}) - E(\mathbf{k}')}$$
$$\sim \frac{|\int \exp[i(\mathbf{k} - \mathbf{k}')\cdot\mathbf{r}]V\,d\tau^2|}{E(\mathbf{k}) - E(\mathbf{k}')} \tag{3.18}$$

The numerator here is zero unless $\mathbf{k} - \mathbf{k}' = \mathbf{q}$, which provides a stringent restriction on the states that may mix together. The interaction energy increases in magnitude as $E(\mathbf{k}) - E(\mathbf{k}')$ decreases. When $E(\mathbf{k}) = E(\mathbf{k}')$, then this treatment is inappropriate and degenerate perturbation theory has to be used. The interaction energy is largest for this case. Several orbital situations may be envisaged and are collected in Figure 3.20. In Figure 3.20(a) there is no change in total energy since both levels $\psi(\mathbf{k})$, $\psi(\mathbf{k}')$ are empty. In (b), if overlap is ignored, then the total energy change is zero, but if overlap is included, then the result is a destabilizing one, just as for the case of the He$_2$ configuration in Figure 1.1(b). Case (c) gives rise to a small distortion since $E(\mathbf{k})$ and $E(\mathbf{k}')$ are not equal, but case (d) gives rise to the strongest stabilization. This is clearly the case when $\mathbf{q}$ connects two degenerate levels, one empty and one full at the Fermi surface. The result of the interaction is to generate a gap between filled and

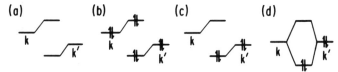

Figure 3.20 Mixing possibilities for bands at different **k** as a result of a perturbation associated with a vector **q**. (a) The two levels that are coupled (at **k** and **k'**) are both empty; (b) they are both full; (c) the lower energy level is full and the upper one empty; (d) as for (c), except that the two levels are degenerate; that is, both lie on the Fermi surface.

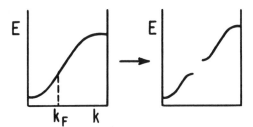

Figure 3.21 The general result of opening a gap as a result of the distortion **q**.

empty states that are initially degenerate just as described in a different way in Section 2.4. Figure 3.21 shows the result. Notice that the Fermi surface has disappeared in the distorted case.

In two or three dimensions not all of the Fermi surface may be nested, and so some of the levels will not mix in the way described previously. In this case some of the Fermi surface may be left behind as pockets, as shown in Figure 3.22(a)–(d). In this case the distorted material is still metallic, but not quite as metallic as before,

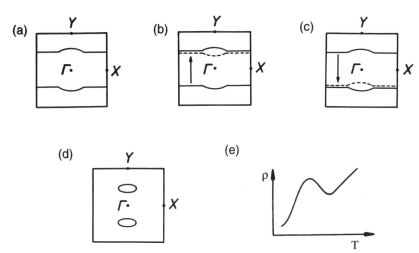

Figure 3.22 The effect of partial nesting (b), (c) of the Fermi surface of (a) to give the small closed loops in the Fermi surface after the distortion **q**. (e) The expected behavior of the resistivity with temperature. The slope at high temperature represents the transport properties of the Fermi surface (a); the (larger) slope at low temperature is associated with conduction via the surface in (d); the intermediate behavior associated with the freezing in of the charge-density wave (CDW). (Adapted from Ref. 82.)

since there are fewer electrons at the Fermi level. An example of this is TaS_2, metallic before and after the distortion. Experimentally the resistivity behavior with temperature that is found often looks like that shown in Figure 3.22(e). Below a critical temperature where the distortion occurs the slope is steeper (less metallic) than above this temperature.

The Peierls distortion of linear chain systems such as polyacetylene may be rephrased in this new language. Here from **3.12** the half-filled energy band leads to $\mathbf{q} = \mathbf{k}_F - (-\mathbf{k}_F) = 2\mathbf{k}_F$. [This Peierls distortion is thus sometimes called a $2\mathbf{k}_F$ distortion. Analogously the distortion of Fig. 2.25(c), which leads to a dimerization even though the band is only one-quarter full, is sometimes called a $4\mathbf{k}_F$ distortion.] In general for a one-dimensional system described by a unit cell of length a, if $\mathbf{k}_F = (1/q)\mathbf{b}$, then the new unit cell will be $2q$ times the old. Polyacetylene, where $1/q = \frac{1}{4}$, is stabilized by a distortion that leads to alternating short and long C–C bonds and thus a doubling of the cell length. The distortion does not necessarily need to be commensurate with the lattice and indeed incommensurate charge density waves (CDW) are found in a variety of materials. Imagine, for example, the one-dimensional chain but one where the band filling is not a regular fraction. In this case q will not be an integer. Thus $K_2Pt(CN)_4Br_{0.3}$ is a metal above 150 K. Below this temperature a static CDW is generated and a band gap opens up. Since the highest occupied (z^2) energy band is now 1.7/2.0 full, $\mathbf{k}_F = 0.425\mathbf{b}$ and thus $2\mathbf{k}_F = \mathbf{q} = 0.85\mathbf{b}$. In Section 2.4 it was shown how the shorter-wavelength distortion led in general to a larger stabilization energy. From **3.13** a shorter-wavelength distortion is given by $\mathbf{q}' = 0.15\mathbf{b}$,

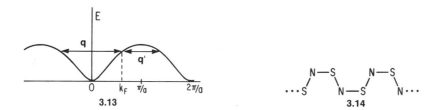

3.13   3.14

or a superlattice with a repeat $1/0.15 = 6.6$ times that expected for the undoped parent.

These nesting ideas suggest a strategy to stabilize an ostensibly one-dimensional system against the CDW distortion predicted from its band filling. If the dimensionality of the material may be increased such that its Fermi surface is converted from that of Figure 3.16(a) to Fig. 3.14(b), then the nesting properties are substantially degraded, and the driving force for the distortion reduced. Electronically this arises from interactions between the chains (Ref. 372). $(SN)_x$ is an example (Ref. 203) (**3.14**) of a "one-dimensional" system, predicted to be distorted from electron count considerations, but stabilized by interchain interactions. Figure 3.23 shows this idea from a different viewpoint. Note how the bands of a half-filled single chain split apart into two when a pair of chains is brought together. The new bands are, respectively, bonding and antibonding between the two. At the simplest level of approximation the Fermi level does not move, but the fractional orbital occupancies of the two bands are now different from 0.5.

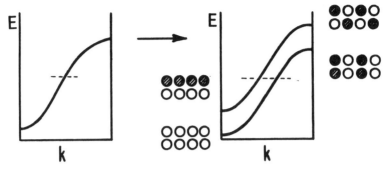

Figure 3.23 The effect on the band structure of a one-dimensional chain when interchain interactions are included. Now k_F is different for the two bands.

3.3 Symmetry considerations

Just as in molecular chemistry (Refs. 100, 209), symmetry considerations determine many of the details of the energy bands of solids. In molecules the point symmetry of the atomic skeleton, determined by the location of the nuclei, sets the allowed symmetry species of the molecular orbitals using a specified atomic orbital basis. Such group-theoretical ideas were used to derive the orbitals of the cyclic polyenes of Section 1.5, the cyclic group of order n (C_n) being used to generate the basic form of the wave functions. In Section 2.1 the analogy was drawn (Ref. 43) between the infinite cyclic group and the translation group (they are isomorphic) in determining the form of the crystal orbitals of the infinite solid. In the cyclic polyenes the actual point symmetry is higher (D_{nh}) than C_n, and indeed in most solids (and especially those that form the basis for textbook examples) there are extra symmetry elements also present that are of importance in determining the band structure. How combinations of point symmetry elements, translations, glide planes (**3.15** shows a

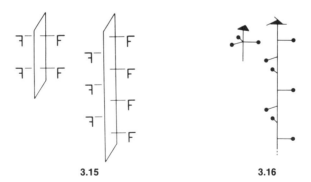

3.15   **3.16**

mirror plane at left and a glide plane at right) and screw axes (**3.16** shows an ordinary rotation axis at left and a screw axis at right) lead to the 230 space groups is not a task for this book, but discussions appear in Refs. 11 and 77. Here the assumption will be made that the reader understands the general ideas behind the theory of group representations as described, for example, in Ref. 100.

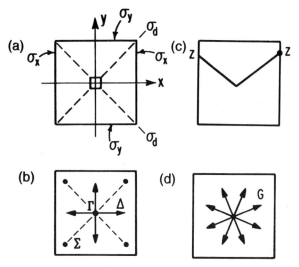

Figure 3.24 (a) The symmetry elements of the square lattice, (b) the vectors at Γ and those that define the lines Σ and Δ, (c) the vector defining the point Z, (d) the vector defining the general point.

Not only, therefore, do solids possess translational symmetry, but also point symmetry elements. Thus it is clear that the square lattice of Figure 3.24(a) itself possesses the symmetry elements $\{E, C_4, C_4^3, C_2, 2\sigma_v, 2\sigma_d\}$, a set that is appropriate for the C_{4v} point group. The $\sigma_v(i)$ plane is the one perpendicular to the ith axis. [In fact, the symmetry of the lattice is higher. There are other symmetry elements, including the plane of the page (σ_h), that are present that lead to the D_{4h} group, but the C_{4v} group will be a sufficient place to start.] Since the label $\mathbf{k}$ used to describe the energy levels of the solid is defined in reciprocal space, it is important to consider the symmetry of the reciprocal lattice. This step is a very simple one. It is easy to show (Refs. 77, 219) that if an operation $\mathscr{R}$ is a symmetry operation of the direct lattice, then its inverse $\mathscr{R}^{-1}$ is a symmetry operation of the reciprocal lattice. Since both $\mathscr{R}$ and $\mathscr{R}^{-1}$ are contained in the symmetry group of the direct lattice (one of the rules of group assembly), then the groups of the two lattices are the same. However, important in the solid are the symmetry properties, not only of the lattice itself, but also the properties of the vector $\mathbf{k}$, called the group of $\mathbf{k}$. For example, take the vector shown in Fig. 3.24(b), which ends at a point labeled Δ. The results of operation of eight elements of the group on $\mathbf{k}$ lead to the four vectors shown. This collection of vectors is called the star of $\mathbf{k}$. Only for the pair of operations $\{E, \sigma_v(y)\}$ is the vector transformed into itself; that is, $\mathscr{R}\mathbf{k} = \mathbf{k}$. Thus this pair of symmetry elements defines the group of Δ. (The group is isomorphic with C_s.) Similarly the group of Σ [Fig. 3.24(b)] is defined by the symmetry elements $\{E, \sigma_d(xy)\}$. This group is also isomorphic with C_s.) The group of Z [Fig. 3.24(c)] shows a new feature. As a result of operation with $\sigma_v(x)$, the vector is shifted from the right to the left boundary of the lattice. However, these two points are translationally equivalent. A translation of the reciprocal lattice vector along x will bring the two points into coincidence. The group of Z is then defined by the elements $\{E, \sigma_v(x)\}$. It too is isomorphic with C_s. These three lines mapped out by the reciprocal lattice vector are called *special lines* since there is a symmetry element other than E that transforms the vector into

Table 3.1 Character tables for some groups of **k**

Γ, M	E	C_2	$2C_4$	$2\sigma_v$	$2\sigma_d$
Γ_1, M_1	1	1	1	1	1
Γ_2, M_2	1	1	1	-1	-1
Γ_3, M_3	1	-1	-1	1	-1
Γ_4, M_4	1	1	-1	-1	1
Γ_5, M_5	2	-2	0	0	0

					Δ	E	σ^y
					Σ	E	σ_d
X	E	C_2	σ^y	σ^x	Z	E	σ^x
X_1	1	1	1	1	Δ_1, Σ_1, Z_1	1	1
X_2	1	1	-1	-1	Δ_2, Σ_2, Z_2	1	-1
X_3	1	-1	1	-1			
X_4	1	-1	-1	1			

itself. The vector shown in Figure 3.24(d) that describes the point G does not lie on any symmetry elements and is thus described by the group containing $\{E\}$ alone. It is called a *general point*.

The points $\Gamma(0,0)$, $X(\frac{1}{2},0)$, and $M(\frac{1}{2},\frac{1}{2})$ are special or symmetry points because they possess symmetry higher than just $\{E\}$. For example, at the X point the C_2 or $\sigma_v(x)$ operation takes the vector at $(\pi/a,0)$ to that at $(-\pi/a,0)$, two points that are translationally equivalent. It is transformed into itself by both E and $\sigma_v(y)$. The result is a group, $\{E, C_2, \sigma_v(x), \sigma_v(y)\}$, isomorphic with C_{2v}. Similar considerations apply at the point Γ that always has the full (holohedral) symmetry of the lattice. The character tables for these groups of **k** are shown in Table 3.1. In the molecular point groups the symmetry labels devised by Mulliken contain information concerning degeneracies (e.g., E or A_1) and symmetry properties with respect to important operations of the group (E_g, E_u or A_2', A_2''). In the solid state the labels used are not informative at all, and the irreducible representations are just numbered consecutively. However, the Q_1 irreducible representation ($Q = \Gamma, X, M, \Delta, \Sigma, Z, G$, etc.) is always totally symmetric.

In molecular problems a question of frequent interest is how the symmetry species of a function under consideration changes during a descent in symmetry. One example is the splitting (Ref. 147) of an atomic ^3F state of an ion (point group K_h) on insertion into an octahedral hole (point group O_h) in a crystal. In the solid such a descent in symmetry occurs on moving away from the symmetry points along the symmetry lines or to a general **k** point. In particular at the points Γ and M degenerate representations are possible (Γ_5 and M_5) that are not possible anywhere else in the zone. No degeneracies (other than those associated with spin) are possible at the general point G, but accidental degeneracies frequently occur, such as at band crossings in one-dimensional systems, for example. One such crossing

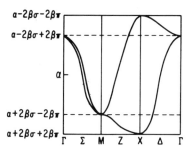

Figure 3.25   The dispersion picture for the x, y orbitals located at the nodes of a square net lying in the xy plane.

is seen in Figure 2.6 and involves the xy and z^2 bands of different symmetry at the cross point.

The behavior of the $p_{x,y}$ orbitals of the square lattice is a nice example showing some of the symmetry consequences of the groups of $\mathbf{k}$. The x orbital (for example) is involved in σ interactions along x, and π interactions along y (**3.17**). The dispersion

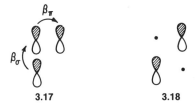

3.17   **3.18**

properties of the two bands, if nearest-neighbor interactions only are included, are simply given by

$$E(\mathbf{k}) = \alpha - 2\beta_\sigma \cos(k_x a) + 2\beta_\pi \cos(k_y a), \qquad \text{for } x \qquad (3.19)$$

$$E(\mathbf{k}) = \alpha - 2\beta_\sigma \cos(k_y a) + 2\beta_\pi \cos(k_x a), \qquad \text{for } y \qquad (3.20)$$

A dispersion picture is presented in Figure 3.25 and shows that the two are degenerate

at Γ and M $[E = \alpha \pm (-2\beta_\sigma + 2\beta_\pi)]$ but not at any other points. This is easy to see from the symmetry properties of the orbitals. The pair transform in the following way under the groups of $\mathbf{k}$.

	Γ	M	X	Δ	Σ	Z	
x, y	Γ_5	M_5	$X_3 + X_4$	$\Delta_1 + \Delta_2$	$\Sigma_1 + \Sigma_2$	$Z_1 + Z_2$	(3.21)

Notice how the degeneracy is split at the lower-symmetry points X, Δ, and Z. Along the Σ line symmetry arguments suggest that the degeneracy should also be lifted. In fact, the two bands are "accidentally" degenerate as the result of the choice of the basis set and inclusion of nearest neighbors only. Inclusion of the next-nearest-neighbor interaction shown in **3.18** leads to off-diagonal elements that connect the

x and y orbitals. The diagonal terms become

$$E(\mathbf{k}) = \alpha - 2\beta_\sigma \cos(k_x a) + 2\beta_\pi \cos(k_y a) - 2\beta' \cos[(k_x + k_y)a] - 2\beta' \cos[(k_x - k_y)a],$$

$$\text{for } x \quad (3.22)$$

$$E(\mathbf{k}) = \alpha - 2\beta_\sigma \cos(k_y a) + 2\beta_\pi \cos(k_x a) - 2\beta' \cos[(k_x + k_y)a] - 2\beta' \cos[(k_x - k_y)a],$$

$$\text{for } y \quad (3.23)$$

The off-diagonal terms, zero in the nearest-neighbor model, are now

$$-2\beta'' \cos[(k_x + k_y)a] + 2\beta'' \cos[(k_x - k_y)a] \quad (3.24)$$

These are zero at Γ and M but nonzero along the Σ line. The degeneracy is thus lifted here by these (smaller) second-nearest-neighbor interactions.

Similar considerations apply to the one-dimensional chain that formed the basis for much of the introductory material of Chapter 2. At the points Γ $(0,(2\pi/a))$ and X $(\frac{1}{2},(2\pi/a))$ the group of $\mathbf{k}$ contains the elements $\{E,i\}$, but for any other $\mathbf{k}$ the group contains $\{E\}$ only. The group that contains $\{E,i\}$ is isomorphic with the set of groups of order two of Table 3.1, and leads to only two different symmetry species, one symmetric (Γ_1) and one antisymmetric (Γ_2) with respect to inversion. The group of G, the general point, which contains $\{E\}$ alone, only contains a single irreducible representation (G_1). Thus at the symmetry points Γ and X valence s and p_x orbitals transform as $\Gamma_1 + \Gamma_2$ and $X_1 + X_2$, respectively, but as $2G_1$ at general points in between. The important result, seen clearly in Figure 2.9, is that s–p mixing can occur only at points away from the two symmetry points. At both zone center and zone edge the crystal orbitals are either pure s or pure p in character.

The problems described so far involve symmorphic space groups. These are groups that do not contain screw axes and glide planes (or where such symmetry elements may be removed by a different choice of origin). These are symmetry elements that are composite operations involving a point symmetry operation (a rotation or reflection, respectively) along with a translation (**3.15, 3.16**). Whereas it is relatively easy to identify the group of $\mathbf{k}$ for symmorphic groups (usually by inspection as for the square lattice above), those for nonsymmorphic groups at the zone boundaries are invariable less readily accessible. Indeed the character tables are often worked out on a case-by-case basis (although a tabulation does exist, Ref. 30). Of the 230 space groups, unfortunately the majority, 157, are nonsymmorphic. As an example, perhaps the simplest that could be imagined, are the groups of $\mathbf{k}$ for the doubled one-dimensional call of **2.13**. Notice that a degeneracy occurs at the zone edge in the electronic description of this problem in Figure 2.19, and that this certainly cannot be understood by using the group of order two employed for the one-dimensional chain already discussed.

The group elements for the doubled cell are readily seen to be $\{e|na\}$, $\{i|2na\}$, $\{e|(2n+1)a\}$, $\{i|(2n+1)a\}$, where $n = 0, \pm 1, \pm 2, \ldots$, etc. This useful Seitz notation specifies a point symmetry operation followed by a translation, $\{\text{point operation}|\text{translation}\}$. The operation $\{e|(2n+1)a\}$ is equivalent to a 2_1 screw axis or glide plane. The distinction is lost in this one-dimensional example.

Table 3.2 Multiplication table

	$\{e\|0\}$	$\{e\|R\}$	$\{e\|\tau\}$	$\{e\|\tau'\}$	$\{i\|0\}$	$\{i\|R\}$	$\{i\|\tau\}$	$\{i\|\tau'\}$
$\{e\|0\}$	$\{e\|0\}$	$\{e\|R\}$	$\{e\|\tau\}$	$\{e\|\tau'\}$	$\{i\|0\}$	$\{i\|R\}$	$\{i\|\tau\}$	$\{i\|\tau'\}$
$\{e\|R\}$	$\{e\|R\}$	$\{e\|0\}$	$\{e\|\tau'\}$	$\{e\|\tau\}$	$\{i\|R\}$	$\{i\|0\}$	$\{i\|\tau'\}$	$\{i\|\tau\}$
$\{e\|\tau\}$	$\{e\|\tau\}$	$\{e\|\tau'\}$	$\{e\|R\}$	$\{e\|0\}$	$\{i\|\tau\}$	$\{i\|\tau'\}$	$\{i\|R\}$	$\{i\|0\}$
$\{e\|\tau'\}$	$\{e\|\tau'\}$	$\{e\|\tau\}$	$\{e\|0\}$	$\{e\|R\}$	$\{i\|\tau'\}$	$\{i\|\tau\}$	$\{i\|0\}$	$\{i\|R\}$
$\{i\|0\}$	$\{i\|0\}$	$\{i\|R\}$	$\{i\|\tau'\}$	$\{i\|\tau\}$	$\{e\|0\}$	$\{e\|R\}$	$\{e\|\tau'\}$	$\{e\|\tau\}$
$\{i\|R\}$	$\{i\|R\}$	$\{i\|0\}$	$\{i\|\tau\}$	$\{i\|\tau'\}$	$\{e\|R\}$	$\{e\|0\}$	$\{e\|\tau\}$	$\{e\|\tau'\}$
$\{i\|\tau\}$	$\{i\|\tau\}$	$\{i\|\tau'\}$	$\{i\|0\}$	$\{i\|R\}$	$\{e\|\tau\}$	$\{e\|\tau'\}$	$\{e\|0\}$	$\{e\|R\}$
$\{i\|\tau'\}$	$\{i\|\tau'\}$	$\{i\|\tau\}$	$\{i\|R\}$	$\{i\|0\}$	$\{e\|\tau'\}$	$\{e\|\tau\}$	$\{e\|R\}$	$\{e\|0\}$

The multiplication table for this group is shown in Table 3.2. Each element has an inverse. These are

$$\{e|0\}^{-1} = \{e|0\} \qquad \{e|R\}^{-1} = \{e|R\} \qquad \{e|\tau\}^{-1} = \{e|\tau'\} \qquad \{e|\tau'\}^{-1} = \{e|\tau\}$$

$$\{i|0\}^{-1} = \{i|0\} \qquad \{i|R\}^{-1} = \{i|R\} \qquad \{i|\tau\}^{-1} = \{i|\tau\} \qquad \{i|\tau'\}^{-1} = \{i|\sigma'\}$$

Here $R = 2a$, $\tau = a$, $\tau' = \tau + R$. The division of the elements into classes proceeds in the normal way. As usual, $\mathscr{A}$ and $\mathscr{B}$ belong to the same conjugacy class if, for all elements of the group, $\mathscr{C}$, $\mathscr{C}^{-1}\mathscr{A}\mathscr{C} = \mathscr{A}$ or $\mathscr{B}$. The element $\{e|0\}$ clearly forms a class on its own, as does $\{e|R\}$. The other elements pair up into three classes, $[\{i|0\}, \{i|R\}]$, $[\{e|\tau\}, \{e|\tau'\}]$, and $[\{i|\tau\}, \{i|\tau'\}]$. This class structure determines the form of the character table. Recall that the number of classes equals the number of different irreducible representations and that the sum of the square of their dimensions, g_i^2, is equal to the order of the group, g. The table that results is shown in Table 3.3. Interestingly it is isomorphic with C_{4v}, but there is no fourfold axis to be seen in the geometrical structure. The most important result for the discussion here is that doubly degenerate representations are possible. Symmetry arguments are even more restrictive than this. Since the wave function must change sign as a result of a translation by R at the point $\frac{1}{2}(2\pi/a)\mathbf{b}$, then the only allowed value of the character, $\chi_i(\{e|R\})$ at this point is $(-1)g_i$, where g_i is the dimension of the ith irreducible representation. The structure of the character table thus requires all functions at this point to

Table 3.3 Character table for the group of $(\frac{1}{2})(2\pi/a)\mathbf{b}_1$

	$\{e\|0\}$	$\{e\|R\}$	$[\{i\|0\}, \{i\|R\}]$	$[\{e\|\tau\}, \{e\|\tau'\}]$	$[\{i\|\tau\}, \{i\|\tau'\}]$
Γ_1	1	1	1	1	1
Γ_2	1	1	1	−1	−1
Γ_3	1	1	−1	1	−1
Γ_4	1	1	−1	−1	1
Γ_5	2	−2	0	0	0

transform as Γ_5 (i.e., to be doubly degenerate). Analogously none of the functions at the zone center, $(0)(2\pi/a)\mathbf{b}$, may be doubly degenerate since here the phase factor is $+1$. In fact, at the zone center the two orbitals have a set of characters

$$\{e|0\} \quad \{e|R\} \quad [\{i|0\}, \{i|R\}] \quad [\{e|\tau\}, \{e|\tau'\}] \quad [\{i|\tau\}, \{i|\tau'\}]$$
$$2 \qquad 2 \qquad\quad 0 \qquad\qquad\quad 0 \qquad\qquad\quad 2$$

which reduce to $\Gamma_1 + \Gamma_4$ and thus the two wave functions shown in Figure 2.21(c), (d).

4 □ The electronic structure of solids

This chapter builds on the results of Chapters 2 and 3 and covers a wide span of bonding types. Pauling (Ref. 275 (1929) summarized early ideas (Ref. 31) concerning the structures of (largely) oxides with his rules based mainly on electrostatic arguments. Hume-Rothery (Ref. 189) showed that electron count was an important consideration and apparently determined the structures of series of metal alloys, which he correspondingly called "electron compounds." The ideas of Lewis (1916) have long been used very effectively in this area (Ref. 274). The structure of many solids are readily understandable at the simplest level because they are effectively giant molecules held together by the two-center, two-electron bond familiar to molecular, and especially organic, chemists. The Zintl phases are examples here, their structural integrity being understood in terms of directional bonding to give arrangements analogous to those found as three-dimensional frameworks (e.g., diamond), two-dimensional sheets (e.g., α-arsenic), slabs (e.g., GaS), or individual molecules (e.g., the halogens) typical of such electronic pictures. A final group of solids, which frequently have structures difficult to understand in detail, are simply condensed collections of gas-phase molecules arranged to give the best packing.

Just described are, of course, the four traditional classes (Refs. 208, 352) of solids, ionic, metallic, covalent, and van der Waals classified by their bonding type. While these may be useful very broad classifications, they should not be taken too dogmatically. There are, for example, whole series of organic solids that are metallic (Ref. 389). The "typically ionic" compound TiO_2 becomes metallic after treatment with a small amount of t-BuLi. The high-temperature superconducting cuprates (Ref. 353) are ceramics, and the observation of metallic behavior in such materials came as a suprise to many. Solid H_2 almost certainly becomes a metal under pressure (Ref. 232), and elemental Ca almost becomes (Ref. 32) an insulator. This chapter will contain discussions of several examples from the solid-state area and will build on the concepts developed earlier. Importantly, the orbital model (Ref. 5) will be seen to be broadly applicable to most of them.

4.1   Oxides with the NaCl, TiO_2, and MoO_2 structures

Many transition metal oxides, nitrides, and carbides are found (Ref. 94, 366) in the rocksalt structure. Shown in Figure 4.1, the structure contains both metal and nonmetal in octahedral six-coordination. It may be regarded as a cubic eutactic (often described as close packed, see Chapter 6) array of anions with the cations occupying

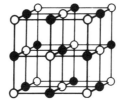

Figure 4.1 The rocksalt (NaCl) structure.

(a) (b)

(c) (d)

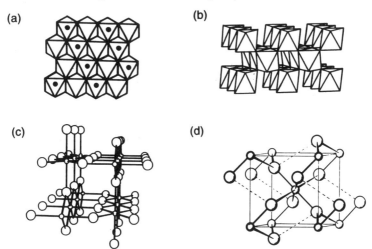

Figure 4.2 The rutile (TiO_2) structure. (a) The relationship to the eutactic structure; (b) the polyhedral connections; (c) the chains of octahedra that share edges; (d) the unit cell. (Reproduced with permission from Ref. 280.)

all the octahedral holes. Another description is as a packing where each MX_6 octahedron shares all of its edges with its neighbors (Ref. 94). The rutile structure shown in several views in Figure 4.2 is found for a series of metal oxides and fluorides. Here one-half of the octahedral holes of a hexagonal eutactic array of anions are filled with cations. A closely related structure is that of MoO_2. As Figure 4.3 shows, it differs from the rutile structure in that the metal atoms form dimers. Both of the structures are distorted away from the ideal structures just described. In the case of TiO_2 the oxygen atom becomes threefold coordinated by metal in a planar geometry. This is a question studied in more detail in Chapter 7.

The energy density of states for a transition metal oxide in the rocksalt structure using the band model is shown (Ref. 56) in Figure 4.4(a). Understanding how his picture originates is best tackled by first using a σ-only model that includes nearest neighbors only. Both metal and oxygen atoms lie in an octahedral environment. Thus

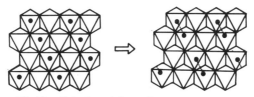

Figure 4.3 The distortion of the rutile structure to that of MoO_2.

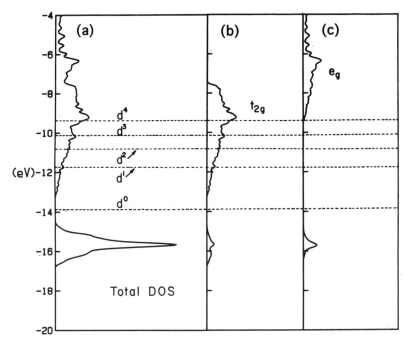

Figure 4.4 (a) The total density of states of TiO in the rocksalt structure; (b) the partial density of states associated with the d orbitals derived from the t_{2g} set of the octahedron; (c) the partial density of states associated with the d orbitals derived from the e_g set of the octahedron. Lines indicating the Fermi level for various d counts are shown.

both species may use $a_{1g}(2s$ or $(n+1)s)$ and t_{1u} ($2p$ or $(n+1)p$) orbitals for σ bonding to nearest-neighbor atoms. In addition, the metal has nd $e_g(x^2, x^2 - y^2)$ orbitals available for σ bonding. Energetically the largest interactions are between the valence s,p orbitals on the two centers, with smaller interactions between the metal d orbitals and the orbitals on the nonmetal. Using this model the transition metal t_{2g} (xz, xy, yz) orbitals remain M–O σ nonbonding. By including interactions between the next-nearest-neighbor metal atoms, however, these orbitals become involved in metal–metal bonding. One of the interactions involved is shown in **4.1**

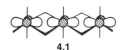

4.1

and is important since the octahedra share edges in this structure. The metal–metal distance $r(M–M)$ is just $\sqrt{2}r(M–O)$ and, for the case where the anion is from the first row, these $M–M$ distances can be quite close. The t_{2g} orbitals generate energy bands via this overlap, which will have metal–metal bonding character at the bottom and be metal–metal antibonding at the top of the band. In addition to these metal–metal interactions the t_{2g} levels are involved in π interactions with filled oxygen donor levels of the correct symmetry. Thus the band structure of the system is expected to be built up as in Figure 4.5.

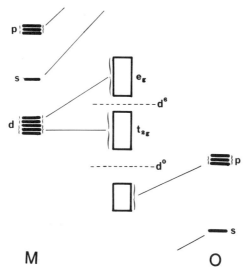

Figure 4.5 Derivation of the band structure of a metal oxide in the rocksalt structure from the orbitals of the octahedrally coordinated metal.

Figure 4.4(a) shows the density of states (DOS) computed (Ref. 56) for rocksalt TiO. (TiO does not, in fact, have the rocksalt structure but an interesting defective arrangement that is discussed in Section 4.4.) The pictures are generated using the method shown in Figure 2.6. Not only can the energy density of states be calculated via this route, but also the overlap populations and the orbital character of the energy bands. In addition, therefore, the total density of states may be decomposed into contributions from any set of orbitals, just as in the case of $B_3N_3H_6$ in Figure 1.9(c). The deepest-lying peak in these plots is from the oxygen p orbitals, and the levels that lie above -14 eV are primarily d in character. In Figures 4.4(b) and (c) are shown, respectively, the t_{2g} and e_g contributions to this band. Notice that the two bands in this region overlap each other. For this particular system then, the band structure may be visualized as in **4.2a**. The case where a gap exists between the e_g

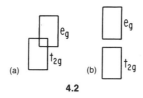

4.2

and t_{2g} bands can readily be imagined (**4.2b**), leading to semiconducting behavior for the d^6 configuration. Which case is found depends on the relative sizes of the M–O σ,π and M–M interactions. It is worth pointing out that the e_g–t_{2g} splitting arises through orbital interactions between the Ti $3d$ and oxygen atom orbitals on this model, just as in molecular transition metal complexes.

This σ-only model suggests that the e_g orbitals are destabilized by M–O nearest-neighbor σ interactions and that the width of this band comes from the extended interactions between metal and oxygen. Analogously, although the t_{2g} band derives

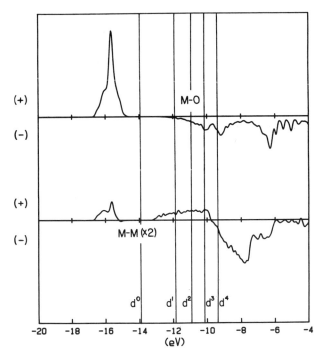

Figure 4.6 Crystal orbital overlap population (COOP) plot for the Ti–O and Ti–Ti bonds in rocksalt TiO. The Ti–Ti plot is scaled by a factor of two.

some of its width via metal–metal interactions, the metal t_{2g} levels are also destabilized by metal–oxygen π bonding, as described previously. The complete bonding picture for these rocksalt oxides is usefully shown in the crystal orbital overlap population (COOP) curves of Figure 4.6. Recall from the discussion of Section 2.1 that when the COOP curve is positive, the orbitals are bonding between a given pair of atoms, and where it is negative, they are antibonding. Notice that the overall Ti–O "bond strength" is best for the d^0 system since here all Ti–O bonding orbitals are occupied. As the d manifold becomes occupied, and especially beyond the d^1 electron count, it decreases since Ti–O antibonding orbitals are now being populated. The COOP curve appropriate for Ti–Ti interactions is positive for electron counts up to a little above d^3, the half-filled point, where all the metal–metal bonding orbitals are filled. However, by comparison with some strongly metal–metal bonded systems, the magnitude of the Ti–Ti overlap population metal–metal bonding is not very large, even at the optimal electron count. For electron counts above d^1 the overlap population results imply that metal–metal bonding is only moderately strengthened but with the cost of filling orbitals that are Ti–O d–$p\pi$ antibonding.

The density of states picture (Ref. 44) for metal oxides in the TiO_2 structure (Fig. 4.7) is quite similar in a general way to that for rocksalt metal oxides. This comes about since the energetic location of the basis orbitals is similar, set by the energies of the atomic orbitals, metal, and oxygen, and by the octahedral metal coordination. Notice TiO_2 itself (d^0) has a gap between filled and empty bands (3.0 eV from experiment for rutile and 3.2 eV for its anatase variant) and that there is also

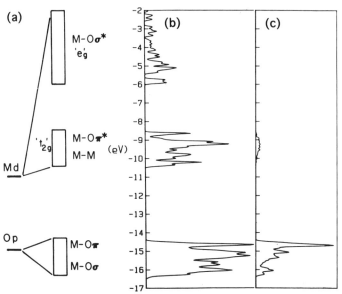

Figure 4.7 Assembly of the band structure of TiO_2. (a) The structure expected from consideration of the local geometry. (b) The calculated density of states. (c) The projection of the p orbital that lies perpendicular to the OTi_3 plane.

a gap between e_g and t_{2g} bands. Accordingly rutile (when pure and stoichiometric) is colorless and an insulator. The picture of Figure 4.7(a) shows that the t_{2g} band derives its width from both $M–M$ and $M–O$ π interactions. In the second panel of Figure 4.7(b) the projection of the oxygen p_z orbital that lies perpendicular to the OTi_3 plane and is involved purely in π interactions is quite wide. In contrast to the situation in rocksalt, in TiO_2 the octahedra are fused only along one direction. This is the direction where the metal atoms pair up in the MoO_2 structure found for VO_2, MoO_2, WO_2, α-ReO_2, and TcO_2. NbO_2 has a related but more complex arrangement.

The effect of the distortion $TiO_2 \rightarrow MoO_2$ on the calculated density of states is shown (Ref. 44) in Figure 4.8(a). Notice a substantially wider band than in the undistorted parent. Its origin is simply uncovered. Figure 4.8(b) shows the decomposition into σ-, π-, and δ-type interactions with respect to the edge-sharing chains. The band gains its width via $M–O$ π interactions, but notice the way the σ and π components split into two on distortion. This may be envisaged as a variant of the Peierls distortion described in Section 2.4 for the MH_4 chain. Roughly speaking, VO_2 (d^1) has a configuration that fills the lower σ band, and MoO_2 (d^2) one that fills both lower σ and π bands. In ReO_2 (d^3) there is some population of the δ band. All of these examples are found in the distorted structure, at least at low enough temperature. (For VO_2, for example, the transition occurs at 340 K.) Interestingly, RuO_2 (d^4) is metallic and has the undistorted rutile structure. In numerical terms some results from tight-binding calculations on TiO_2 itself are shown (Ref. 44) in Figure 4.9. These are "rigid-band" calculations, ones where the behavior of systems other than the one directly under consideration are obtained by adding to or removing electrons from it. No change is made in orbital or geometrical parameters. It is important to see, irrespective of the choice of the geometrical

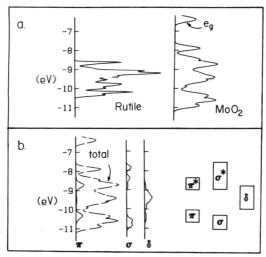

Figure 4.8 (a) Behavior of the "t_{2g}" bands of rutile on distortion to the MoO_2 structure. (b) Decomposition into σ, π, and δ contributions.

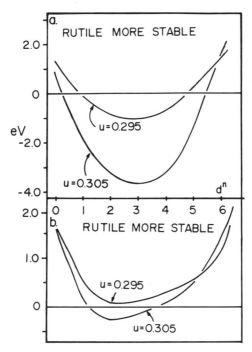

Figure 4.9 Calculated energy differences between the rutile and MoO_2 structures for TiO_2 for two different values of the u parameter, which defines the oxygen atom locations in the structure. (a) and (b) The results of calculations with orbital parameters appropriate for Ti and Fe, respectively.

95

parameters, that the d^0 configuration is resistant to distortion. One way to view this is as the result of the elastic forces of the underlying structure described by Eq. (5.1). Note that such a rigid-band calculation gives the wrong result for RuO_2. The general shape of the energy difference curve as a function of d count shows the importance of the generation of stronger metal–metal bonds on distortion.

4.2 The diamond and zincblende structures

Diamond is perhaps the quintessential "covalent" solid, and it is instructive actually to write down (Refs. 90, 160, 224) the form of the Hamiltonian matrix, which, when solved at different **k** points, will give the band structure of the material. There are other ways to get a qualitative feel for the electronic structure (Refs. 228, 298, 299), one of which is described in Chapter 5. Zincblende, or sphalerite (the two labels are in common use), is the simplest derivative diamond structure with Zn and S atoms alternately substituted for the carbon atoms of this net. There are many diamond like structures that are more complex derivatives such as chalcopyrite ($CuFeS_2$) and stuffed variants such as those of the MgAgAs type, where the Ag and As atoms occupy the Zn and S sites of zincblende and the Mg atoms the interstices.

The structure of diamond is shown in Figure 4.10(a). It may be regarded as a simple four-connected net, or alternatively as a face-centered cubic packing of carbon atoms with extra carbon atoms occupying one-half of the tetrahedral holes. [This latter description is the usual one for the simplest derivative structure of diamond, zincblende, or sphalerite (ZnS), where the atoms of one type occupy the holes in the lattice formed by atoms of the opposite type.] Alternatively it may be regarded as being composed of two interpenetrating face-centered cubic lattices shifted by $(\frac{1}{4},\frac{1}{4},\frac{1}{4})$. The unit cell as usually drawn is cubic and contains eight atoms. The smallest, primitive unit cell is the rhombohedral one that contains two atoms. This is shown in Figure 4.10(b), where for clarity only the atoms of the face-centered cubic packing are shown. This eight-orbital problem is the most complex one to be tackled analytically in this book. The entries for the Hamiltonian matrix are given (Ref. 90) in Table 4.1 and are constructed in exactly the same way as shown for graphite in Section 3.1. In anticipation of extension to the diamond derivative structure of zincblende, the orbitals of the two atom types are given anion (A) and cation (C)

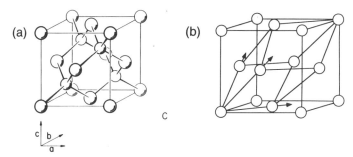

Figure 4.10 (a) The structure of diamond. (Reproduced with permission from Ref. 280.) (b) The rhombohedral cell connecting the atoms of the face-centered cubic lattice.

labels. [The meaning of the interaction parameters are given in Eqs. (4.3) and (4.5).] The Brillouin zone is shown in **4.3**.

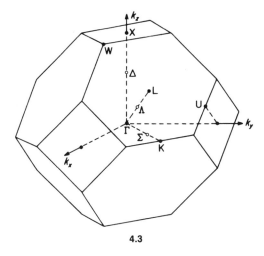

4.3

It is interesting to break the problem down into three parts, the s and p bands by themselves and then the interaction between the two. In the last step the approach will be a more complex (three-dimensional) version of the assembly of the band structure of the sp-bonded chain. The 8×8 matrix of Table 4.1 reduces to 6×6 and 2×2 matrices once $s-p$ interactions are dropped. The s band is then given by a very simple expression

$$E(k) = e_s \pm \beta_{ss} |g_0(k)| \tag{4.1}$$

A dispersion picture and the corresponding density of states is shown (Ref. 90) in Figure 4.11. The p band, more complex, of course, is shown in Figure 4.12 and requires numerical solution. Note the use of symmetry labels to describe the levels. Recall that the labels L_3, Γ_{25}, etc. are just the solid-state analog of the Mulliken symbols

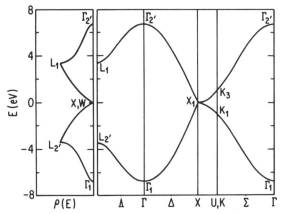

Figure 4.11 The density of states and band structure of the s band of diamond, with no $s-p$ mixing. (Adapted from Ref. 90.)

Table 4.1   Hamiltonian matrix elements for the diamond structure

	$s(C)$	$s(A)$	$p_x(C)$	$p_y(C)$	$p_z(C)$	$p_x(A)$	$p_y(A)$	$p_z(A)$
$s(C)$	$e_s(C)$	$V_{ss}g_0$	0	0	0	$V_{sp}g_1$	$V_{sp}g_2$	$V_{sp}g_3$
$s(A)$	$V_{ss}g_0^*$	$e_s(A)$	$-V_{sp}g_1^*$	$-V_{sp}g_2^*$	$-V_{sp}g_3^*$	0	0	0
$p_x(C)$	0	$-V_{sp}g_1$	$e_p(C)$	0	0	$V_{xx}g_0$	$V_{xy}g_3$	$V_{xy}g_1$
$p_y(C)$	0	$-V_{sp}g_2$	0	$e_p(C)$	0	$V_{xy}g_3$	$V_{xx}g_0$	$V_{xy}g_1$
$p_z(C)$	0	$-V_{sp}g_3$	0	0	$e_p(C)$	$V_{xy}g_1$	$V_{xy}g_2$	$V_{xx}g_0$
$p_x(A)$	$V_{sp}g_1^*$	0	$V_{xx}g_0^*$	$V_{xy}g_3^*$	$V_{xy}g_1^*$	$e_p(A)$	0	0
$p_y(A)$	$V_{sp}g_2^*$	0	$V_{xy}g_3^*$	$V_{xx}g_0^*$	$V_{xy}g_2^*$	0	$e_p(A)$	0
$p_z(A)$	$V_{sp}g_3^*$	0	$V_{xy}g_1^*$	$V_{xy}g_1^*$	$V_{xx}g_0^*$	0	0	$e_p(A)$

$$\left.\begin{aligned}\mathbf{d}_1 &= [111]a/4\\ \mathbf{d}_2 &= [1\bar{1}\bar{1}]a/4\\ \mathbf{d}_3 &= [\bar{1}1\bar{1}]a/4\\ \mathbf{d}_4 &= [\bar{1}\bar{1}1]a/4\end{aligned}\right\}$$

$$\left.\begin{aligned}g_0(\mathbf{k}) &= e^{i\mathbf{k}\cdot\mathbf{d}_1} + e^{i\mathbf{k}\cdot\mathbf{d}_2} + e^{i\mathbf{k}\cdot\mathbf{d}_3} + e^{i\mathbf{k}\cdot\mathbf{d}_4}\\ g_1(\mathbf{k}) &= e^{i\mathbf{k}\cdot\mathbf{d}_1} + e^{i\mathbf{k}\cdot\mathbf{d}_2} - e^{i\mathbf{k}\cdot\mathbf{d}_3} - e^{i\mathbf{k}\cdot\mathbf{d}_4}\\ g_2(\mathbf{k}) &= e^{i\mathbf{k}\cdot\mathbf{d}_1} - e^{i\mathbf{k}\cdot\mathbf{d}_2} + e^{i\mathbf{k}\cdot\mathbf{d}_3} - e^{i\mathbf{k}\cdot\mathbf{d}_4}\\ g_3(\mathbf{k}) &= e^{i\mathbf{k}\cdot\mathbf{d}_1} - e^{i\mathbf{k}\cdot\mathbf{d}_2} - e^{i\mathbf{k}\cdot\mathbf{d}_3} + e^{i\mathbf{k}\cdot\mathbf{d}_4}\end{aligned}\right\}$$

for molecules. The last step, assembly of the complete diagram by combining the s and p interactions, follows along the same lines as that for the assembly of any molecular orbital diagram. Here it is done at some symmetry points in Figure 4.13. The final dispersion diagram is then assembled by connecting levels of the correct symmetry and is shown for the lowest four bands in Figure 4.14. Perhaps the most important s–p interaction has occurred at X, where the strong interaction between

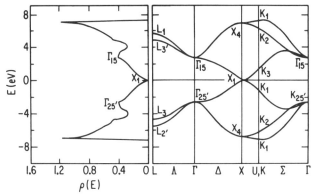

Figure 4.12   The density of states and band structure of the p band of diamond, with no s–p mixing. (Adapted from Ref. 90.)

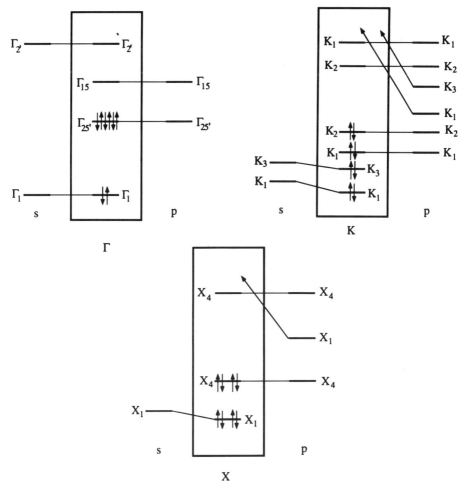

Figure 4.13 Crystal orbital interaction diagrams at three **k** points that show how the *s* and *p* bands mix together. Although at *K* all the levels of K_1 symmetry may in principle mix together, the largest interaction is with the middle level of the set. How this occurs is readily understandable from Figure 4.12, where the correlation between the levels at *K* and *X* is shown. The strong interaction with the middle level of K_1 symmetry reflects the strong interaction with its parent at *X*.

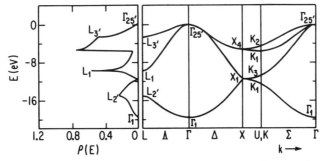

Figure 4.14 The density of states and band structure of the occupied bands of diamond. (Adapted from Ref. 90.)

the levels labeled X_1 has ensured that a gap exists between the filled and empty levels for the diamond electron count. The dispersion diagrams that are generated for the Group 14 elements in this way are quite good but may be improved by the addition of next-nearest-neighbor interactions. It is too much to hope for an analytic expression for all of the energy bands, but at some of the symmetry points it is possible to write the energies (Ref. 90) in closed form. So at Γ

$$
\left.
\begin{aligned}
E(\Gamma_1) &= e_s + \beta_{ss} \\
E(\Gamma_2) &= e_s + \beta_{ss} \\
E(\Gamma'_{25}) &= e_s + (e_p - e_s) - V_{xx} + U_{xx} \\
E(\Gamma_{15}) &= e_s + (e_p - e_s) + V_{xx} + U_{xx}
\end{aligned}
\right\}
\tag{4.2}
$$

where

$$
\left.
\begin{aligned}
V_{xx} &= \tfrac{1}{3}\beta_{pp\sigma} + \tfrac{2}{3}\beta_{pp\pi} \\
V_{xy} &= \tfrac{1}{3}\beta_{pp\sigma} - \tfrac{1}{3}\beta_{pp\pi}
\end{aligned}
\right\}
\tag{4.3}
$$

and at X

$$
\left.
\begin{aligned}
E(X_4) &= e_s + (e_p - e_s) \pm V_{xy} - U_{xx} \\
E(X_1) &= e_s + \tfrac{1}{2}(e_p - e_s + U_{xx}) \pm \sqrt{(e_p - e_s + U_{xx})^2 + (2V_{sp})^2}
\end{aligned}
\right\}
\tag{4.4}
$$

where

$$
V_{sp} = (-1/\sqrt{3})\beta_{sp\sigma}
\tag{4.5}
$$

The next-nearest-neighbor interactions, labeled U_{ij}, are defined in an analogous way. Equation (4.4) shows the general quadratic expression for $E(X_1)$ and, via the term $2V_{sp}$, the strong effect of s–p interactions at this point. A schematic, showing the way the s and p bands interact to give the final band structure, is shown in Figure 4.15 along with a simple orbital diagram for methane for comparison.

Diamond, therefore, is an insulator with the density of states shown in Figure 4.14.

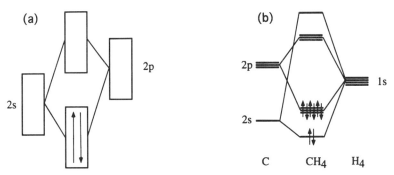

Figure 4.15   A comparison between (a) the band structure of diamond and (b) the molecular orbital diagram for methane.

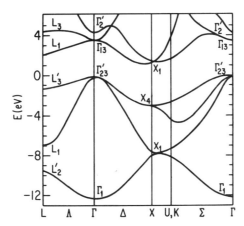

Figure 4.16 The valence and conduction bands of Si. (Adapted from Ref. 91.)

Compare, however, the density of states shown in Figures 4.11 and 4.12 for the s and p bands separately, with that for graphite of Figure 3.5(a). They are in fact quite similar, consisting of "bonding" and "antibonding" bands that just touch. It is the strong s–p mixing between the two that is possible by symmetry in diamond, but not possible in graphite, that leads to the insulating character of diamond and the semimetallic nature of graphite.

The band gaps in some materials are direct ones; that is, the highest energy level in the filled (valence) band lies at the same point in **k** space as the lowest energy level of the empty (conduction) band. The situation is more complex here, and the gap is an indirect one, as shown in a more complete band picture for silicon (Ref. 91) in Figure 4.16. The two levels concerned are those labeled Γ_{25}' and X_1. The band gap is thus a complex function of V_{xx}, U_{xx}, V_{sp}, and $e_p - e_s$, the difference between the energies of the atomic levels. Table 4.2 shows values for the various parameters for C, Si, and Ge. Notice that the largest change on moving from carbon down the group is associated with the drop in V_{sp}.

The picture for the ZnS, zincblende structure is quite similar. The Hamiltonian matrix of Table 4.1 is readily modified by introducing different values for e_p and e_s, and two values for V_{sp}. The most important result that comes from the calculation is the splitting off of the "s" and "p" bands, as shown in Figure 4.17 for ZnSe. Both bands are more heavily located on the more electronegative Se atom, as expected from the simplest electronegativity considerations such as those described for HHe in Section 1.3. The energy levels at some symmetry points may be written in closed

Table 4.2 Matrix elements for the group 14 elements

	V_{ss}	V_{sp}	V_{xx}	V_{xy}	U_{xx}
C	−15.2	10.25	3.0	8.30	—
Si	−8.13	5.88	1.71	7.51	−1.46
Ge	−6.78	5.31	1.62	6.82	−1.0

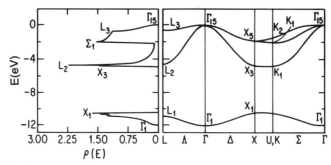

Figure 4.17   The density of states and band structure of the occupied bands of ZnSe. (Adapted from Ref. 90.)

form. At Γ and X

$$
\left.\begin{aligned}
E(\Gamma_1) &= (e_s + e_{s'})/2 \pm \sqrt{[(e_s - e_{s'})/2]^2 + \beta_{ss}^2} \\
E(\Gamma_{15}) &= (e_p + e_{p'})/2 \pm \sqrt{[(e_p - e_{p'})/2]^2 + V_{xx}^2} \\
E(X_1) &= (e_s + e_{p'})/2 \pm \sqrt{[(e_s - e_{p'})/2]^2 + V_{sp}^{2'}} \\
E(X_3) &= (e_{s'} + e_p)/2 \pm \sqrt{[(e_{s'} - e_p)/2]^2 + V_{sp}^{''2}}
\end{aligned}\right\} \tag{4.6}
$$

where

$$
\left.\begin{aligned}
V_{sp}' &= (-1/\sqrt{3})\beta_{sp'\sigma} \\
V_{sp}'' &= (-1/\sqrt{3})\beta_{s'p\sigma}
\end{aligned}\right\} \tag{4.7}
$$

4.3   "Localization" of "Delocalized" Orbitals in Solids

Perhaps one of the most confusing pairs of terms in the area of chemical bonding is that of "localization" and its antonym "delocalization" (Refs. 101, 111, 156–158, 225–227, 162). The two words were used in Chapter 1 to characterize the nature of mathematical form of the wave function that describes the electronic state in H_2. There is was shown that whether the "localized" (H–L) or "delocalized" (M–H) wave function was appropriate depends upon the ratio of two- to one-electron terms in the energy, $\kappa = U/4\beta$. In a different sense in Figure 1.3, within the delocalized molecular orbital model, one would say that the electrons in the HHe^+ molecule were largely localized on the more electronegative helium atom. In a third usage of this term the bonding electrons in the CH_4 molecule, in diamond, in BeH_2, or in the silicate of **4.5** might be described as being localized in pairs between the connected atoms in a Lewis sense. Thus the conventional localized description of BeH_2 would be as shown in **4.4**, where two pairs of electrons occupy localized Be–H bonding orbitals constructed from the central atom s and p orbitals. In the discussions for solids so far in this book the delocalized band model has been used throughout. Here the wave function (crystal orbital) contains, in principle, contributions from every

H:Be:H   ⬭▷Be , Be◁⬭
4.4

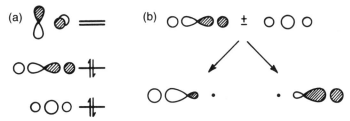

4.5

atom in the solid. How is such an extended electronic description related to the Lewis structure of the silicate chain of **4.5**, for example?

The simplest molecular example is that of BeH_2. Figure 4.18(a) shows the molecular orbital diagram for this species. Notice that in both occupied molecular orbitals the wave function is spread over all three atoms. Now a judiciously chosen linear combination of these orbitals produces two functions, shown in Figure 4.18(b), which are simply mirror images of each other (Ref. 214), and are a perfectly respectable way to describe the electron density in the molecule. These are frequently called bond orbitals (Ref. 5). However, they are not stationary states of the Schrödinger equation and will not be of use in understanding photoelectron spectra, for example. In fact, they will only be a valid descriptor of the system for what are called "collective" properties (Ref. 111), namely those for which all of the electrons need to be included. Two examples would be the total energy and the electron density. As a general result N pairs of electrons in N bonding orbitals may be localized in this sense to produce N two-center, two-electron bonds. However, where there are fewer pairs of electrons in bonding orbitals than prejudice wants to make "bonds," then the delocalized picture has to be used (unless one wants to create "bond orbitals" that are spread over more than two atoms). The circle conventionally drawn inside the hexagon of benzene is an example. Here there are only three π-bonding pairs of electrons and six C–C close contacts. How well "localized" are these functions that describe two-center, two-electron bonds is a good question to ask.

For linear BeH_2 molecular orbital theory leads (Ref. 214) to a pair of canonical valence orbitals which after Figure 4.18 may be written

$$\left. \begin{aligned} \psi_1 &= c_1\phi_s + c_2\phi_1 + c_2\phi_2 \\ \psi_2 &= c_3\phi_p + c_4\phi_1 - c_4\phi_2 \end{aligned} \right\} \tag{4.8}$$

where $\phi_{1,2}$ are the hydrogen $1s$ orbitals and $\phi_{s,p}$ are the central atom $2s$, $2p$ orbitals. Now the total wave function, the antisymmetrized product wave function used to describe the electronic state, remains invariant by replacing $\psi_{1,2}$ with the two

Figure 4.18 Generation of localized orbitals in BeH_2 from the σ_u^+ and σ_g^+ delocalized orbitals. (a) A part of the molecular orbital diagram for linear BeH_2. (b) The construction of the localized orbitals.

equivalent functions $\psi'_{1,2}$, where

$$\left.\begin{array}{l}\psi'_1 = \dfrac{1}{\sqrt{2}}(\psi_1 + \psi_2) = \dfrac{c_1}{\sqrt{2}}\phi_s + \dfrac{c_3}{\sqrt{2}}\phi_p + \dfrac{(c_2 + c_4)}{\sqrt{2}}\phi_1 + \dfrac{(c_2 - c_4)}{\sqrt{2}}\phi_2 \\[4mm] \psi'_2 = \dfrac{1}{\sqrt{2}}(\psi_1 - \psi_2) = \dfrac{c_1}{\sqrt{2}}\phi_s - \dfrac{c_3}{\sqrt{2}}\phi_p + \dfrac{(c_2 + c_4)}{\sqrt{2}}\phi_1 + \dfrac{(c_2 + c_4)}{\sqrt{2}}\phi_2\end{array}\right\} \quad (4.9)$$

If the interactions of the central atom s and p orbitals with the hydrogen $1s$ orbitals are equal, then $c_1 = c_3$ and $c_2 = c_4$, so that the two equivalent MOs become

$$\left.\begin{array}{l}\psi'_1 = (c_1/\sqrt{2})(\phi_s + \phi_p) + c_2\sqrt{2}\phi_1 \\[2mm] \psi'_2 = (c_1/\sqrt{2})(\phi_s - \phi_p) + c_2\sqrt{2}\phi_2\end{array}\right\} \quad (4.10)$$

These, of course, are the two "localized" orbitals formed by interaction of each sp hybrid on the Be atom with the corresponding hydrogen $1s$ orbital. Some idea of how "localized" such functions are may be obtained by considering the overlap between the two hybrids of Eq. (4.9). It contains terms such as $(c_1^2 - c_3^2)$ and $(c_2^2 - c_4^2)$ and will be small if the s and p orbitals are as equally bonded as possible to the hydrogen $1s$ orbitals. This will, in fact, only be the case if the s and p orbitals on the central atom have a radial extent that is very similar. This is only true for first row atoms. It will also depend upon the strength of the s and p orbital interactions relative to the s–p energy separation. On moving down the periodic table the s-p mismatch becomes larger, and so the concept of the localized bond becomes much less appropriate. Such a mismatch is shown, for example, by the change in the difference between the s and p pseudopotential radii of Table 6.2. It is fortunate that much of organic chemistry involves first row atoms where such localized ideas are most valid.

These molecular ideas may be extended (Ref. 47) to the infinite solid-state array. The backbone of the silicate shown in **4.5** can be used as an example. Such AB_4 chains of vertex-sharing tetrahedra occur in silicates such as the pyroxenes (SiO_3^{2-}), phosphates such as Maddrell's salt, $NaPO_3$, and indeed in several Zintl salts such as Ca_3GaAs_3 ($GaAs_3^{6-}$). The structure is understood quite well in traditional terms by using the two-center, two-electron bond picture of **4.5**. Each silicon uses its four electrons to form two bonds to the oxygen atoms of the chain backbone, and two bonds to two pendant O^- units. Although a qualitative picture will be described, there is no reason why it could not be derived in a more quantitative way using the theoretical tools already described. The first step is the derivation of the band structure of a linear chain of alternating silicon and oxygen atoms, each containing one s and one p orbital to model the backbone of the $(SiO_3^{2-})_n$ chain. Slightly more complex results are found, which only differ in detail if the proper geometry of the chain is used instead.

Figure 4.19 shows construction of the energy bands of a homoatomic linear chain that carries both s and p orbitals. The simplest approach is to take the diagram for the sp-bonded chain of Figure 2.9, [Fig. 4.19(a)], fold the band in two for a two-atom

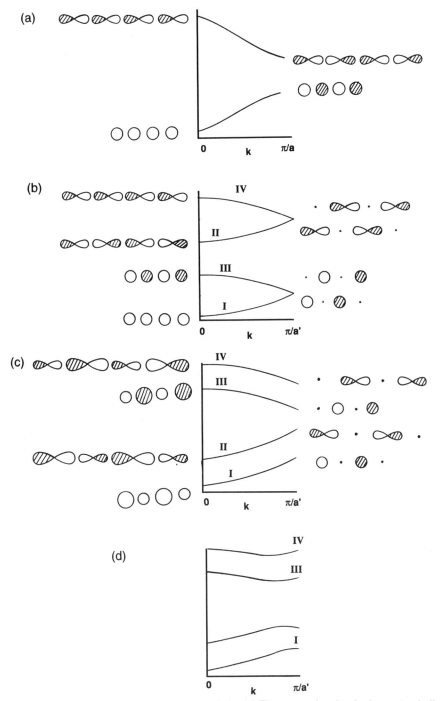

Figure 4.19 The band structure of a linear *sp* chain. (a) The energy bands of a homoatomic linear chain that carries both *s* and *p* orbitals. (b) How the band is folded in two for the two-atom cell. (c) The result of an electronegativity perturbation. (d) The effect of *s–p* mixing to give the final band structure.

cell [Fig. 4.19(b)], and then apply an electronegativity perturbation [Fig. 4.19(c)]. Finally s–p interaction is switched on (**4.6**) to give the final picture [Fig. 4.19(d)].

$$(I) + \mu(IV)$$

$$(II) + \lambda(III)$$

4.6

The result is a largely oxygen $2s$ band (I), which lies lowest in energy, followed by an oxygen $2p$ band (II). Both of these bands contain two electrons. Higher in energy lie the two silicon located bands, one composed largely of silicon $3s$ (III) and the other silicon $3p$ (IV).

Notice that the levels of Figure 4.19 are delocalized throughout the solid in much the same way as the molecular orbitals of BeH_2 are over the molecule in Figure 4.18(a). A completely analogous technique may be used to localize these extended solid-state orbitals as performed for the molecule in Figure 4.18(b). The localization process should be averaged over the whole zone to be really correct, but the central features of the result are obtained from localization at the zone center and the zone edge. The result, shown in Figure 4.20, leads to occupied bond orbitals (two-center, two-electron bonds) that hold the skeleton together. This important result shows in quite a striking way the relationship between two very different ways to regard solid-state bonding. The argument is similar to those used in molecules, but here the result has a stronger impact since the delocalized orbitals extend throughout the whole solid. The validity of this localized picture is determined by exactly the same arguments as used for BeH_2; the valence atomic s and p orbitals need to be localized in the same region of space. An exactly analogous approach is applicable to many other systems, of course.

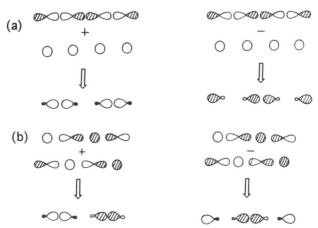

Figure 4.20 Localization of the delocalized orbitals of Figure 4.19. Judicious choice of the mixing coefficients leads to localized orbitals at (a) the zone center and (b) the zone edge.

These results may be formalized by introducing the idea of the Wannier function, a route towards the generation of localized functions in solids. Some of their properties will be quite useful when viewing electronic conduction. If $\mathbf{r}_n$ is a lattice point, then the Wannier functions (Refs. 16, 162, 338, 394, 395) $W(\mathbf{r} - \mathbf{r}_n)$ are defined as

$$W(\mathbf{r} - \mathbf{r}_n) = N^{-1/2} \sum_k \exp(-i\mathbf{k} \cdot \mathbf{r}_n)\psi(k, \mathbf{r}) \tag{4.11}$$

Such functions are orthogonal about different lattice points; that is,

$$\int W^*(\mathbf{r} - \mathbf{r}_n) W(\mathbf{r} - \mathbf{r}_m) \, d\tau = 0, \qquad n \neq m \tag{4.12}$$

The functions are peaked about the lattice sites. So if $\psi(k, \mathbf{r})$ is written as a simple Bloch function, as a product of a plane wave, $e^{i\mathbf{k} \cdot \mathbf{x}}$, with $\zeta(\mathbf{x})$, a function with the periodicity of the lattice [i.e., $\zeta(\mathbf{x}) = \zeta(\mathbf{x} - \mathbf{x}_t)$], then

$$W(\mathbf{x} - \mathbf{x}_n) = \zeta(\mathbf{x}) \sin[\pi(x - x_n)/a]/[\pi(x - x_n)/a] \tag{4.13}$$

This is shown in Figure 4.21. Notice that it is not possible to "localize" the system completely on one center. Only if $(x - x_n)$ is very large will contributions from sites other than $x = x_n$ be zero. This corresponds to the trivial case of a collection of isolated, noninteracting atoms. If overlap is included in the calculation, then the tails of the Wannier function become more important.

Wannier functions of other bands are in general more difficult to generate. There is no magic formula, the form of the localization following one's chemical prejudices (Refs. 394, 395). Thus in BeH_2 the delocalized functions were mixed together to generate bond orbitals in the "correct" place. In solids the mixing coefficients of the delocalized levels at different $\mathbf{k}$ points can be determined in the same way. Figure 7.45 shows an example in MoS_2, where the band is a mixture of metal z^2, xy, and $x^2 - y^2$ orbitals, as well as containing some sulfur character. This "localized" orbital introduces another language problem. The orbital contains two electrons but is spread out over three atoms. In this sense this "localized" orbital is "delocalized" just as for the bonding orbital of cyclic H_3^+. In principle, this is not really a problem. One is quite used to finding orbitals strongly localized in various parts of the solid.

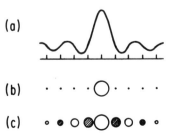

Figure 4.21   (a) The Wannier function for a one-dimensional chain with no overlap included in the normalization of the Bloch orbitals. (b) Another view of (a). (c) The Wannier function when overlap is included in the normalization.

For example, the orbitals in solid methane are well localized on individual CH_4 molecules. The case in Figure 7.45 shows one orbital that is well localized in one part of the solid, but there are other energy bands in the same material that are not. This diversity of the orbital structure is of frequent occurrence in solids.

Further insight into this problem comes from the treatment (Ref. 162) of the many-electron problem in solids. Consider the electronic description of a filled band of electrons. The Pauli Principle sets strong restrictions on the form of the total wave function. It must be antisymmetric with respect to electron exchange, as in the Slater determinant for n electrons:

$$
\Phi_{\text{total}} \sim
\begin{vmatrix}
\phi_{k1}(1) & \phi_{k1}(2) & \cdots & \phi_{k1}(n) \\
\phi_{k2}(1) & \phi_{k2}(2) & \cdots & \phi_{k2}(n) \\
\cdots & \cdots & & \cdots \\
\phi_{kn}(1) & \phi_{kn}(2) & \cdots & \phi_{kn}(n)
\end{vmatrix}
\tag{4.14}
$$

$\phi_{ki}(j)$ is the Bloch spin orbital at the point k_i in the Brillouin zone occupied for the jth electron. The Slater determinant may in fact be constructed using Wannier functions instead. Recalling the definition of Eq. (4.11), the expression for Φ_{total} using this set of functions turns out in fact to be identical (except for a phase factor) to that obtained by using the Bloch orbitals. The two are just related by a simple unitary transformation. Thus for the full band (and only for the full band) the electronic description is identical, irrespective of whether the "delocalized" Bloch orbitals are used, or whether the "localized" Wannier functions are employed. The electronic description of the partially filled band is different. It is still appropriate to use that subset of the ϕ_{ki} that are occupied in the partially filled band to construct a Slater determinant, but now the correspondence with the Wannier description has disappeared. All the Bloch orbitals of a given band are needed to represent all of the Wannier orbitals, as Eq. (4.11) shows. So if some of the Bloch orbitals are not occupied (partially filled band), then this equivalence is lost. (In fact, the Bloch determinant has to be written as a sum of Wannier determinants for this case, as discussed more fully in Ref. 162.) There are some similarities to the situation in molecules where the C–H bonds in CH_4 are often regarded as being "localized" but the π system of benzene as being "delocalized." In solids this important result for the filled band [the gapped insulator of Fig. 2.10(e)] applies not only to the band filled with electron pairs, but also to the case of the ferromagnetic insulator (**2.6**), where the band is filled with electrons of one spin only.

Before concluding this section, it should be pointed out once again that the energy levels associated with the delocalized model are the ones that have to be used to make comparisons with the photoelectron spectra of molecules and solids. Only for collective properties (Ref. 111) are the localized bond orbitals appropriate. Thus the photoelectron spectrum of BeH_2 shows two peaks in accord with the molecular orbital model, rather than one peak corresponding to the picture of two identical hybrids. Similarly the photoelectron spectrum of the one-dimensional chain of Figure 2.4 has a finite width (the dispersion of the band) and does not represent excitation from an infinite set of states identical to that of Figure 4.21.

4.4   The structure of NbO

NbO has the unique structure shown in Figure 4.22. It may be described as based on the rocksalt structure but with a quarter of both metal and oxygen atoms removed. This is easy to see in Figure 4.22(b) but less easy in Figure 4.22(a). Although the structure is then a defect rocksalt arrangement, since the "defects" are ordered, it is best to think of it as a different structure type. The structure is quite unusual since both Nb and O occupy sites of square-planar coordination within a simple four-connected three-dimensional net. Such square-planar metal coordination is a very unusual feature for a d^3 metal. Such geometries are typical of low-spin d^8 metal complexes. The observation of oxygen in a square-planar environment is unusual, too. The theoretical challenge here is then to understand these two observations. Defect rocksalt structures are frequently found for early transition metal oxides and chalcogenides (Ref. 310), and thus the problem is more general than appears at first sight. Stoichiometric TiO also has an ordered, defect rocksalt structure, but here one-sixth of the metal and oxygen atoms are missing. An investigation of the electronic structure of NbO with the aim of understanding why this structure is found for this particular d count presents another opportunity to show (Ref. 56) how use of the theoretical tools described in Chapter 2 leads to insights into chemical bonding.

   The electronic structure and density of states for the rocksalt oxide was described in Section 4.1. A calculation shows in quite a dramatic way why the NbO structure is favored over rocksalt for the d^3 case. Figure 4.23 shows the density of states calculated for this structure. Notice that the region between metal and oxygen bands, clear for the rocksalt case shown in Figure 4.4(a), now contains a large density of states. The Fermi level has dropped considerably, and although some of the oxygen levels appear to have gone up in energy, overall there has been a strong stabilization of the system compared to the rocksalt parent. Figure 4.24 shows the COOP curves, and it is clear that these new levels are strongly metal–metal bonding.

   Understanding the origin of this stabilization is in fact straightforward. The removal of two oxygen atoms from the octahedral environment of the metal atom leads to a square-planar geometry with the level diagram shown in Figure 4.25(a). Probably the most striking feature of this diagram is the strong stabilization of the z^2 orbital relative to the octahedron. This orbital of course points into the space left by the ejected oxygen atoms. One of the reasons it lies to low energy is its mixing in the lower-symmetry with the metal s orbital in a similar way to that described in Section 7.7. The presence of this low-lying orbital, generated by oxide vacancies, is vital for understanding the change in metal–metal bonding. A numerical calculation shows (Ref. 57) that the z^2–z^2 overlap (**4.7**) with a neighboring metal is larger than the t_{2g}-type xz–xz overlap shown in **4.1**. Along with the fact that the z^2–z^2 overlaps

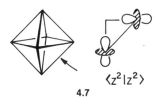

$\langle z^2 | z^2 \rangle$

4.7

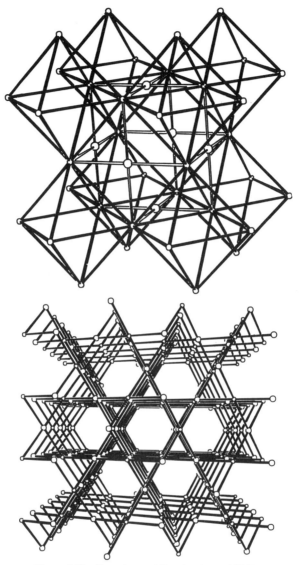

Figure 4.22   Two views of the structure of NbO.

are associated with eight nearest neighbors but the xz–xz overlaps only four, these z^2–z^2 interactions are extremely important and give rise to the low-lying metal–metal bonding orbitals in the density of states picture of Figure 4.23. In Figure 4.26 the width of the "z^2" band is an indicator of this good metal–metal interaction.

The oxygen environment is also important for the stability of the solid. Recall that metal atom loss leads to the creation of square-planar oxide ions. Figure 4.25(b) shows how the oxygen orbital energies change during this process. The z, and to a lesser extent the s, orbitals are destabilized as a result of the loss of M–O σ-bonding interaction. This electronic situation is one of the unusual features of the NbO structure. Four-coordinate oxygen is usually tetrahedral or nearly so since the filled

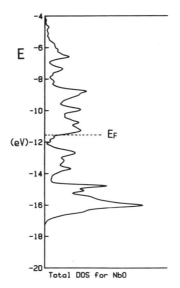

Figure 4.23 Total density of states for NbO. The Fermi level corresponds to a d^3 electron count.

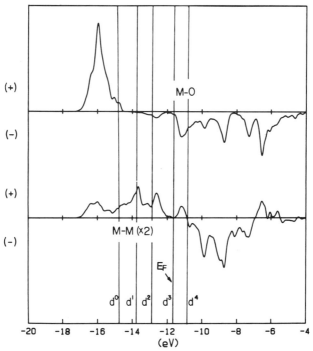

Figure 4.24 COOP curves for the Nb–O and Nb–Nb bonds in NbO. The Fermi level for NbO itself corresponds to a d^3 electron count. The Nb–Nb plot is scaled by a factor of two.

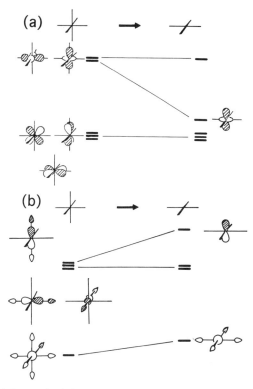

Figure 4.25 Effect on (a) the octahedral metal d levels and (b) those of the octahedral oxygen atom, on removing two trans ligands to go to a square-planar structure.

oxygen z orbital is considerably stabilized in this geometry. If this orbital can be stabilized by interaction with the metal, then this would provide some compensation for the loss of σ bonding. The argument in fact is the same as that used in Section 7.8 to rationalize the three-coordinate and planar environment of the oxygen atom in rutile. The π interactions between metal d orbitals and oxygen z shown in **4.8** provide a stabilization of the occupied orbitals for d^3 NbO. M–O π bonding and

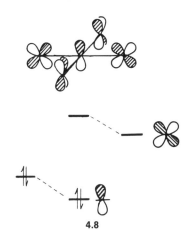

4.8

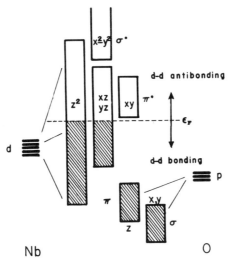

Figure 4.26 Schematic level diagram for NbO. The occupied levels are shaded. The d^3 electron count leads to an $(xz)^1(yz)^1(z^2)^1$ configuration.

π antibonding levels result (Fig. 4.26). Figure 4.27 shows the oxygen p_z character in the energy bands. The picture for the p_z orbital is very similar to that of **4.8**. However, filling the metal d orbitals that are M–O π antibonding will destabilize this arrangement. Notice that the Fermi level for NbO lies just beneath this point and so maximizes M–O π bonding.

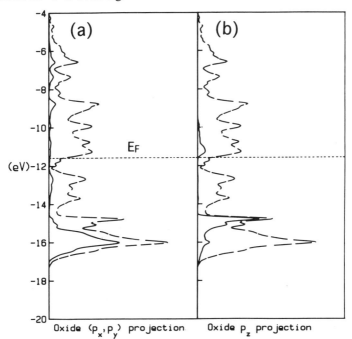

Figure 4.27 Projections of the oxide p orbital character from the density of states (dashed lines) for NbO; (a) x, y, (b) z.

Table 4.3 Madelung constants (A_M) for various monoxide candidates

Structure[a]	A_M	Density (formula units/R_O^3)
Rocksalt (NaCl)	1.7476	0.5
Observed NbO	1.5043	0.375
Observed TiO	1.5762	0.416
Alternative NbO	1.6219	0.375
Alternative TiO	1.6759	0.416
Sphalerite (ZnS)	1.6381	0.325
Cooperite (PdO)[b]	1.6102	0.334

[a] In each structure, R_O, the metal–oxygen distance, remains fixed.

[b] The Pd–O–Pd angle in PdO was taken as 98°.

The electronic structure of this system is therefore rather interesting. The largely oxygen $2p$ levels are M–O bonding, in both σ and π senses, but have lost some stabilization due to the loss of two ligands. The energy region between oxygen $2p$ and metal t_{2g}, empty in rocksalt, now contains orbitals that are involved in metal–metal bonding. This has come about by removing oxygen atoms such that metal orbitals, M–O σ bonding in rocksalt, now become effectively involved in M–M bonding. Notice the important result from Figure 4.24 concerning the location of the Fermi level for d^3, the d count for NbO. For this electron count no strongly M–O antibonding orbitals are occupied, but most of the M–M bonding levels are occupied.

It is interesting to ask if the stability of the NbO structure may be rationalized along electrostatic lines. In fact, such arguments fail completely (Ref. 57). As Table 4.3 shows, Madelung energies for NbO and TiO are unfavorable not only with respect to rocksalt, but also with respect to some possible alternatives with as many defects. The comparison with rocksalt is easy to appreciate. Electrostatic arguments always favor the more densely packed structure with the largest number of ions since the energy is proportional to Nq_1q_2/r_{12}, where N is the number of interactions between pairs of charges q_1, q_2 separated by a distance r_{12}. The electrostatic energy will always be lower per formula unit for the rocksalt structure than for the more open NbO and TiO structures with a given metal–oxygen distance. The ionic model also fails in predicting the experimentally observed arrangement of these defects within the NaCl structure. Table 4.3 shows the Madelung constants for some structures with the same percentage of vacancies as found in NbO but with a different arrangement. Notice the more favorable Madelung energy for the structural alternatives. In fact, of all the structures listed, the NbO and TiO arrangements actually observed have the smallest Madelung constants. It could be argued that the defect structure shrinks with respect to structural competitors to increase its Madelung energy and thus stability. However, there is no obvious reason for this to be the case for one defect ordering pattern and not the other of similar density.

The observed structure of NbO is thus one that is well described by the band model (Ref. 57) and is the one that best adapts its environment to maximize metal–metal bonding given the number of d electrons present. If the number of d electrons is different, as in TiO, then similar arguments should hold, but the details of the structure should be different. Indeed the TiO arrangement, although a defect rocksalt structure, has one-sixth of the metal and oxygen atoms missing.

4.5   Chemical bonding in ionic compounds

Figure 4.28(a) shows how the band structure of NaCl would be generated using the ideas of earlier chapters. Chlorine is more electronegative than sodium, and so the chlorine levels lie deeper in energy. The band gap should then be given approximately by the difference in the free atom energies $|e_p(\text{Cl}) - e_s(\text{Na})|$. Some figures are given in Table 4.4. The chlorine bandwidth will be controlled both by Na–Cl and Cl–Cl interactions, although, since the Na and Cl levels are energetically far apart, the width (less than 2 eV from photoelectron spectroscopy) arises largely from the latter. Notice that the top of the band is destabilized more than the bottom is stabilized, a direct result of the inclusion of overlap into the orbital problem. Sodium chloride, however, is often regarded as the prototypical ionic compound held together by electrostatic forces. On this model there are no bands at all. How can the two points of view be reconciled? Figure 4.28(b) shows how this is done (Refs. 103, 250, 354) for the general case of an AX system, where A is more electropositive than X. At the far left are the energies of the A and X atoms. Ionization and electron attachment, respectively, shift the energies of the A^+ and X^- states by the ionization potential (I) and electron affinity (A_e), respectively, of the two species. The atomic A level is stabilized in this

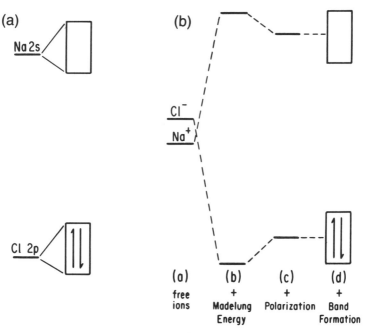

Figure 4.28   (a) Band structure expected for NaCl in the rocksalt structure. (b) The shifts in the atomic valence levels of sodium and chlorine in the generation of solid NaCl on the ionic model.

Table 4.4 Electronic parameters for comparison with alkali halide band gaps[a]

	Li	Na	K	Rb	Cs
F	13.6	11.6	10.7	10.3	9.9
	(11.5)	(11.9)	(12.8)	(13.1)	(13.4)
Cl	9.4	8.5	8.4	8.2	8.3
	(6.8)	(7.2)	(8.1)	(8.4)	(8.8)
Br	7.6	7.5	7.4	7.4	7.3
	(5.7)	(6.1)	(7.0)	(7.3)	(7.6)
I	—	—	6.0	6.1	6.2
	(4.5)	(4.8)	(5.8)	(6.0)	(6.4)

[a] Shown are the experimental band gaps, and in parentheses values of $e_s(C) - e_p(A)$, where A and C are anion and cation. (Adapted from Ref. 160.) Units are eV.

process, and the atomic X level is destabilized. On bringing the ions together to form the solid the lattice energy, dominated by the electrostatic terms, leads to a stabilization of the system. The result is stabilization of the anion state since it is surrounded by cations and a destabilization of the cation site since it is surrounded by anions. The shift in the energies of the levels, the site potentials, are determined by A_M, the Madelung energy. There are other effects, such as those associated with polarization forces between the ions that reduce this electrostatic contribution. These represent the response of the environment to the presence of the charge. However, the important result is that the effect of the Madelung interaction is to shift the levels in the solid close to the values expected for the neutral atoms. As a numerical example in CuO, the ionization potential of Cu^{2+} is 36.8 eV, the site potential at copper is -25.3 eV, and thus the copper level is found at -12.5 eV. The purely ionic model would have energy bands of vanishingly narrow width, but in NaCl, for example, much of the bandwidth arises through like-atom [Cl$\cdots$Cl (small) and Na$\cdots$Na (large)] interactions.

In the simplest model the separation between the A and X levels in the solid should depend inversely on the anion–cation distance. A comparison with experiment is given in Figure 4.29(a). Notice that this correlation works best really only for the most "ionic" materials, namely the fluorides. Here, orbital overlap between the anions and between anion and cation is small. Figure 4.29(b) shows the correlation with $|e_p(X) - e_s(A)|$. In general terms this separates the series of halides, one from another, but is clearly only a part of the picture. Contributions from overlap interactions, such as those behind the finite width of the energy bands in NaCl (especially the broad sodium band, around 5 eV), are also important. Both anion–anion and anion–cation interactions increase with the size of the halide.

Even though the anion band in NaCl derives its width mainly through Cl$\cdots$Cl interactions, the electrons may be regarded as being localized on the chlorine atoms since the band is full. In this sense, even though there are energy bands of finite

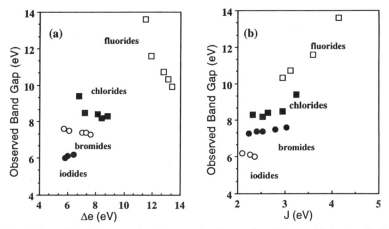

Figure 4.29 (a) Correlation between band gap and anion–cation distance (d) in the alkali halides. J is proportional to $1/d$. (b) Correlation between Δe, the difference in the atomic s and p levels for cation and anion, respectively, $|e_p(X) - e_s(A)|$, and the band gap in the alkali halides. (Data from Ref. 160.)

width, with states delocalized through the solid, the point charge model may be applicable. As the orbital interactions between anion and cation become stronger, however, the band gains width from this route as well. The electronic picture changes, too. The orbital character of the "anion" band is not purely anionic in character any more, and its description in terms of localized functions will be in terms of a mixture of anion and cation contributions.

The change in the qualitative picture for the AX solid on moving from an electrostatic (ionic) model to an orbital ("covalent") one is then quite smooth. The general picture is one where narrow levels, located energetically close to the relevant atomic values (and that may have a finite width from like-ion interactions), and broadened by anion–cation interactions. In the simplest model, as the electronegativity difference between the atoms decreases, orbital interactions increase and the point charge model starts to lose its applicability. The Madelung energy is reduced, since now there is less than a full charge on either A or X and this leads to a reduction in the band gap of Figure 4.28(b). However, the orbital interactions between A and X now become important so that the A and X states interact to form bands that are $A–X$ bonding and $A–X$ antibonding.

Although the discussion in this book centers around orbital models, it should not be forgotten that the "ionic model" found in basic texts is extremely useful in the study of a wide range of materials. Applications range from the qualitative to the quantitative (Refs. 85, 198, 345). One such study is described in Section 6.5.

4.6   The transition metals

The transition elements are found (Refs. 21, 114) in the crystal structures depicted in Figure 4.30 and listed in Table 4.5. The close-packed and bcc structures are in fact geometrically much more similar than obviously apparent from these pictures. **4.9**

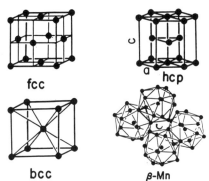

Figure 4.30   The fcc, hcp, bcc, and β-manganese structures. (Adapted from Ref. 221.)

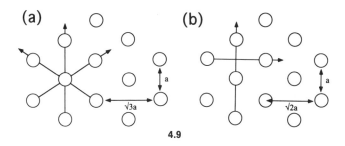

4.9

shows a comparison between a (111) plane of the fcc structure [or a (0001) plane of the hcp structure] and the (110) plane of bcc. The latter is sometimes called the pseudo-close-packed plane to highlight this similarity. Other than the first row elements, iron and manganese, it is clear that location in the periodic table (or d electron count) determines crystal structure. These materials will be treated (Ref. 9) as most metals are in this book, namely, as covalently bound materials with partially filled energy bands as in Figure 2.10 (Refs. 130, 149). The band model will be quite

Table 4.5   Crystal structures of the transition elements[a]

3	4	5	6	7	8	9	10	11
Sc	Ti	V	Cr	Mn	Fe	Co	Ni	Cu
hcp	hcp	bcc	bcc	α-Mn[b]	bcc[b]	hcp[b]	fcc[b]	fcc
fcc			hcp					
Y	Zr	Nb	Mo	Tc	Ru	Rh	Pd	Ag
hcp	hcp	bcc	bcc	hcp	hcp	fcc	fcc	fcc
La	Hf	Ta	W	Re	Os	Ir	Pt	Au
dhcp[c]	hcp	bcc	bcc	hcp	hcp	fcc	fcc	fcc

[a] Low-temperature and low-pressure forms
[b] Magnetic at room temperature
[c] dhcp close packing is a structure intermediate between hcp and fcc (dhcp = double hexagonal closest packing).

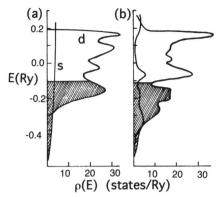

Figure 4.31 The density of states of an elemental transition metal. (a) A narrow *d* band crossed by a wider, free-electron-like *s* band. (b) Hybridization between the two. (1Ry = 13.6 eV) (Adapted from Ref. 9.)

sufficient to understand their electronic and geometrical structure. The density of states for a typical transition metal is shown in Figure 4.31(a). It consists of a relatively narrow *d* band that is crossed in energy by a much wider *s* or *sp* band (Refs. 172, 229, 255). It turns out that since the width and location of the two bands are element dependent, there is approximately one *s* electron present for each transition element (Ref. 333). Thus the maximum in the cohesive energy plot of Figure 2.13 appears for the "six-electron" Group 6 elements. The bands do mix together to produce the picture of Figure 4.31(b).

For the first row elements the bandwidth is frequently narrow enough to be able to support a ferromagnetic state where some of the electrons at the top of the band are located in higher-energy orbitals but with parallel spins (Fig. 4.32). This is of particular interest in terms of physical properties for iron, one of the elements that does not fit into the regular classification of Table 4.5. The hcp structure would be expected by analogy with its neighbors, but the bcc structure is actually found. Figure 4.33 shows the calculated density of states (Ref. 283) for a bcc metal. Notice the narrow part of the band, or peak in the density of states, at the electron count (8) appropriate for iron. Section 1.6 described the balance between high- and low-spin molecules in terms of the ratio ($|P/\Delta|$) of the pairing energy (P) and orbital separation (Δ). If $|P/\Delta| > 1$, then a high-spin state results. In solids a more appropriate but related measure is the product of the density of states at the Fermi level and the

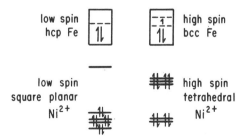

Figure 4.32 The analogy between nonmagnetic and ferromagnetic iron and high- and low-spin nickel complexes. Notice in both cases there is a change in structure with spin state.

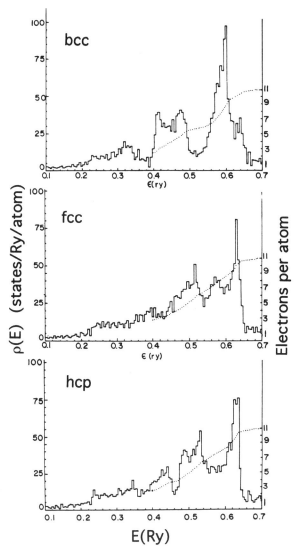

Figure 4.33 Calculated densities of states for the three most common structures found for the transition metals, bcc, fcc, and hcp. The scale at the right-hand side is to be used with the dotted line, which gives the number of electrons per atom. Notice the peak in the density of states for the bcc structure for 8 electrons per atom. (1 Ry = 13.6 eV) (Adapted from Ref. 283.)

pairing energy. If $P\rho(E_F) > 1$, then a high-spin, ferromagnetic state is possible. This is known as the Stoner criterion (Ref. 159). Such changes in structure with spin state have analogies to those of the d^8 molecules described in Section 1.6 and are shown side by side in Figure 4.32. Such magnetic behavior is not found for the elements of the second and third rows, where bandwidths are larger due to larger overlap, and thus $\rho(E_F)$ smaller.

The dependence of structure on electron count of Table 4.5 suggests that a straightforward explanation exists for these observations. Figure 4.34(a) shows the

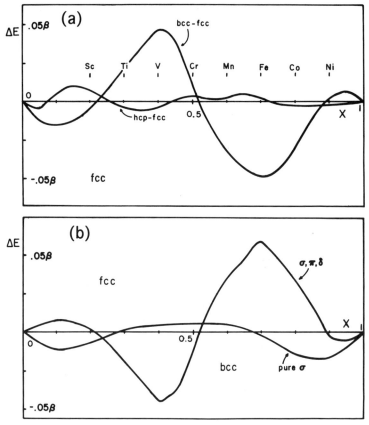

Figure 4.34   (a) Computed energy difference curves between the bcc, fcc, and hcp structures for the transition metals using a d-orbital-only model. The variable x measures the extent of band filling. The location of the individual transition metals on this plot allows for one electron to be effectively located in a metal $4s$ orbital. (b) The fcc versus bcc plot. Two curves are shown, one that contains σ, π, and δ interactions and the other σ interactions alone.

results of a set of band structure calculations using the Hückel approximation with just the d orbitals alone (Refs. 63, 118, 284). The results are qualitatively in good agreement with experiment. The hcp–bcc–hcp–fcc sequence is reproduced. The actual electron count that should be used for a particular element to compare this plot with experiment, is from the preceding discussion, around d^{n-1}. Thus the model that emerges from the calculation is one where d orbital overlap and electron filling determine the structure of a particular element. All three calculations, for the hcp, fcc, and bcc structures, used a unit cell containing two atoms, and thus this is a ten-atom problem. Although evaluation of the interaction integrals for the orbital set is not a trivial task, there are only three Hückel parameters involved. These are just β_σ, β_π, and β_δ of **4.10**. Numerical evaluation of these for a typical system

4.10

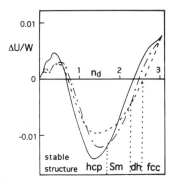

Figure 4.35 Computed energies of the hcp (solid line), double hexagonal closed packed, dh, (dashed line), and Sm-type (dash-dot line) structures relative to that of fcc for low *d* counts. The predicted structures are shown for an ideal value of *c/a*. Actual comparison with experiment requires the knowledge of the number of *f* and *s* electrons. (Adapted from Ref. 284.)

leads to a ratio of 1:0.8:0.1. Perhaps surprising, compared to the relative strengths of σ and π bonding in main group chemistry, is the importance here of π bonding. Figure 4.34(b) shows the rather different energy difference plot (Ref. 63) that comes from a calculation which only included σ-type interactions. It is clearly not successful in reproducing the observed structural trends at all. Thus the structures of the transition elements are determined by σ and π bonding of the *d* orbitals in a way that, in principle, is no more complex than Hückel's rules for cyclic polyenes. (Notice that the traditional concept of the "metallic bond" has not made an appearance here.) Similar considerations allow (Refs. 123, 284, 294) an explanation for the more complex set of structures found for the lanthanides. These are based on variants of the hcp and fcc building algorithm, with mixed stacking of close-packed sheets. Figure 4.35 shows the energy differences calculated as a function of *d* count. For these systems the actual correlation with experiment is not straightforward. Both the effects of *sd* hybridization and the partitioning of electrons between *d* and *f* levels is important. However, the agreement with experiment is quite good (Ref. 123).

It is pertinent to ask at this point why the transition elements have these "close-packed" structures. A more complete discussion is reserved for Chapter 7, but it is important to point out here that close-packed arrays of atoms are those that produce the largest concentration of three-membered rings. From the discussion of Hückel's rule in Section 1.5, an important result stemming from the method of moments is that three-membered rings are particularly stable at small values of *x*, the fractional band occupancy (Ref. 45). As far as the *s* orbital occupancy is concerned there is approximately one electron of this type present per transition metal atom. The *s,p* band is therefore approximately one-eighth full. The narrow *d* bands may have only a small effect in determining whether the structure is an open or close-packed one. Here, though, the values of β are sufficient to distinguish between fcc, hcp, and bcc structures. An alternative viewpoint is that when there are insufficient electrons to form two-center, two-electron bonds, then the energetically favored alternative is to form structures containing three-membered rings where, just as in H_3^+ of Section 1.5, there are two electrons but three "bonds." The closo boranes are molecules that also contain a high density of three-membered rings. In this electronic

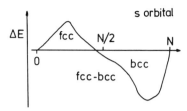

Figure 4.36 The results of a calculation using an *s*-orbital-only model; the energy difference curve between the fcc and bcc structures.

language these close-packed structures are not only close-packings of atoms but also of three-membered rings. The basic structure of these elements therefore arises from the electronic demands of an electron-poor *s,p* manifold of orbitals, with the finer details controlled by the *d* orbital interactions. Notice that these structures, contrary to the picture of "close-packed" oxides of Section 6.1, are probably well described as close packed, since here the number of attractive interactions is maximized. An interesting calculation using an *s*-orbital-only model is shown in Figure 4.36. Here the energy difference curve for the fcc and bcc structures shows clear third-moment character in accord with these ideas.

A simple model was used in Section 2.2 to show that the cohesive energies of these elements varies with the number of electrons, as $W(N)N(10-N)$, where $W(N)$ is the bandwidth, a result in qualitative agreement with experiment. In terms of moments the bandwidth is in general given by $W \sim (\mu_2 - \mu_1^2)$, or if the value of $\mu_1 = 0$ (to set the zero of some energy scale), then it is proportional to the second moment only, related as in Eq. (2.15) to the coordination number. The same type of model may be used to view the parabolic dependence on N of the surface tension of these metals. The surface tension may be envisaged as the energy involved in generating a bare surface from the bulk metal, namely, the process of cleaving the solid into two. Bonds are broken, converting the Γ coordinate bulk atoms into Γ' coordinate ones at the surface. The variation in surface tension with *d* count (Fig. 4.37) should look very similar to the variation in cohesive energy [Fig. 2.13(a)]. It does (Refs. 7, 217).

Extension of the model to the case of *AB* alloys of these elements showed a dependence [Eq. (2.14)] on $(\Delta N)^2$, which from Figure 2.15 is clearly only a part of

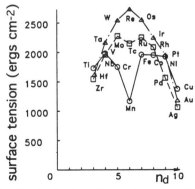

Figure 4.37 The surface tension for the transition metal series. (Adapted from Ref. 217.)

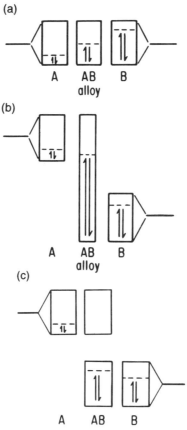

Figure 4.38   Models for transition metal–transition metal alloys. (a) The case where A and B have the same electronegativity and the same bandwidth. (b) The common band model. (c) The ionic model.

the picture. There are two other contributions (Refs. 285–287, 291, 293, 355) to the heat of formation that need to be taken into account.

The first contribution to the heat of alloy formation is a repulsive term that arises from the difference in the interactions between the A and B atoms (β_{AB}) and the corresponding values for the pure components (β_{AA} and β_{BB}). This arises since the bandwidth depends on N, as noted. The second contribution involves a more detailed discussion of the electronic model. Figure 4.38(a) shows the model used in the generation of Eq. (2.14). The band filling shown is appropriate for the case of an alloy between early and late metals, where the heat of formation is largest. The bandwidth used for the AB alloy is equal to that used for the pure A and B end members. However, the real state of affairs will correspond to that of Figure 4.38(b), where the electronegativity difference between the two is nonzero. The interactions are not of the ionic type where electrons move from the pure "A" band to the pure "B" band as might be expected [and shown in Fig. 4.38(c)], since the orbital interactions that control the bandwidth are between A and B located orbitals. Thus the band that is generated is a common band (Ref. 361), one that has both A and

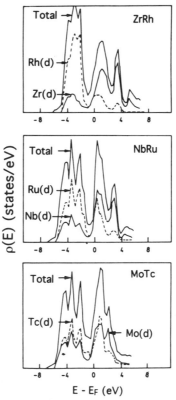

Figure 4.39 Calculated total and partial densities of states for some transition metal–transition metal alloys. The total density of states does not appear to change much, but the partial densities of states change in accord with the variation in electronegativity difference ($\Delta\chi$), which follows the change in Z. $\Delta\chi$ becomes smaller on moving down the diagram. (Adapted from Ref. 9.).

B character, just as the bonding and antibonding orbitals in the HHe molecule have orbital character from both H and He. The results of a calculation are shown in Figure 4.39. One could study this problem by fixing the second moments of the constituent bands and the alloy band (Refs. 105, 107, 117); however, there is an added, important consideration for the case of metals. These are materials where perfect screening occurs, a property that implies that at the atomic level there is no charge transfer from A to B during the formation of the alloy. In orbital terms the character of the molecular orbitals of HHe depends upon the interaction integral and the energy mismatch between the two starting orbitals. Thus the only way to ensure that such local charge neutrality holds using the orbital model is to allow the energy separation of the A and B atomic orbitals to change from the values used for the pure end members. Such an energy change is said to arise from the "renormalization" of the atomic energy levels. [Such renormalization of atomic energy levels is commonplace in solids (Ref. 385). In molecular calculations the energy separation between unperturbed nd and $(n + 1)s$ orbitals is usually large. For the solid metals the H_{ii} values used in the tight-binding calculation need to be adjusted to mimic properly the electronic structure (Refs. 362, 363, 385). The renormalization in this case lowers the s orbital energy closer to that of the d.]

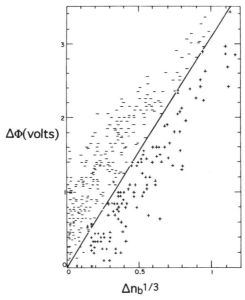

Figure 4.40 The correlation with experiment for Eq. (4.15). A minus sign indicates that there is at least one stable binary compound between the elements, that is, that $\Delta H < 0$. A plus sign indicates that there is no stable binary compound known and also that there is limited solubility ($< 10\%$) of the two metals. (Adapted from Ref. 240.) Values of ϕ and $n_b^{1/3}$ are assigned each element.

The effect of these two extra (positive) contributions to the heat of alloy formation is shown in Figure 2.15. The overall result is a heat of formation that is negative for alloys between early and late metals but positive elsewhere. The correlation with "experiment" comes from the heats of formation calculated from Miedema's scheme (Refs. 26, 27, 239–242).

Miedema used a "macroscopic atom" model to describe the heat of formation of these alloys via the following equation. The parameter $f(c)$ labels the stoichiometry, P and Q are constants, and each metal atom is assigned two indices, ϕ and $n_b^{1/3}$.

$$\Delta H = f(c)[-P(\Delta\phi)^2 + Q(\Delta n_b^{1/3})^2] \qquad (4.15)$$

The agreement with experiment is shown in Figure 4.40 and is certainly very good indeed, in fact virtually perfect. It is, however, difficult to understand the microscopic basis behind the scheme using the orbital language of this book. It has been suggested that the main driving force for alloy formation is the transfer of electrons from the states associated with the more electropositive metal to those of the more electronegative atom. In this sense the model is more like the "ionic" picture of Figure 4.38(c) rather than the common band model used previously to generate the heat of formation of the alloy. One of the atomic indices, ϕ, is in fact the electronegativity of the element. As a first approximation to ϕ the metal work function is used, although it is refined on the scheme to give a better agreement with experiment. The contribution to the heat of formation from this source is proportional to $-(\Delta\phi)^2$. The second index is positive and represents an energy cost. Miedema argued that while the electron density is continuous across the boundaries of the atomic cells in the pure A and pure B systems, a readjustment is necessary to ensure the same is true of the AB alloy. The energy penalty is argued as being proportional to $+(\Delta n_b^{1/3})^2$, where Δn_b is the electron density discontinuity at the boundary between A and B cells. The origin of this second term is much more difficult to envisage using the chemical ideas described in this book. However, when the heat of formation is written as in Eq. (4.18), it becomes clear that use of the two indices for each element i, ϕ_i, and $n_{bi}^{1/3}$ just leads to two terms with which to mimic the heat of formation, one positive and one negative, just as in the orbital model.

4.7   The free- and nearly-free-electron models

The ideal of the metallic bond where the electrons move freely under the rather weak influence of the nuclei has not been used so far in this book to describe chemical bonding in solids. It is, however, an idea that has been widely used since the band structures of many materials, especially those of the sp-bonded elements, are well described by the nearly-free-electron (NFE) model (Refs. 9, 38, 98, 159, 199, 254). However, although the model gives insights into such physical properties such as heat capacity, thermal and electrical conductivity, and the magnetic susceptibility of metals, it fails dramatically when it comes to distinguishing between metals, semimetals, semiconductors, and insulators themselves.

At its most basic, the model imagines the "metallic" electrons (mass, m) of the solid moving in a box (of side a) defined by the location of the atoms. This "free-electron" picture of the metal is modified by allowing a weak periodic perturbation provided by the nuclei (Refs. 167, 171). It starts off, therefore, from exactly the opposite premise of the LCAO approach, where the nuclei and the orbitals they hold are very much the starting point of the electronic structure problem. The free-electron model itself appears to be particularly appropriate where the energy bands derived from the orbitals on a given atom are very broad and overlap considerably with each other. [The electronic situation in the transition elements of Fig. 4.31 are often described in terms of a narrow d-band crossed by a free-electron-like $(s + p)$ band.] In such a case the atomic identity of the levels begins to disappear, and the energy of the levels themselves are given by the particle-in-the-box expression

$$\left.\begin{aligned} E &= (n_x^2 + n_y^2 + n_z^2)h^2/8ma^2 \\ &= r^2 h^2/8ma^2 \end{aligned}\right\} \tag{4.16}$$

using the three quantum numbers, n_x, n_y, and n_z. The wave functions are just the plane waves $\psi_k(r) \sim e^{i\mathbf{k}\cdot\mathbf{r}}$, where the propagation vector $\mathbf{k}$ is given by

$$\mathbf{k} = (2\pi/a)(n_x\mathbf{x} + n_y\mathbf{y} + n_z\mathbf{z}) \tag{4.17}$$

The energy is just

$$E = |\mathbf{p}|^2/2m = (1/2m)\hbar^2|\mathbf{k}|^2 \tag{4.18}$$

In addition to energy quantization, the momentum $\mathbf{p}$ is also quantized via $\mathbf{p} = \hbar\mathbf{k}$.

Let $M(E)$ be the total number of states with an energy less than E. $M(E)$ is equal to the number of levels contained in one octant of a sphere of radius r, namely $\frac{1}{8}$ $\frac{4}{3}\pi r^3$, so that

$$M(E) = \frac{2}{3}\pi r_{max}^3$$
$$= (4\pi/3)(2mE/h^2)^{3/2}a^3$$
$$= (V/6\pi^2)(2m/\hbar^2)^{3/2}E^{3/2} \tag{4.19}$$

The density of states $\rho(E)$ is just $\partial M(E)/\partial E$ or

$$\rho(E) = (V/4\pi^2)(2m/\hbar^2)^{3/2}E^{1/2} \tag{4.20}$$

This needs to be divided by V to describe the number of states per unit volume and multiplied by 2 to take spin degeneracy into account. This function is shown in **4.11**. Figure 4.41 shows the computed density of states from calculations on some elemental

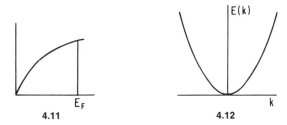

4.11 **4.12**

metals. Notice the strong resemblance to **4.11** for the alkali metals. The description is less good for some of the other metals or for the alloys shown (Ref. 288) in Figure 4.42, where there is a difference in nuclear charge.

Knowledge of the number of electrons in the solid and its density enables a

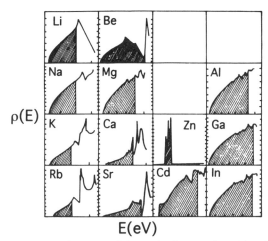

Figure 4.41 Densities of states for some of the sp (main group) metals, showing the similarity for many to that expected from the free-electron model. (Adapted from Ref. 288.)

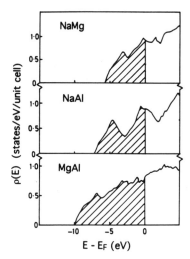

Figure 4.42 Densities of states for some alloys of the sp (main group) metals. The poorest agreement with the free-electron parabola is seen for NaAl, where there is the largest difference in Z and hence electronegativity. (Adapted from Ref. 288.)

straightforward determination from Eq. (4.19) of the energy of the highest occupied level, E_F, an observable that comes from photoelectron spectroscopy. Given a total number of electrons N, then the Fermi level is calculated as

$$
\left.
\begin{aligned}
N &= 2 \int_{-\infty}^{E_F} \rho(E)\, dE = \tfrac{2}{3}(V/4\pi^2)(2m/\hbar^2)^{3/2} E_F^{3/2} \\[2mm]
E_F &= (3N\pi^2/V)^{2/3}(\hbar^2/2\,m)
\end{aligned}
\right\}
\tag{4.21}
$$

N/V comes from the density and the number of electrons per atom. Calculated values of E_F for Na, Mg, and Al are 3.2, 7.2, and 12.8 eV, to be compared with their experimentally determined counterparts of 2.8, 7.6, and 11.8 eV. In the free-electron model the Fermi surface for three-dimensional systems is a sphere.

The Fermi energy is an important parameter in the free-electron model since the total energy depends directly upon it.

$$
E_T = 2 \int_{-\infty}^{E_F} \rho(E) E\, dE = \tfrac{2}{5}(V/4\pi^2)(2m/\hbar^2)^{3/2} E_F^{5/2}
\tag{4.22}
$$

Using E_F from Eq. (4.21),

$$
E_T = \tfrac{3}{5} N E_F
\tag{4.23}
$$

From Eq. (4.20), $E(\mathbf{k})$ takes on the parabolic form shown in **4.12**. It is interesting to compare the form of this dispersion equation with that from the tight-binding model of Section 2.1. Expansion of Eq. (2.2) leads to

$$E(\mathbf{k}) = \alpha + 2\beta \cos ka = (\alpha + 2\beta) - \beta(ka)^2 + \cdots \qquad (4.24)$$

to be compared with the free-electron model result

$$E(\mathbf{k}) = V_0 + (\hbar k)^2/2m \qquad (4.25)$$

where a constant potential has been added to Eq. (4.18). At the bottom of the energy band, therefore, for small $\mathbf{k}$,

$$V_0 = \alpha + 2\beta \quad \text{and} \quad -\beta(ka)^2 = (\hbar k)^2/2m \qquad (4.26)$$

so that

$$-\beta = \hbar^2/2ma^2 \qquad (4.27)$$

Since a is fixed, in order to match tight-binding and free-electron models, the electron mass is adjusted to fit β, such that

$$m^* = -\hbar^2/2\beta a^2 \qquad (4.28)$$

Thus a small effective mass, m^*, describes wide energy bands, and a large effective mass, narrow energy bands. This effective-mass description is one in widespread use, especially in the physics community.

The effect of the nuclei may be added to the model as a perturbation assuming a weak periodic potential. Thus the nearly-free-electron model is generated. If the nuclear potential is written as $V = \sum_g V(g) \exp(i\mathbf{g} \cdot \mathbf{r})$, where $\mathbf{g}$ is a reciprocal lattice vector, then the second-order perturbation theory expression [just as in Eq. (3.16)], is zero unless $\mathbf{k} - \mathbf{k}' = 2\pi\mathbf{g}$.

$$\Delta E = \frac{|\int \psi_k^* \psi_k V \, d\tau|^2}{E(\mathbf{k}) - E(\mathbf{k}')} \qquad (4.29)$$

$$\Delta E = \frac{2m}{\hbar^2} \sum_G \frac{|V(g)|^2}{\mathbf{k}^2 - |\mathbf{k} + \mathbf{g}|^2} \qquad (4.30)$$

This expression starts to become large as $2\mathbf{g} \cdot \mathbf{k}$ approaches $\mathbf{g}^2$, when degenerate perturbation theory needs to be used. The secular determinant for this case is just

$$\begin{vmatrix} E - E(\mathbf{k}) & V(g) \\ V(g) & E - E(\mathbf{k} - \mathbf{g}) \end{vmatrix} = 0 \qquad (4.31)$$

The result is shown in Figure 4.43, where gaps open up [$2|V(g)|$ in magnitude] at

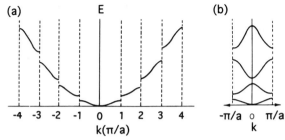

Figure 4.43 (a) The result of the perturbation associated with the nuclear potentials on the free-electron levels. Gaps are opened up at $\pm n\pi/a$. (b) The bands using the extended zone scheme of (a) may be folded into the first zone. Note the resemblance of the behavior of the deepest-lying pair of bands to the tight-binding picture of Figure 2.29.

$\pm n\pi/a$. The form of the lowest-energy band of the "nearly-free-electron" model is quite similar to that of the tight-binding band of Figure 2.2(c). Indeed if the first two nearly-free-electron bands of Figure 4.43(a) are folded into the first Brillouin zone, then the similarity between the resulting picture [Fig. 4.43(b)] and that of the folded one-dimensional chain with s and p_z bands of Figure 2.9 is striking. (This is in fact a result that may be readily proven algebraically; p. 161 of Ref. 210).

Within the nearly-free-electron theory the adoption of one particular structure over another is determined (Refs. 167, 199, 254), by noting the energetic disadvantage of filling the upper band where the Fermi surface touches a zone face as shown in Figure 4.43(a). In three dimensions the parameters determining this condition are somewhat more complex, since the zones themselves are not quite as simple. (See, for example, the fcc zone of **4.3.**) Thus in Figure 4.44 the Fermi surface does not touch any of the zone faces, and for this electron count the structure is stable. The electron count for which the two surfaces do touch will differ from one solid to another, and the system will search out alternative geometrical arrangements where the structure of the zone is such that no high energy levels need to be filled for that particular electron count. One area where the method was used with apparent success is in understanding the structures of the Hume–Rothery electron compounds (see Chapter 8). The structures of the brass alloys depend, as shown in Table 4.6, on the number of electrons per atom. Just by calculating the number of electrons that just fills the set of lower levels for the four structures shown, Jones (Ref. 199) was able to determine the critical electron count at which that particular structure started to become unstable. The agreement with experiment is reasonably good. However, there are some real problems for this explanation. The results for copper, for example, cannot be correct (Refs. 12, 336). The metal has the fcc structure, yet both from

Figure 4.44 A spherical Fermi surface lying inside the zone faces of the fcc lattice.

copper

Figure 4.45   The Fermi surface of copper. (Adapted from Ref. 210.)

Table 4.6   Experimental and calculated values of the number of electrons per atom at which one Hume–Rothery phase becomes less stable than the next in the series

Phase	N_{expt}	N_{calc}
α, fcc	1.38	1.36
β, bcc	1.50	1.48
γ, complex	1.62	1.54
ε, hcp	1.75	1.69

experiment and theory more exact than that of Jones, its Fermi surface, and those of the other Group 11 metals, have actually passed through a zone face for one $(s + p)$ electron per atom as in Figure 4.45. Thus the calculated critical electron counts of Table 4.6 are incorrect. The general idea of structural stability being controlled by the perturbation provided by the atomic potential on the free-electron states is revived in Section 8.5, but the best explanation of the structures of the Hume–Rothery phases is a tight-binding one and is given in Section 8.3.

4.8   Compounds between transition metals and main group elements

Section 4.1 presented a discussion of the electronic structure of some transition metal–main group compounds, and the structures of several other systems of this type are discussed elsewhere in the book. The electronic structure of many of these systems may be described by first allowing for the expansion of the metal lattice and then studying the effect of insertion of the main group atoms. Three broad categories may be envisaged (Ref. 136): that where the metal d levels lie much deeper in energy than those of the main group atom (for example, PdLi); the case where the opposite is true (for example, PdF); and the intermediate case where the two are of similar energy (for example, PdB). Figure 4.46 shows a decomposition of the electronic densities of states calculated for the three systems (Ref. 136). In the first two cases there is a rather weak interaction between the two sets of orbitals. The palladium d band is somewhat narrower than in the pure metal, simply as a result of stretching the metal lattice to accommodate the nonmetal. In the PdF case the situation is quite ionic, involving charge transfer to the deep-lying

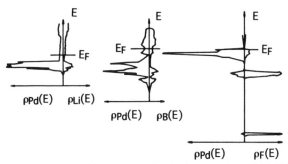

Figure 4.46 Calculated partial densities of states for the three systems PdLi, PdB, and PdF. (Adapted from Ref. 136.)

fluorine-located band. Notice the drop in the Fermi level as a result of this donation of one electron to fluorine compared to that for PdLi, where an electron has been added to the metal.

The case of PdB is the most interesting, since this corresponds to the situation where strong mixing occurs between the two sets of levels. Figure 4.47 shows the density of states in more detail and the assignment of the levels in terms of their Pd–B bonding, nonbonding, and antibonding characteristics. For these borides the dependence of the heat of formation on electron count should have a similar form to that for the metals of Figure 2.13. As metal–boron bonding orbitals become filled, the heat of formation should rise to a maximum, beyond which it should decrease as metal–boron antibonding orbitals become occupied.

The heats of formation for a number of borides have been experimentally determined (Ref. 237). For the transition metal $M B_2$ systems with the AlB_2 structure, the experimental results are shown in Figure 4.48(a). Figure 4.48(b) shows the results of a calculation (Ref. 49) of the variation in the heat of formation, and the forms of the two curves are quite similar. (These are tight-binding calculations and so the absolute values indicated by the energy scale are not to be directly identified with the scale for the experimentally determined values.) Some predictions can be made on this model. If the electronegativity of the main group element is larger than that of boron so that the "metal" levels are always antibonding between metal and

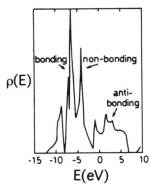

Figure 4.47 The density of states for PdB showing the bonding, nonbonding, and antibonding levels. (Adapted from Ref. 136.)

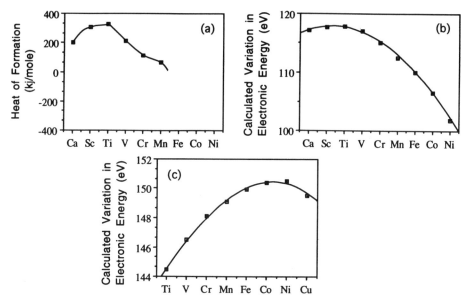

Figure 4.48 (a) Experimental values of the heat of formation for MB_2 borides, where M is a first row transition metal. (b) Calculated heat of formation for MB_2. (c) Calculated heat of formation for MC_2.

nonmetal, then there will not be a maximum in the plot of the heat of formation as a function of the transition metal identity. If the nonmetal electronegativity is increased, as on moving from boron to carbon, then the peak in the heat of formation is expected to move to higher d counts, as shown for calculations (Ref. 50) on MC_2 species (also with the AlB_2 structure) in Figure 4.48(c).

4.9 The nickel arsenide and related structures

A view of the structure of NiAs is shown in Figure 4.49. It consists of a hexagonal $(\cdots ABABAB \cdots$ or $\cdots hhhhhh \cdots)$ eutactic array of arsenic atoms with metal atoms occupying each octahedral interstice. The structure is therefore the hexagonal analog of rocksalt $(\cdots ABCABC \cdots$ or $\cdots cccccc \cdots)$. Each arsenic atom is coordinated by metal atoms in a trigonal prismatic fashion, whereas in rocksalt both atoms are in octahedral coordination. In rocksalt the nearest-neighbor metal atoms are located across an edge of the coordination octahedron; in nickel arsenide the closest metal–metal interaction is closer, being across a face. The TiP structure is exactly intermediate between the two $(\cdots ABCBAB \cdots$ or $\cdots hchchc \cdots)$. As shown in Figure 4.50, here the metal–metal contacts across the octahedral faces occur in pairs. These close cation–cation contacts are important in determining the stability of the structure of nickel arsenide relative to that of rocksalt. If there is an attractive potential between the metal atoms, then it will be stabilized; if a repulsive one, then destabilized. Clearly one way to generate an attractive metal–metal potential is via metal–metal bonding. Variants of this structure occur in different guises. Thus β-ZrCl$_3$ (d^1) has a defect nickel arsenide structure where analogous close Zr–Zr contacts are found. The TiP structure is found for d^1 systems. This is the electron

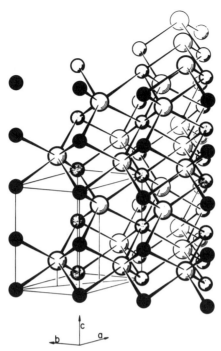

Figure 4.49 The structure of NiAs. (Reproduced with permission from Ref. 280.)

count where there are just the right number of electrons to form a single metal–metal bond between pairs of metal atoms across the octahedral face of Figure 4.50.

A structure related to that of NiAs is that of WC. It consists of a simple hexagonal lattice ($\cdots$AAA$\cdots$) of both metal and nonmetal. Both types of atoms are trigonally prismatically cooordinated by atoms of the opposite type. Figure 4.51 compares the local arrangements of the WC and rocksalt types. In WC the layers of metal atoms are eclipsed (simple hexagonal packing), leading to close metal–metal contacts across a shared face of trigonal prisms of anions analogous to the close metal–metal contacts made across a shared face of octahedra of anions in rocksalt. Table 4.7 shows the structures adopted by the transition metal carbides. Interestingly there is a sharp dividing line between rocksalt and WC types. Figure 4.52 shows the results (Ref. 384) of a band calculation on a metal carbide in the three structures: rocksalt, NiAs, and WC. Notice the perfect agreement with the results of Table 4.7, namely the change in structural preference with metal d count between 5 and 6. The systems where metal–metal bonding is possible along the z direction are stabilized as the electron

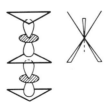

Figure 4.50 Metal–metal interactions through a common face of a coordination octahedron.

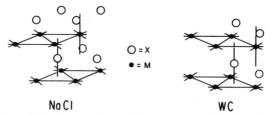

NaCl   WC

Figure 4.51   A comparison of the structures of NaCl and WC.

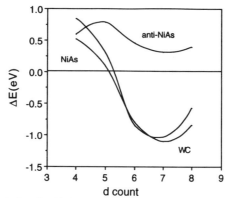

Figure 4.52   Calculated variation (for MnP) in total energy of the anti-NiAs, NiAs, and WC structures relative to that of NaCl with metal d count. A positive energy indicates that the NaCl structure is more stable. (Adapted from Ref. 384.)

count increases. This is also shown in the variation in the metal–metal overlap population as a function of d count shown in Figure 4.53. As a nice test of this viewpoint the figures also show some results for the anti-NiAs structure (found, in fact, for one form of NbN). Here it is the nonmetals with the close contacts across the shared face of the octahedra. Notice that it behaves similarly to rocksalt in terms of metal–metal overlap populations.

Table 4.7   Examples of carbides and nitrides with the rocksalt or WC structure

Structure type	number of valence electrons				
	8	9	10^a	11	12
NaCl	TiC	VC	VN		
	ZrC	NbC	MoC_{1-x}	CrN	
	HfC	TaC	WC_{1-x}		
WC			MoC	MoN	RuC?
			WC	WN	OsC?
			TaN		

^{a} There are series of ternary systems such as $Ti_{0.7}Co_{0.3}N$ (10.5 electrons) and $Mo_{0.8}Ni_{0.2}N_{0.9}$ (11.3 electrons) that have the WC structure.

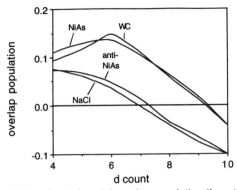

Figure 4.53 Variation in calculated metal–metal overlap population (from the calculations in Figure 4.52) with metal d count. (Adapted from Ref. 384.)

The NiAs structure is found for a number of transition metal chalcogenides and phosphides. Table 4.8 shows how for the phosphides there is an interesting electron count dependence associated with the stability of the parent structure compared with those of two distorted variants, the MnP and NiP structures. The MnP structure (Fig. 4.54) contains a distorted metal atom framework that leads to the formation of a zigzag chain of metal atoms along [010] and distortions of the phosphorus network also to give chains. In the NiP structure (Fig. 4.55) this zigzag network has distorted further to give Ni–Ni pairs. Transitions between the NaAs and MnP structures may be induced by temperature, pressure, and composition, especially in the $Mn_{1-x}M_xP$ series, where M is another transition metal. A similar electron count dependence exists for the sulfides, but many of these are nonstoichiometric and the trends difficult to see. However, for the first row systems, both TiS and CrS are found in the NiAs structure, but VS in the MnP structure. Interestingly at ~550°C it undergoes a transition to the NiAs structure. In terms of a pair-potential model, if an isotropic potential is used (i.e., one that only depends on distance), then it may

Table 4.8 Phosphides with the NiAs, MnP, and NiP structure

Structure type	number of valence electrons						
	9[a]	10[a]	11	12	13	14	15
NiAs	TiP ZrP[b] HfP[b]	VP NbP[c] TaP[c]					
MnP			CrP MoP[c] WP	MnP	FeP RuP	CoP IrP	
NiP							NiP

[a] ScS, YS, HfS have the rocksalt structure.
[b] TiP structure type.
[c] Structures related to the WC type.

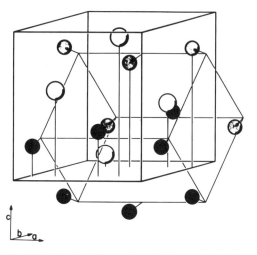

Figure 4.54 The MnP structure (Reproduced with permission from Ref. 280.)

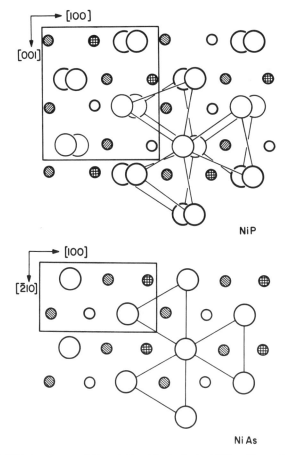

Figure 4.55 The NiP structure and its relationship to that of NiAs. (Reproduced with permission from Ref. 280.)

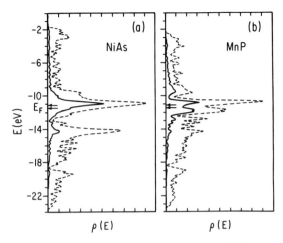

Figure 4.56   The densities of states for (a) the NiAs and (b) the MnP structures. The Fermi levels shown correspond to those of CrP and VP. (Adapted from Ref. 350.)

be shown (Refs. 151, 152) that the MnP distortion is one of only two possible ways to distort the NiAs structure. It is then not much of a surprise to find the zigzag chains that are generated during such a distortion for many other systems. They are found for $MoTe_2$, $NbTe_2$, and $TaTe_2$, as discussed in Section 7.9, and other systems such as ZrI_2, Nb_2Se_3, Cr_3S_4, Mo_2As_3, Nb_3Se_4, and structures related to them. An analogous structural relationship in terms of the geometrical changes involved is found between the Ni_2In structure and that of Co_2Si. They are just the filled-up analogs of NiAs and MnP.

Figure 4.56 shows the computed density of states (Ref. 350) for the NiAs and MnP structures. It is clear to see that there are not only changes in the density of states in the region associated with the metal d levels, but also with the deeper-lying levels located largely on the phosphorus atoms. During the distortion NiAs → MnP there are then important energetic changes associated with both P–P and $M–M$ interactions, and it is difficult to separate the relative importance of the two. Notice, however, the opening of a gap in the in-plane density of states around the point between VP and CrP, where the crucial electron count for the NiAs → MnP transition occurs. Thus this transition is driven by electronic factors involving metal–metal interactions.

4.10   Molecular metals

The fact that many solids typically regarded as "molecular" are metallic, and often superconducting, is perhaps at first sight somewhat surprising. Hower, as shown in Section 2.5, if the interaction between the units is large enough and the highest occupied band is partially occupied, then metallic character is indeed possible (Refs. 161, 243, 312). Although there is a sense that materials of this type are held together by weak van der Waals forces that leave the properties of the individual molecules intact, in fact the interatomic separations that result are close enough for important

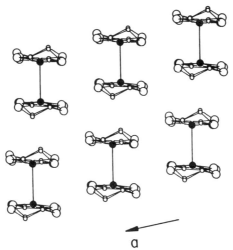

Figure 4.57   Close metal–metal contacts in the structure of the Pd[M(dmit)$_2$]$_2$ materials. (Adapted from Ref. 81.)

orbital interactions to develop between the molecular units. Section 5.1 contains a description of the consequences of the orbital interactions between carbon atoms on adjacent sheets in graphite, for years a prototypical "van der Waals solid." Partially filled bands may be generated by doping as in the TTF series of Figure 2.27 and the fullerenes of Section 5.6, to give a stable charge-transfer salt of the radical anion or cation type, which are fequently metallic. In this section two systems will be compared that develop these ideas further. These are the salts formed between the electron acceptors M(dmit)$_2$ ($M = $ Ni, Pd) (where dmit is a C_3S_5 ligand) and donors such as $X = $ TTF, NMe$_4$, and Cs (Refs. 81, 115).

The structures of all the $X[M$(dmit)$_2]_2$ materials are similar and consist of [M(dmit)$_2$] dimers arranged in infinite slabs separated by the donors X. Figure 4.57 shows close metal–metal contacts that are possible in these pairs. The Ni/TTF salt is metallic at ambient pressure to 3 K, but the Pd analog undergoes a CDW distortion at 220 K, leading to a semiconductor. It becomes superconducting at 1.6 K under 20 kbar of pressure. The calculated band structures of the two systems are dramatically different (vide infra), but why? The HOMO and LUMO of the M(dmit)$_2$ unit are shown in **4.13**, and are constructed from the same basic dmit orbital.

HOMO   **LUMO**

4.13

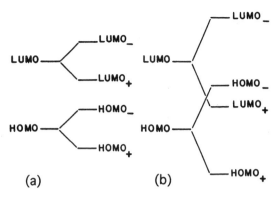

Figure 4.58 The results of molecular orbital calculations on the $[M(\text{dmit})_2]_2$ dimer, (a) weak and (b) strong dimerization. (Adapted from Ref. 81.)

Although the metal xz orbital may in principle mix into the LUMO, its nodal properties are such that admixture is small. No metal d orbital can mix into the HOMO. The two lie close together in energy (about 0.4 eV from calculation). Thus the difference between the Ni and Pd systems cannot lie with the metal–ligand interactions in HOMO and LUMO. An important difference between the two is revealed (Fig. 4.58) when calculations are performed on the dimer itself (Ref. 81). Whereas simple reasoning would suggest that the extra electron in the $[M(\text{dmit})_2]_2^-$ unit goes into an orbital derived from the LUMO of the $M(\text{dmit})_2$ molecule, this is only true in fact for $M = \text{Ni}$. The HOMO–LUMO separation in the monomer is small, for reasons just outlined, and if the intermolecular interaction is large enough, then the nature of the HOMO of the dimer may be dramatically altered. This is the case for the $M = \text{Pd}$ system, where the HOMO of $[M(\text{dmit})_2]_2^-$ is the out-of-phase combination of the HOMO of the monomer. Thus, as shown in Figure 4.59, the band structures of the Ni and Pd compounds are quite different, and as a result their transport properties also differ.

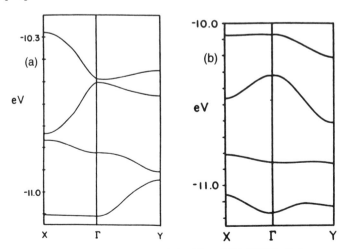

Figure 4.59 The computed band structures of the Ni and Pd $X[M(\text{dmit})_2]_2$ compounds. (Adapted from Ref. 81.)

The key to understanding their different behavior lies in the arrangement of the two molecules of $M(\text{dmit})_2$ in the dimer. For $M = $ Pd the two molecules are almost eclipsed (**4.14**), but for $M = $ Ni the two are staggered (**4.15**). The interactions between

4.14

4.15

the dimers are similar in the two cases. Similar structures are found for all of the $M = $ Pd materials irrespective of the nature of X, suggesting that it is the metal itself that controls the local geometry in the dimer, but how? The eclipsed monomer arrangement found for the dimer in $[\text{Pd}(\text{dmit})_2]_2-$ leads to maximum repulsions between the sulfur atoms at short distances. These show up electronically in orbitals that lie deeper than those shown in Figure 4.58. Such eclipsed arrangements are generally not found in solids, since the balance of attractive and repulsive forces tend to avoid such direct interactions. However, the LUMO does contain some metal character and thus an extra stabilization in ψ_{LUMO}^+ from metal–metal overlap. Thus two electrons in ψ_{LUMO}^+ will stabilize the eclipsed arrangement. If the stabilization from this source is large enough at distances where the S–S repulsions are not too fierce, then this eclipsed structure can be favored. The metal wave functions for the $M = $ Pd case are much less contracted than for the case of the first row transition metal (see Section 8.6), and so significant $M \cdots M$ interactions develop at longer interunit distances. For the more contracted d orbitals in nickel, these interactions are only significant at shorter distances where nonbonded repulsions are now overwhelming.

This example highlights the changes in electronic structure triggered by rather subtle changes in the geometrical arrangements of the molecules in the solid. Such effects are widespread. Changes in intermolecular interactions, and hence electronic structure, can be modulated by the size of the cations and by the actual crystal structure adopted (Refs. 386, 387, 389). Thus the β form of $(\text{ET})_2\text{I}_3$ is metallic all the way down to 1.5 K when it becomes a superconductor. [ET is an abbreviation for BEDT–TTF or bis(ethylenedithio)tetrathiofulvalene.] The α phase of the same stoichiometry undergoes a metal–insulator transition at 135 K.

4.11 Division into electronic types

The introduction to this chapter presented the traditional division of bonding types in solids (Refs. 2, 216, 376). The use of one of them, the "metallic bond," has been much curtailed during the course of the chapter. In its nearly-free-electron format, although of use in the study of the alkali metals, for structural and electronic problems elsewhere in the periodic table it can be complementary but is invariably inferior to the tight-binding approach. It is shown in Chapter 5 that even for the electronic picture of elemental calcium, a case where the nearly-free-electron model is often used, the tight-binding approach leads to an insightful picture of the geometrical and electronic structure using the orbital language of the chemist. The discussion of the pressure dependence of its conductivity in terms of the orbital model is a very useful one. Just as there are advantages to using the same orbital picture in diverse areas of molecular chemistry [organic polyenes, hypervalent main group molecules, and organometallic complexes, for example (Ref. 5)], so the orbital picture has a wide area of applicability in the area of solids. Some solids (such as the Zintl compounds) may be described in terms of localized Lewis structures in the same way as some molecules (such as CH_4). These materials are equally well described by the delocalized picture, the one that has to be used to understand spectroscopic properties. Some solids (such as the transition elements) have to be described in terms of delocalized orbital pictures, just as some molecules (such as the boranes) have to use such a model. There is really no electronic distinction between the two. As shown in Section 2.2, diamond is an insulator because it has a gap between filled and empty bands. The delocalized band model was used for this material, but in addition it can be described in terms of four localized two-center, two-electron bonds, just as in the organic chemists' description of methane.

In the area of solids traditionally regarded as "ionic," it is clear that the general electronic descriptions of these materials are not given well by simple electrostatic considerations. [the cohesive energies of such systems, however, are very well matched by such a model, (Refs. 198, 345).] The importance of the generation of energy bands from combinations of orbitals that are largely localized (located) at the anion or cation sites is now well established. Similarly the presence of the filled oxygen band of rutile leads to a description in terms of electrons largely localized on the oxygen atoms. In the organic metals of Section 4.10 the bands are generated via interactions of orbitals localized (spatially) on individual molecules. The strength of the crystal orbital model is that it is able to bridge these areas. Thus there is no built in requirement that any of these three descriptions has to occur. The transition from "ionic" to "covalent" interactions proceeds smoothly as the wave functions describing the levels of the bonding band take on additional anionic character. Whether the electrons of the organic metal are localized on individual molecules or delocalized through the solid depends on the strength of the inter- and intraunit interactions. Metallic behavior is simply the result of a partially filled energy band and does not occur through any special bonding mechanism. This topic will be taken up again in the next chapter.

5 □ Metals and insulators

Certainly one of the most interesting areas from both the chemical and physical points of view is the connection between metals and insulators. What decides whether a particular solid is a conductor of electricity, how well it does it, and how external events such as pressure and temperature control these physical properties are important questions to ask (Refs. 4, 47, 109, 112, 133, 144–146, 182, 192, 197, 202, 212, 232, 252, 253, 282, 328, 344, 397). There are many interesting physical effects associated with metals and insulators and the transition between them. The temperature dependence of the resistivity and the role of magnetism are just two. This chapter concentrates on chemical aspects. In recent years the study of the metal–insulator transition has received an added stimulus from the discovery (Refs. 87, 330) of the high-temperature superconductors. Here, with small changes of chemical composition, $La_{2-x}Sr_xCuO_4$ changes from being an insulator ($x = 0$) to a superconductor (with a maximum T_C at $x \sim 0.15$) to a "normal" metal ($x > 0.25$). This is one of several systems (Ref. 310) where physical properties change dramatically on doping. In this chapter several of the electronic situations encountered have readily identifiable molecular analogs, but many do not. In several places in this chapter the discussion will come close to the limits of present theoretical understanding.

5.1 The importance of structure and composition

Metallic conduction is a consequence of a partially filled energy band. Figure 5.1(a) shows, within the framework of free-electron theory, how in the absence of an applied field each filled state with a momentum k has a partner with momentum $-k$. As a result, there is no net flow of current. Figure 5.1(b) shows the case where there is an applied field. The field raises the energy of the k states at the expense of the energy of the $-k$ states, yielding a net current. This can only occur if there are empty states just above the unperturbed Fermi level; in other words, there is no band gap. The situation for the filled band is shown in Fig. 5.1(c), (d). Although the picture of Figure 5.1 employs the free-electron model, a similar viewpoint comes from the tight-binding picture. Here the wave function has an equal amplitude in each cell so that it is delocalized to an extent depending on intercell overlap. Of course, even when there is a band gap, the wave function is still delocalized in this sense, even though it is an insulator. This should be distinguished from the case where small intercell overlap leads to localization of the wave function and hence produces an insulator because of a low transition probability for interatomic hopping. These ideas will be expanded later.

Since the electronic structure of a solid is controlled by its geometrical structure, in principle one ought to be able to make comments concerning the electrical

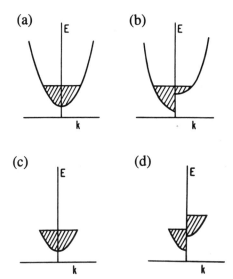

Figure 5.1   The microscopic behavior behind electronic conduction. The levels at $\pm k$ (a) without and (b) with an applied field for a partially filled band. (c) and (d) the same but for a filled band.

properties of a material from the way its atoms are arranged. Although the use of pseudopotential theory is introduced in Section 8.4, for most systems of chemical interest the tight-binding, orbital model is the best one for this purpose. Perhaps the electronic structures of diamond and graphite are the best places to start. Graphite is a semimetal (or a zero-gap semiconductor), as is apparent from the way its band structure was derived in Section 3.1. Diamond, however, is an insulator with a large gap between filled and empty orbitals. It is the difference in structure between the two that makes the difference. It was shown in Section 4.2 how this is determined purely by the possibility of s–p mixing between the s and p bands of diamond, absent by symmetry for the π system of graphite.

Diamond is an example from a much larger group of solids, the Zintl compounds (Refs. 211, 319). These are compounds, often between an electropositive atom and atoms from Groups 13–17 of the Periodic Table, that satisfy the octet rule around each electronegative atom. Thus $CaIn_2$ contains Ca^{2+} ions stuffed inside a hexagonal diamond network of In^- (isoelectronic with a Group 14 element). The structure of $GaGa_2$ (Fig. 5.2) might be called a Zintl compound since it contains Ca^{2+} ions sandwiched between graphite-like sheets of Ga^- ions. (The actual structure type is that of AlB_2.) In its more extended use of the idea, Zintl clusters are those that obey Wades's rules (Ref. 359). These rules set the number of skeletal bonding electron pairs at $(n + 1)$ for an n-vertex deltahedron. They are of great utility. For the most basic type of Zintl compound, however, if there are N_b obvious two-center, two-electron bonds and N_n obvious lone pairs, then, from the discussion of Section 4.3, there will be $N_n + N_b$ filled deep-lying bands (b = bonding, n = nonbonding), separated from N_b empty antibonding bands (a = antibonding) in the solid as shown in **5.1**. Thus GaSe (Fig. 5.3) is a semiconductor possessing Ga–Ga and Ga–Se two-center, two-electron bonds and Se lone pairs. Zintl phases such as KSnSb, where the tin and antimony atoms form a derivative structure of α-arsenic (Fig. 5.4), are

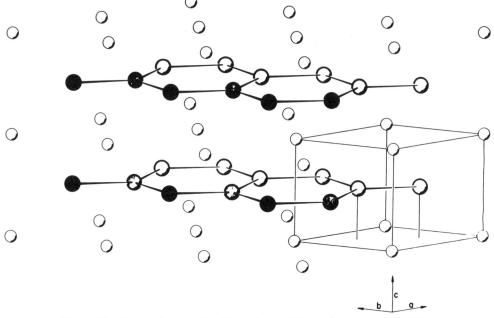

Figure 5.2 The structure of AlB$_2$. (Reproduced with permission from Ref. 280.)

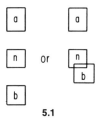

5.1

also semiconductors, with three two-center, two-electron bonds linking each tin or antimony atom and one lone pair at each center. From the picture of **5.1** there is an obvious way to generate a metal, namely arrange for the empty and filled bands to overlap as in **5.2**. This may be achieved by increasing the pressure on the solid. Iodine,

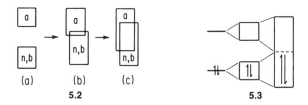

for example, forms a molecular solid, but the band structure is similar to that in **5.1**, since there are significant interactions between the orbitals on adjacent atoms. As pressure is applied the band gap smoothly drops to zero (~ 170 kbar), and at the same time the solid becomes metallic. A similar effect is predicted (Refs. 383) to occur for hydrogen, a process shown in **5.3**. Here the critical pressure is certainly much

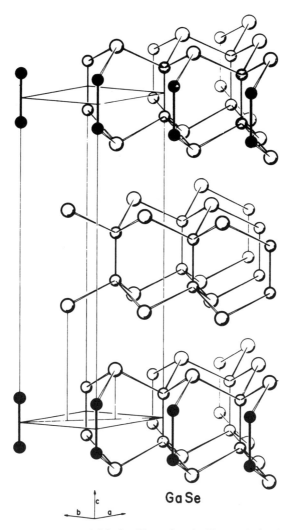

Figure 5.3   The structure of β-GaSe. (Reproduced with permission from Ref. 280.)

higher (4 Mbar or higher) (Ref. 232). The bonding and antibonding orbitals of the gas-phase molecule start to broaden into bands as the interaction between adjacent molecules increases. Eventually at very high pressures bonding and antibonding bands coalesce. This is just the reverse of the process described in **2.19** and **2.20**. Band overlap of a different type occurs under pressure for materials such as CsI (Ref. 313). Here a filled p band and empty d band located on the anion broaden and overlap under pressure. A metal results, that is, Figure 2.10(c) → (b).

Such a structurally induced metal–insulator transition can occur at ambient pressure by just changing the chemical identity of the material. The structure of α-arsenic, also found for antimony and bismuth, is shown in Figure 5.4. It emphasizes the three bonds around each atom. However, there are relatively close contacts between the sheets so that the bands are broadened enough by intersheet interactions

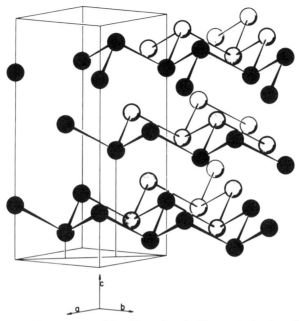

Figure 5.4 The structure of As. (Reproduced with permission from Ref. 280.)

so that there is a small overlap (**5.2b**), leading to metallic character. As the elements of this group becomes heavier, the ratio (d_2/d_1) of the "nonbonded" to "bonded" distances decreases. Table 5.1 shows these values for the heavier elements of Groups 15 and 16. The relationship of this structure to the simple cubic arrangement, where this is unity, is shown in Figure 3.11. Figure 5.5 shows the structures of the heavier Group 16 elements, and the correlation between the spiral structure found for the lighter elements and the simple cubic structure found for α-polonium, where $d_2/d_1 = 1$. As this ratio increases, then the bands of **5.1** broaden and the overlap between them increases. In general, the weaker driving force for distortion on moving down the groups of the periodic table leads to larger band overlap and the generation of metallic conductors. This process is sometimes called metallization (Ref. 160), and is increasingly important as the mass number of the atomic constituents of the solid increases.

Table 5.1 The ratio of nearest-neighbor to next-nearest-neighbor distances in the elements of groups 15 and 16

Element	d_1	d_2	d_2/d_1	Element	d_1	d_2	d_2/d_1
P	2.19[a]	3.88	1.77				
As	2.51	3.15	1.25	Se	2.32	3.46	1.49
Sb	2.87	3.37	1.17	Te	2.86	3.46	1.31
Bi	3.10	3.47	1.12	α-Po			1.00

[a] Average value.

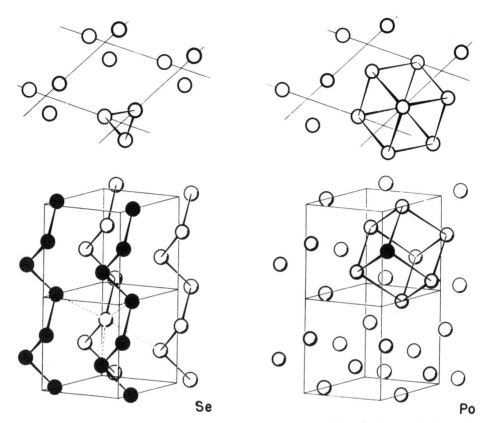

Figure 5.5 The structure of Se and the simple cubic structure of Po. At the top is shown the relationship between the two structures. (Reproduction with permission from Ref. 280.)

The compounds $Sr[Sn_2As_2]$, $K[SnSb]$, and $K[SnAs]$ contain arseniclike sheets as befits their description as Zintl compounds (Ref. 321). The Group 14 and 15 atoms alternate in the structure. All three are semiconductors, in contrast to the behavior of grey arsenic. For $Sr[Sn_2As_2]$ the gap is smaller than in the other two compounds. The reason is easy to see. In the potassium-containing compounds these ions are arranged in layers that alternate with the SnAs or SnSb sheets. In the strontium-containing compound, these metal ions are only found between every other pair of sheets. The results in the presence of one set of nonbonded close contacts between adjacent SnAs sheets in $Sr[Sn_2As_2]$, which are absent in the potassium-containing compounds. In grey arsenic each sheet has close contacts on both sides. Thus the electropositive metal atoms in the Zintl compounds acts as spacers that hold the sheet apart so that intersheet interactions are reduced.

A similar type of behaviour occurs (Ref. 326) in the high-pressure form of LiGe. The low-pressure form consists of a three-connected net for the germanium atoms, as would be expected for the Ge^- species isoelectronic with phosphorus. However, the high-pressure form has a more complex structure, which at first sight ought to be a semiconductor, too. Its structure, shown in Figure 5.6 has both two- and four-coordinate atoms, and simple electron counting suggests that the octet rule is

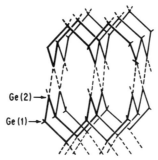

Ge(2) →

Ge(1) →

Figure 5.6 The structure of the high-pressure form of LiGe. (Adapted from Ref. 326.)

satisfied at both centers. Indeed the results of a band structure calculation [Fig. 5.7(a)] on a single slab of the structure (with Ge–Ge distances of 2.7 Å) show a nice gap between filled and empty bands. However, shown dashed in Figure 5.6 are some longer Ge···Ge contacts (3.04 Å) between these slabs. These are sufficient to broaden the bands of the slab such that the gap disappears (Ref. 326) [Fig. 5.7(b)].

Metals are often generated under pressure as a result of a change in structure. Thus arsenic and black phosphorus transform under pressure (Ref. 196) to the simple cubic structure and metallic conductivity is found under these conditions. This is the reverse of the process shown in Figure 3.11. The behavior of the conductivity with applied pressure is somewhat different from that described for iodine. If there is a change in structure under pressure, then the onset of metallic conductivity is much more abrupt.

The prediction concerning generation of metallic behavior for semiconductors under pressure appears to be quite a general one. The related picture is that of materials that are metallic, becoming insulators under "negative pressure." This may only be realized in practice by studying liquid or gaseous systems where the density may be reduced relative to the normal conducting structure. So from Figure 2.11 the prediction can be made that calcium becomes an insulator because of band separation at low densities.

5.1 gave a readily visualized electronic picture of the Zintl compounds. There are some other rather straightforward considerations that allow predictions to be made for other solids. The general statement can be made that the simple cubic structure will be metallic for elemental boron, carbon, nitrogen, and their congeners. The reason for this is quite simple. Since there is only one atom per unit cell, the dispersion for any orbital will be continuous in $\mathbf{k}$ space. Partial occupancy, therefore, will always lead to metallic behavior as long as the band model is appropriate. Elemental polonium with this structure under ambient conditions is an example. The metallic nature of arsenic and phosphorus under pressure (as the structure changes to the simple cubic one) was noted previously.

With more than one atom (and thus more than one set of orbitals) per cell, no statements may be made without recourse to the geometry and the nature of the atoms themselves, except the obvious result that an odd number of electrons per cell is a metal in the band model. So, as discussed, graphite, with two atoms per cell, is semimetallic since π and π^* bands touch at the point K in the Brillouin zone, but isoelectronic BN is a white insulator. Figure 7.17 shows (Ref. 43) a similar picture for

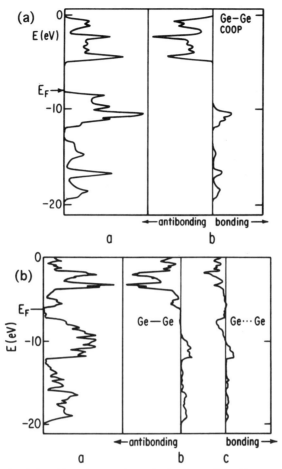

Figure 5.7 (a) Calculated densities of states for a single sheet of the high-pressure LiGe structure. The COOP curve for the Ge–Ge bonds is also shown. (b) As for (a), but now for the full three-dimensional structure that includes interactions between the sheets. Notice that the COOP curve for the intersheet Ge···Ge interactions (panel c) shows occupation of both bonding and antibonding parts of the band formed by intersheet overlap. (Adapted from Ref. 326.)

the NaCl structure. When both of the atoms are chemically identical the simple cubic structure results. Increasing their electronegativity difference eventually leads to the insulating rocksalt structure for eight electrons. The factors influencing the electronic preference for the rocksalt over the sphalerite structure are discussed in Section 7.5.

A well-discussed electronic question is the origin of the band gap in diamond. In Section 4.2 the band structure was derived by brute force solution of a secular determinant and s–p mixing identified as a vital ingredient. However, the general form of the band structure of a tetrahedral solid, such as diamond, may be understood in qualitative terms in the simple way shown (Ref. 160) in Figure 5.8. First s and p orbitals on the same atom are combined to form sp^3 hybrid orbitals (a), pairs of which are linked together to give bonding and antibonding combinations (b). Since these hybrids are not orthogonal (as described for BeH_2 in Section 4.3), they broaden

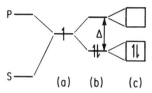

P ——

S ——

(a)   (b)   (c)

Figure 5.8 (a) Combination of *s* and *p* orbitals on the same atom to give sp^3 hybrid orbitals. (b) Pairs of hydrids are linked together to give bonding and antibonding combinations. (c) The hybrid orbitals broaden into bands through on-site interactions.

into bands as shown in (c). In quantitative terms the energy of the sp^3 hybrid is $E_{hy} = \frac{1}{4}(e_s + 3e_p)$ where $e_{s,p}$ are the energies of the atomic levels. They then interact to give bonding and antibonding pairs separated in energy by $\Delta = 2\beta_2$. Finally, by allowing each of these combinations to interact with similar hybrids on the same atom [an energy of $\beta_1 = \frac{1}{4}(e_s - e_p)$] leads to the formation of the two energy bands. Since the bandwidth of this type of band from Figure 2.1 is four times the interaction energy of the basis orbitals, the width of each of the bands is approximately $|(e_s - e_p)|$. The band gap, E_g, is then roughly equal to $\Delta - |(e_s - e_p)|$. It is interesting to see how it varies down the Group 14 series. The term $(e_s - e_p)$ does not change dramatically, but Δ varies inversely as the square of the internuclear distance. The largest change (Table 5.2) is from carbon to silicon. In qualitative terms it is easy to see why carbon is an insulator, and silicon, germanium, and grey tin are semiconductors. Smaller orbital interactions lead to smaller gaps between bonding and antibonding or between nonbonding and antibonding bands, leading to larger overlap between occupied and unoccupied bands. This general picture of increasing "metallization" on moving down this group is also applicable to the elements of the other main group.

This process may be performed in a more analytical way as shown (Ref. 39) in Figure 5.9 by solving the band structure for the four-coordinated net assuming that there are just the two hopping integrals, $\beta_{1,2}$, which define, respectively, the on-site and intersite interactions in the structure. The band structure that results [Fig. 5.9(b)] has most of the features of the more accurate one of Figure 4.16. It may be readily shown that a band gap exists if $|\beta_1/\beta_2| < \frac{1}{2}$.

An interesting result comes from comparison of this discussion and that for the Peierls distortion of Section 2.4. One of the criteria for the generation of an

Table 5.2 Single bond distances in the group 14 elements

Element	Single bond distance, d (Å)	$1/d^2$
Carbon	1.54	1.00^a
Silicon	2.34	0.43
Germanium	2.44	0.40
Tin	2.80	0.30
Lead	2.88	0.29

^{a} Scaled relative to the value for carbon.

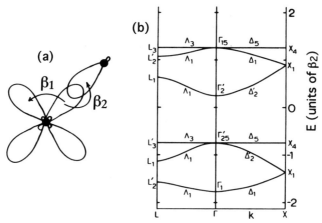

Figure 5.9 The band structure of diamond (b) using the simplest model with hybrid orbitals (a) that are linked together by on-site and intersite interactions. (Adapted from Ref. 39.)

insulator for the half-filled band situation is that there are two different interaction integrals between the orbitals of the problem. In the case of the alternating chain (Peierls distortion), these are β_1 and β_2, and in the diamond case, the different inter- and intra-atom interaction integrals between the atomic hybrids. Of course, polyacetylene can be converted in its distorted geometry into a metal simply by doping the upper band of the two with electrons.

Returning to the differences between graphite and diamond, it is certainly true that a single Lewis structure may not be written to describe the bonds in graphite. Two resonance structures are shown in **5.4**. Similarily two resonance structures

5.4

5.5

(**5.5**) need to be written to describe the electronic arrangement in polyacetylene (assuming that all the C–C distances are equal). In diamond, a single Lewis structure with "localized" two-center, two-electron bands is quite adequate. In terms of the possibilities for localization discussed in Section 4.3, the N filled delocalized orbitals for diamond may be arbitrarily localized to give N filled orbitals that exactly account for all of the N bonds between adjacent carbon atoms. As described in Section 4.3, both localized and delocalized pictures are in principle appropriate for diamond, but in symmetrical polyacetylene and graphite this is not the case. Just as in benzene, where there are three π electron pairs and six linkages, a delocalized π picture is all that is possible. Within the band model this results in the presence of a partially occupied band and thus metallic behavior. Thus partially filled bands demand an electronic description "delocalized through the crystal" and thus metallic character. The filled bands of the Zintl compounds mean that the electrons can be "localized" into Lewis-type two-center, two-electron bonds, and thus an insulator results. The two cases are often described as containing itinerant or localized electrons, respectively. (There are other mechanisms for localization that will be described in Section 5.4.)

The insulating character of the rocksalt structure for NaCl is due to the large electronegativity difference between Na and Cl, which leads to a substantial energy gap between the Na and Cl located bands. Similar effects occur in many other solids irrespective of their structure. One of the reasons most silicate and phosphate minerals are insulators is because the bonding bands, largely oxygen located, lie much deeper than their antibonding (largely Si- or P-located) analogs (Fig. 4.20). Also, as shown in Section 5.4, many metal ions found in such mineral structures, which in principle could form delocalized d bands, actually contain localized electrons.

TiO_2 of Figure 3.7 is another example. The energy gap between the O-located bonding bands and Ti-located antibonding bands is nonzero (about 3 eV from experiment). The bonding in the solid-state structure of rutile leads to energy bands, but the d^0 solid is an insulator because of the large band gap. The calculated density of states for the rocksalt MO compound shows overlap of the e_g and t_{2g} bands of the metal (Fig. 3.4). If the bands do not overlap, as in the calculation for rutile in Figure 3.7, then semiconducting behavior is possible for the low-spin d^6 configuration (**4.2b**). In a related system that contains octahedral metal coordination, the perovskite $LaRhO_3$ is nonmetallic with a gap of ~ 1.6 eV. $LaCoO_3$ with the same structure is also nonmetallic, but only just so. At higher temperatures a whole series of transformations take place triggered by the small e_g–t_{2g} gap. Thus the chemical identity of the system is vital here. Unlike the case of the Zintl compounds, where electronic saturation usually leads to semiconducting behavior, the e_g–t_{2g} gap in transition metal systems is frequently nonexistent and the state of affairs system dependent.

The electronic picture for graphite is in fact a little more complex than suggested in Section 3.1. The idea of filled π and empty π^* bands that touch is one that comes from the treatment of the isolated 6^3 sheet. It does, however, need to be modified to describe the real three-dimensional structure of the material. Although the sheets are usually described as being held together by van der Waals forces, there is a non-negligible orbital interaction between the two. This is readily visualized from the geometrical structure of the solid shown in **5.6**. Notice that the mode of sheet stacking leads to two types of carbon atom, those (A) that form a linear chain of atoms along the c direction and those (B) that lie in the middle of the six-membered rings of sheets above and below. Of course, the interaction along the chain is small, since the intersheet separation is large (3.35 Å), but it is large enough to change the orbital picture subtly. How the sheets interact is seen quite nicely experimentally from results using the scanning tunneling microscope (STM). This instrument probes the Fermi

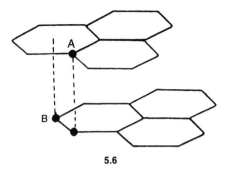

5.6

level of the surface density of states of metallic solids, and thus atoms where there is no such electron density are not imaged. The experimental STM picture of graphite is shown (Ref. 342) in **5.7**, and interestingly only half of the atoms of the sheet may

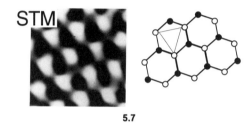

5.7

be imaged giving the appearance of a 3^6 rather than 6^3 net. (The picture was kindly provided by Professor S. G. Louie.)

The dispersion associated with the single (and hence two-dimensional) sheet is shown in Figure 3.5. Since each $p\pi$ orbital contains one electron, these bands are half-full and the Fermi level lies at the point K. This point is one where the two bands are nonbonding and degenerate with energies of α. The pair of orbitals that will be most useful in the description of the levels are shown in Figure 3.6(c). The electron density for each of the two carbon atoms of the unit cell should be identical at the Fermi level for this single-sheet model. Although the distance between adjacent sheets is about 3.3 Å, the value of H_{ij} representing carbon $p\sigma$ interaction (β') is about 0.5 eV, so that there may be significant dispersion perpendicular to the sheets. The dispersion along $\mathbf{a}_3$ for $p\sigma$ orbitals on the B atoms is zero since there is no interaction between orbitals of this type on adjacent sheets. Calculation of the dispersion along the line K $(\frac{1}{3},\frac{1}{3},0)$ to H $(\frac{1}{3},\frac{1}{3},\frac{1}{2})$ (Fig. 3.2) leads to the very simple energetic dependence on $\mathbf{k}$ appropriate for a one-dimensional band with one orbital per unit cell, $E = \alpha \pm 2\beta' \cos(k_z c/2)$. The dispersion curves for the two types of $p\pi$ orbital are shown (Ref. 342) in Figure 5.10(a). The result is that some of the levels associated with the A atoms drop below the Fermi level of the isolated sheet. The larger elcctron density at the Fermi level is now found for the type B atoms. In terms of the electrical properties such overlap leads, from the picture of Figure 5.10(b), to a higher number of carriers at the Fermi level. So the three-dimensional structure is a better metal than predicted for a single sheet, where the bands exactly touch.

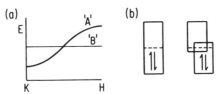

Figure 5.10 (a) Dispersion of the π bands of graphite ($K \to H$) as a result of intersheet interaction. Only the orbitals located on the A atoms have any interaction along $\mathbf{a}_3$. (b) The overlap between π and π^* bands that such interaction allows.

5.2 The structures of calcium and zinc

As noted previously, since the energy bands of solids broaden as the hopping integrals increase with a decrease in interatomic separation, a general prediction is that all materials should become metallic under pressure. However, the Group 2 elements Ca–Ba, with the s^2 configuration, behave rather differently in that their conductivity decreases with an increase in pressure. For the case of Yb, a semiconductor is generated under pressure. Other s^2 metals, zinc and cadmium, with a d^{10} core ($d^{10}s^2$) show an unusual structural distortion. Both elements crystallize in the hcp structure, but with c/a ratios (1.856, 1.886, respectively) quite different from the ideal value of 1.633. Thus the structure is stretched along the c direction. Although the calculations described in Section 8.1 correctly predict that for twelve electrons the preferred structure is this distorted (or η-hcp) phase, there are some interesting electronic ideas behind (Refs. 32, 252) the behavior of these d^0s^2 and $d^{10}s^2$ metals.

The key to understanding the behavior of these materials lies in the extension to three dimensions of the discussion of the avoided crossing of the s and p bands of the one-dimensional chain of Section 2.1. As a result of the difference in parity of the two orbitals, the dispersions of the s and p bands are expected to be of opposite sign. However, if the atomic s–p orbital separation is small compared to the magnitude of the interaction between them, then the two curves, expected to cross somewhere along $\mathbf{k}$, will be involved in an avoided crossing. The important result of the avoided crossing is that the top of the lower-energy band is, by symmetry, the in-phase bonding arrangement of the valence p orbitals, and the bottom of the upper band is the out-of-phase antibonding arrangement of the valence s orbitals. Without such a crossing the top of the lower-energy band is the out-of-phase antibonding arrangement of the valence s orbitals, and the bottom of the upper band is the in-phase bonding arrangement of the valence p orbitals, as shown schematically in Figure 5.11.

This difference is very important. Elemental zinc corresponds to case (a), but elemental calcium to case (b). For the former notice that on this model the electronic situation is a little like that for He$_2$ of Figure 1.1(b). Bringing together two filled s bands should lead to a repulsion between the atoms. The result for zinc and cadmium is an increase in metal–metal distance along c. Changing the internuclear separation should have very different effects on the relative energies of the top of the lower band and bottom of the upper band of Figure 5.11(a), since there will be controlled by the bonding or antibonding nature of the orbitals concerned. For the case of calcium, a reduction in internuclear separation leads to a drop in energy of the levels at the top of the lower band (since they are bonding) and an increase in energy of the levels at

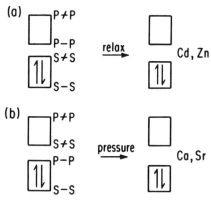

Figure 5.11   Behavior of the two types of orbital situations shown in Figure 2.9(a). No crossing of the s and p levels with **k**. The top of "s" band is s–s antiboding, and, as in Zn and Cd, the structure relaxes to give a larger than ideal c/a ratio. (b) Crossing of the s and p levels with **k**. The top of the "s" band is now p–p bonding and, thus under pressure in Ca and Sr, drops to lower energy.

the bottom of the higher-energy band (since they are antibonding). Thus as the metal–metal distance becomes smaller, the gap between the two bands increases. This result is behind the pressure effect on conductivity observed in elemental calcium. In order to see this in its proper perspective, recourse must be made to the full band structure of the material. This is shown (Ref. 32) in Figure 5.12. The point in **k** space where the avoided crossing occurs is the point L. The Fermi level is shown by a solid horizontal line, and as may be seen the electronic picture is quite similar to that of Figure 2.10(b) in that filled and empty bands overlap. Notice that on reducing the internuclear separation (to simulate compression) then the levels at L move in just the way predicted. There are now fewer electrons at the Fermi level for this compressed structure, and the material is a poorer conductor of electricity.

5.3   Geometrical instabilities

As described in Sections 2.4 and 3.2, partially filled bands are often unstable to some extent to a Peierls (Refs. 82, 282), or charge-density wave (CDW), distortion, which lowers the energy of the occupied orbitals and raises the energy of the unoccupied ones. If large parts of the Fermi surface may be translated by a unique vector **q** so as to be superimposable on other parts of the surface, then the Fermi surface is said to be strongly nested. The nature of the distortion is given by the vector **q**. In a simple way the extent of nesting measures how many of the levels in **k** space will be stabilized by the distortion. Such distortions are most likely in electronically one-dimensional systems since all of the Fermi system is nested, whereas in more than one dimension some levels will be stabilized during the distortion, but others may go up in energy, a result quantified by the extent of the nesting. There are exceptions to this statement. As described in Section 3.1, the distortions away from the simple cubic structure for the group 15 elements arise as a result (Refs. 60, 282) of an electronic structure that is effectively that of three almost independent perpendicular one-dimensional systems. The Fermi surface is thus a cube. The half-filled band of the CuO_2 sheet of the cuprate

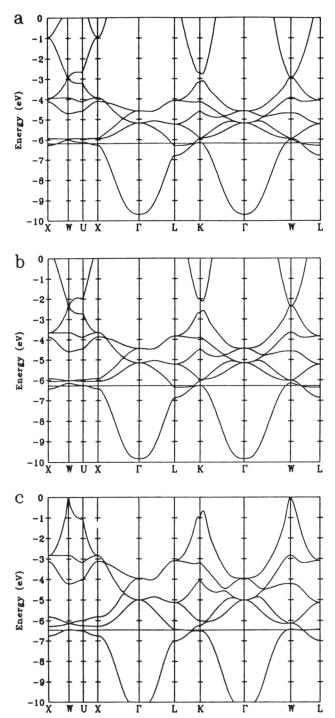

Figure 5.12 The calculated band structure of elemental calcium for three different values of the cell volume V relative to that at ambient pressure, V_0, (a) $V/V_0 = 1.0$, (b) 0.86 and (c) 0.60. (Adapted from Ref. 32.)

superconductor (Fig. 5.38), if the band model were a valid electronic description, should be susceptible to a two-dimensional distortion that is simply the sum of two perpendicular one-dimensional ones.

It is interesting to explore what governs the magnitude of the distortion. It should be recognized in this context that the driving force for the geometrical distortion comes from those occupied energy levels that lie at highest energy. Thus, resisting the distortion, driven by these electrons, are the "elastic" forces of the underlying electronic structure. If the energy of the latter as a function of the distortion is written using Hooke's Law as $2V = k(\delta x)^2$, where δx is the displacement away from the equilibrium geometry, then for the asymmetric distortion of **5.8** the energy per two

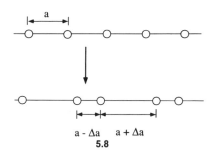

5.8

linkages is

$$2V = k(a + \Delta a)^2 + k(a - \Delta a)^2 = 2[ka^2 + k(\Delta a)^2] \tag{5.1}$$

This is a function obviously minimized for $\Delta a = 0$, the undistorted arrangement. In general, therefore, a solid composed of a collection of balls and springs and obeying Hooke's law has its minimum energy at the most symmetrical structure. The same result comes from a model where the interatomic potential is given by one of the Lennard–Jones type. If $V = -A/r + B/r^n$, then the total energy at equilibrium is $\propto -Aa[1 - (1/n)]$. On distortion as before by $\pm \Delta a$, then the total energy is now $\propto -A/a[1 - (1/n)] + nA(\Delta a/a)^2$, a function minimized for $\Delta a = 0$. (Parenthetically it should be noted that extending this idea to three-dimensional systems leads to insights into why there are crystals. A given motif repeats regularly in space.) Thus the energetics of the CDW distortion need to overcome the inherent stiffness of the underlying electronic framework in order to convert a metal into a semiconductor.

It is also important to understand why the application of pressure frequently reverses the stability of the parent and Peierls-distorted structures. Clearly from the discussion of Section 2.4 the distorted structure will always be the more stable whatever the pressure. This result remains true even if second-nearest-neighbour interactions are included. Figure 5.13 shows the qualitative results of a slightly more sophisticated model, one that includes the repulsive part of the interatomic potential (B/r^n of the previous paragraph). The difference in repulsive energy for the two structures is proportional to $(2a)^{-n}x^2$, the repulsion being less fierce for the symmetrical structure. The result is an energy minimum that lies to smaller a for the symmetrical structure than for the asymmetrical one. Thus, under conditions where

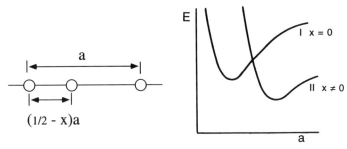

Figure 5.13   The dependence of the energy of the symmetrical (I) and Peierls-distorted (II) chains on cell volume.

it is the repulsive part of the potential which dominates (as a becomes smaller), the symmetric structure becomes more stable. Increasing the pressure can therefore lead to a reversal of the Peierls distortion.

The variation in the magnitude of the driving force from system to system is an interesting one to explore. **5.9** shows a van Arkel–Ketelaar (Refs. 208, 352) diagram

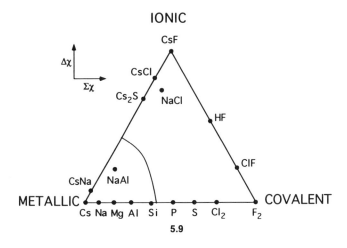

used traditionally to separate compounds into "covalent," "ionic," and "metallic" regions. These diagrams may be constructed quantitatively (Ref. 8) using as a horizontal coordinate the sum of the electronegativities of the constituents, and as a vertical axis their difference. [The electronegativity scale used comes from multiplet-averaged values of the experimental ionization energies using the scheme of Allen or that of Martynov and Batsanov (Ref. 234) described in Section 6.2.] Clearly the electronegativity difference is a good measure of ionic character, but moving horizontally across the diagram using the electronegativity sum leads to a good separation of "metallic" and "covalent" systems using the van Arkel–Ketelaar language. Thus the process of metallization occurs in systems with a smaller than critical electronegativity sum. These are compounds whose constituent elements have diffuse wave functions such that small changes in nearest-neighbor interactions do not lead to large changes in the energy. Also, through the Wolfsberg–Helmholz approximation of Section 1.1, the interaction integrals β_{12} between orbitals on adjacent atoms are set proportional to $(\alpha_1 + \alpha_2)$, a function that scales directly with electronegativity as described in Section 6.2. Thus the magnitude of the nearest-neighbor interactions and the driving force for distortion is expected to be smallest in those systems with the smallest electronegativity. The electronic description of a metal is one of the formation of energy bands by orbital overlap, but with a small electronic driving force for the CDW distortion, unable to overcome the elastic forces of the underlying structure. In general, one expects that the larger the distortion, the larger the band gap that is opened up. One numerical definition of metallicity (Ref. 160) does involve the magnitude of the band gap in these systems. In this scheme, the smaller the gap, the larger the metallicity. Nowhere in the discussion here is the term *metallic bonding* used. It should be dropped from use.

The driving force for the CDW distortion also depends upon the magnitude of the distortion vector. If, for the one-dimensional system, $q = 2k_F = (1/j)(2\pi/a)$, then the new unit cell is j times the old. It may readily be shown (Section 2.4) how the electronic driving force for the distortion decreases as j increases. As a general result, a static long-wavelength distortion, expected from the presence of a small amount of electron doping into an empty band does not usually occur.

Temperature is an important factor in determining CDW instabilities. A simple model will show why this is so. At absolute zero the orbital occupancy of the levels below E_F is 1, and the occupancy of levels above E_F is 0. However, as the temperature is raised, the higher energy levels become thermally populated and the lower energy levels depleted. The statistics of such behavior in terms of the fractional occupancy, $f(E)$, of the levels with energy E is described (Ref. 16) by the Fermi–Dirac distribution function

$$f(E) = 1/[1 + \exp(E - \mu)/kT] \tag{5.2}$$

where μ is the chemical potential (equal to E_F at $T = 0$). As the temperature is raised, $f(E) < 1$ for levels just below μ, and $f(E) > 0$ for the levels just above μ. Thus, as shown in Figure 5.14, this tends to reduce the driving force for the distortion by population of some levels that increase in energy on distortion and loss of occupation of those levels that are most stabilized on distortion. As the material is cooled, a critical temperature will be reached where the thermally excited population is insufficient to stop the distortion from taking place. Table 2.1 shows some one-dimensional examples where this occurs. $K_2Pt(CN)_4Br_{0.3}$ is another. It is a metal above 150 K just as VO_2 is a metal above 340 K. (More complex systems will be described in the following.)

In one dimension all of the Fermi surface is nested, and a gap opens up on distortion. The result is a metal–semiconductor, or metal–insulator, transition. The Fermi surface has disappeared, there being no place in $\mathbf{k}$ space where filled and empty

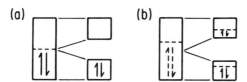

Figure 5.14   Effect of temperature on CDW distortions, (a) at 0 K, (b) at finite temperature. Lowering the temperature favors the structure where a gap is opened in the density of states.

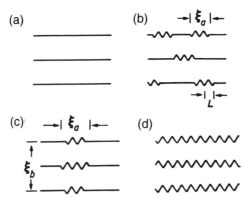

Figure 5.15 Locking in of the CDW. (a) Temperature above the critical temperature for CDW formation. (b) The temperature is lower than in (a), and CDWs are formed dynamically at various parts of the individual chains. (c) Lowering the temperature further leads to coherence between the CDW segments in neighboring chains. (d) At the lowest temperature the CDW distortion is locked in, in both dimensions. (Adapted from Ref. 82.)

levels are degenerate. In two and three dimensions this is usually not true. As shown in Figure 3.19, the incompletely nested two-dimensional surface of the undistorted structure leads to the generation of pockets in the distorted structure. Since there is still a Fermi surface, the system is metallic, but clearly it is not such a good metal as before. The temperature dependence of the resistivity of such a system often takes the form shown in Figure 3.19(c).

Figure 3.19(c) shows (Ref. 82) that the metal–insulator transition that arises from the CDW distortion is not sharp but sets in over a finite temperature range. Imagine, for example, the set of one-dimensional, weakly coupled chains of Figure 5.15(a). As the temperature is lowered, CDWs are formed dynamically at various parts of the individual chains. Lowering the temperature further leads to coherence between the CDW segments in neighboring chains, followed eventually at the lowest temperature by the locking in of the CDW distortion in both dimensions. The various stages of this process may be followed by diffuse x-ray scattering. At the lowest temperature [Fig. 5.15(d)], superlattice spots, separated by $\mathbf{q}$ in reciprocal space, are seen, which immediately give an experimental determination of the wavelength of the CDW (Ref. 249).

As an example, consider the properties (Ref. 373) of the rare earth molybdenum bronze, $La_2Mo_2O_7$. This has a simple structure, but the electronic principles behind its structural and electrical behavior are very similar (Ref. 82) to those determining the properties of a variety of metal oxides including the red, purple, and blue bronzes and the Magnélli phases. The material consists of infinite chains (**5.10**) of stoichiometry $Mo_2O_7^{6-}$ with La^{3+} ions inserted between them. The material is metallic down to 125 K, where it undergoes a phase transition. The Mo atoms have a d^2 configuration, and so study of the deepest-lying d bands will be sufficient. The Mo–Mo distance across the shared octahedral edge of the structure is 2.478 Å, and thus the lowest bands are expected to be strongly involved in metal–metal interactions. The calculated band structure is shown in Figure 5.16 and shows a Fermi level that crosses two bands. The corresponding Fermi surfaces are shown in Figure 5.16. There is an obvious choice of inter- or intraband nesting here. The interband nesting is involved

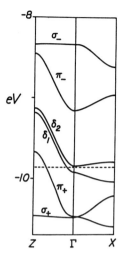

Figure 5.16 The "t_{2g}" bands for the $Mo_2O_7^{6-}$ slab of $La_2Mo_2O_7$. Here $\Gamma = (0, 0)$, $X = (\mathbf{b}_1/2, 0)$, $Z = (0, \mathbf{b}_3/2)$. The dashed line is the Fermi level. (Adapted from Ref. 82.)

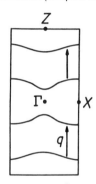

Figure 5.17 Combined Fermi surfaces for the $Mo_2O_7^{6-}$ slab of $La_2Mo_2O_7$. (Adapted from Ref. 82.)

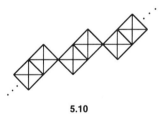

5.10

with a larger area of the Fermi surface and so should be the more important of the two. In Figure 5.17 the vector $\mathbf{q} \sim (0, 0.27\mathbf{b}_3)$. Thus the prediction is made that the CDW with this wave vector is the one responsible for the 125 K phase transition.

5.4   Importance of electron–electron interactions

The influence of the interactions between electrons on adjacent atoms on metallic or insulating properties is perhaps easier to appreciate at the most basic level (Refs. 103,

104) than the effect of structure of the previous section. Imagine a solid composed of atoms each carrying one electron per site. Electrical conduction occurs by movement of an electron from one atom to the next, but if the Coulombic repulsion of two electrons located at the same site (U) is large, then there may be a prohibitively large barrier to passage of current and an insulator results. This system is thus one where the electronic description is a localized one, the electrons individually constrained to lie on individual atomic centers because of the energetic penalty on moving to an adjacent atom. Electronically the picture is very similar indeed to that described for the H_2 molecule in Section 1.2. At large internuclear separation (small β) the lowest-energy electronic situation is best described as $H\cdot + H\cdot$, states such as that described as $H^+ + H^-$ lying higher in energy by the sum of ionization potential (to give H^+) and negative of the electron affinity (to give H^-). Since the overall result of the process $2H\cdot \rightarrow H^+ + H^-$ is to place two electrons on the same site, the energy penalty may be identified with U, the Coulomb repulsion between the two electrons.

The picture described by the Hubbard model of Eq. (1.9) is in fact directly applicable to the solid-state problems (Refs. 142–146, 182, 397). For the case of one electron per center, the right-hand side of Figure 5.18(a) contains two levels described by Heitler–London wave functions and with energies of 2α and $2\alpha + U$, and therefore separated by an energy U. At the left-hand side of the figure, the Mulliken–Hund, or band, model is appropriate, and the result is a half-filled energy band. This corresponds to the case where $\kappa = |U/4\beta|$ is small. Starting off from the insulating side, as the ratio κ decreases, these two Hubbard bands broaden until a critical point is reached where they touch. Hubbard calculated (Ref. 182) this to be for $\kappa = 2/\sqrt{3}$. To the right-hand side of this point, the solid is described as an antiferromagnetic insulator, the two bands separated by a correlation gap. At the left-hand side the system is a metal with a half-filled band. Thus the important parameters of this problem are the Coulomb repulsion term U and β. Although a partially filled band

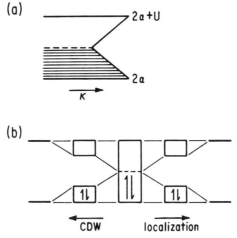

Figure 5.18 (a) Change in the electronic description of electrons as a function of $\kappa = |U/4\beta|$. For small κ the band model is appropriate, but for large κ, or large on-site Coulombic repulsion, U, then a localized model is appropriate. (b) Composite picture for the half-filled-band situation, showing generation of an insulating state either by localization (to the right) or via a CDW (to the left).

Table 5.3 The electronic factors influencing types of metallic and insulating behavior

System	Stability	Stability relative to metallic state
Metallic	$2\alpha - W/\pi + (U + 4V - 2J)/4$	O
Ferromagnetic	$2\alpha + 2(V - J)$	$W/\pi - (U - 4V + 6J)/4$
Peierls distortion	$2\alpha - W/\pi + (U + 4V - 2J)/4$ $+ \Delta_{dist} + \delta^2(U - 4V - 2J)/4$	$\Delta_{dist} + \delta^2(U - 4V + 2J)$
Antiferromagnetic	$2\alpha - W/\pi + (U + 4V - 2J)/4$ $- \Delta_{he} - \delta^2(U - 2J)$	$-\Delta_{he} - \delta^2(U - 2J)$

Source: Refs. 169, 368, and 371.

in principle will lead to a metal, remember that within the band model there are geometrical instabilities that have to be taken into consideration before a metal is formed. Specifically the half-filled band may split into two as a result of a Peierls or CDW distortion.

Figure 5.18(b) shows the two effects side by side. The extreme form of the CDW is shown at the far left-hand side of the diagram, where the levels are associated with isolated units. In very general terms in order to generate a metal, wide bands are needed (small U/β or small U/W, where W is the bandwidth), and those electron counts should be avoided that lead to stable CDWs, and destroy conductivity via a geometrical distortion. A third consideration is associated with ferromagnetism (**2.6**) and the energetic competition with the antiferromagnetic metal (**2.5**). Hubbard's calculations (Ref. 182) using the simplest model showed that this state was always less favorable than the other possibilities. This is not true in practice, of course, but the energetic factors that determine its occurrence [bandwidth, W, or $\rho(E_F)$ and the pairing energy] are closely related (Refs. 369, 371) to κ. Table 5.3 shows the form of the energy differences between the metallic and other states in terms of the one- and two-electron terms in the energy. U and V are the on-site and intersite Coulomb repulsion parameters; J is the intersite exchange interaction. Δ_{dist} is the gain in the one-electron part of the energy as a result of the CDW distortion, and Δ_{he} the loss in one-electron energy on formation of the antiferromagnetic insulator. δ is a system-dependent weight, equal to 0 for the metal and $\frac{1}{2}$ for the antiferromagnetic insulator. The prediction of the behavior of a given system is difficult, set as it is by the balance between the one-electron and many-electron contributions to the energy. These are the many-body effects of the physicist (Ref. 192). The antiferromagnetic insulating state for the half-filled band represents the simplest case of a spin-density wave. Here the spins are located alternately up and down on adjacent centers. More complex possibilities are readily visualized.

Slater (Ref. 328) proposed, prior to Hubbard's model, that antiferromagnetic interactions, rather than electron–electron interactions, were responsible for splitting the band into two. The splitting is proportional to the antiferromagnetic exchange interaction. The effect as it turns out is usually not very large and is only included in systems where the gap is small.

Of interest is a comparison between the antiferromagnetic insulator of this section, the ferromagnetic insulator of **2.6**, and the gapped insulator of Figure 2.10(e). All three are associated with localization of electrons. Recall that for the ferromagnetic and gapped insulators of Section 4.3, a localized description could be invoked because the secular determinant assembled by using Wannier functions is identical (to within a phase factor) of that obtained by using Bloch functions instead (Ref. 162). There is thus a duality of descriptions available for this electronic state of affairs. By way of contrast, the localization that is behind the generation of the antiferromagnetic insulator arises through a special part of the physics, namely the presence of strong on-site Coulombic repulsions. The two types are experimentally distinguishable. Both the ferromagnetic and gapped insulators will have a bandwidth associated with the occupied states that is set in the normal way by the magnitude of the relevant hopping integrals. The bandwidth of the antiferromagnetic insulator could be very narrow, especially if it lies to the far right-hand side of Figure 5.18(a). At this, the simplest level, the two types of localization occur through different mechanisms.

The RENiO$_3$ perovskites (RE = La, Pr, Nd, and Sm) are an interesting series that show how the details of the structure of the solid control the metal–insulator properties (Ref. 215). For RE = La the material is a metal, but the other three compounds show (Fig. 5.19) metal–insulator transitions whose transition temperatures increase with decreasing size of the RE atom. Very few compounds described as perovskites actually have the symmetrical structure shown in Figure 5.34(a). Most examples are distorted (Ref. 258). Figure 5.20 shows the GdFeO$_3$ structure, one distorted form of the perovskite structure adopted by RENiO$_3$, and how adjacent pairs of NiO$_6$ octahedra bend at the common oxygen atom. In this way the structure distorts to improve the oxygen coordination around the rare earth atom, the extent of bending set by its size. Thus the Ni–O–Ni angle decreases in the series $\sim 180°$ (La), 158.1° (Nd), 156.8° (Sm) as the rare earth ion becomes smaller. There are only

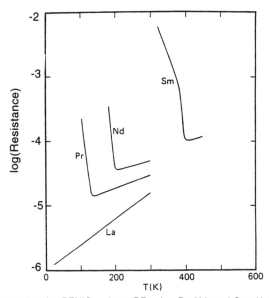

Figure 5.19 Resistance data for RENiO$_3$, where RE = La, Pr, Nd, and Sm. (Adapted from Ref. 215.)

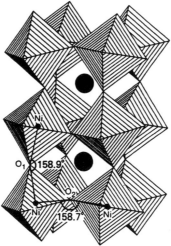

Figure 5.20   The GdFeO$_3$ structure, a distorted variant of the perovskite structure. The large Gd atoms are shown as solid circles. The bond angles shown as those appropriate for PrNiO$_3$. (Adapted from Ref. 215.)

small changes in the Ni–O distances; 1.940(8), 1.943(9), and 1.955(6) Å for Pr–Sm, and so these changes may be ignored. Of particular interest in this series is how the bandwidth of the e_g orbitals varies with the structure of the material. Some comments will be made about the width of the bands in this structure in the next paragraph, but **5.11** shows how the overlap of the oxygen $2p$ orbital with one of the metal σ

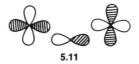

5.11

orbitals ($x^2 - y^2$ is shown) varies with this angle, larger angles leading to greater overlap. Since this overlap directly controls β, proportional to the width of the Ni–O bands, then assuming that U remains constant across the series, $|U/W|$ should increase in the order La ≪ Pr < Nd < Sm. In order to understand the data of Figure 5.19, temperature needs to be added to the model in a qualitative way. Using the model shown in Figures 5.13 and 5.17(b), higher temperatures favor the system where there is no gap, leading to the expectation, therefore, that the temperature of the metal–insulator transition scales with the value of $|U/W|$. This is in accord with experiment. A different explanation of this behavior is given in Section 5.5.

A similar structural control of the balance between metals and insulators is found by comparing the series of oxides $A(I)BO_2$ with the cadmium halide structure and their $A'(III)BO_3$ perovskite analogs (Ref. 143). For example, LiNiO$_2$ is an antiferromagnetic insulator, but LaNiO$_3$ is metallic. One of the reasons behind the difference is the smaller bandwidth in the cadmium halide structure compared to that for the perovskite. The true state of affairs is somewhat more complex and includes the effect of the oxygen $2s$ orbitals, but the basic electronic underpinnings of this result are easiest to see in the p-orbital-only model of the two types of one-dimensional chains of which the full structure is composed. The important point is that by symmetry

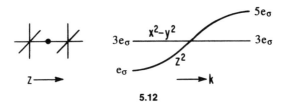

5.12

there is no oxygen p contribution at the very bottom of the perovskite-like chain (**5.12**) of Section 3.1 for the z^2 band, but there is at the top. Thus the band is quite wide. In the cadmium halide case there are oxygen p contributions both at the zone center and at the zone edge (**5.13**). (The energies depicted come from the angular overlap

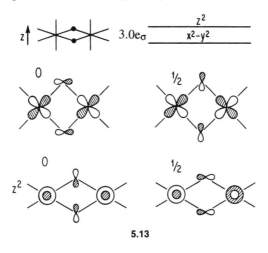

5.13

model.) As a result, the "e_g" bandwidth in this case is smaller than that of the perovskite and thus there is a greater tendency for localized behavior and generation of an antiferromagnetic insulator for the half-filled band.

The picture of a balance between one- and two-electron terms in the energy via the parameter $\kappa \sim |U/W|$ in determining the metallic properties of a solid enables broad generalizations to be made concerning the properties of transition metal containing materials in much the same way as used for the molecular d^8 systems in Section 1.6 (Fig. 5.21). Large values of W will be found at the left-hand side of the first transition metal series and for the second and third row metals. U is larger for first row than for second or third row transition metals. Thus TiO (a defect rocksalt structure) is metallic (itinerant electrons), but NiO is an insulator (localized electrons). Indeed, the electronic spectrum of the latter system is very similar indeed to that of $Ni(H_2O)_6^{2+}$, which dramatically emphasizes this point. Similarly, ruby, Cr^{3+} doped into Al_2O_3 (white sapphire), shows textbook electronic excitations that are analogous to those for $Cr(H_2O)_6^{3+}$. Since W increases with the extent of covalent bonding, metallic behavior should be favored for compounds with the heavier elements from Groups 13–15, localized behavior should be favored for the lighter halides, and the oxides and sulfides should lie in between. Important contributions to the bandwidth may arise though metal–metal bonding, as described in detail for NbO in Section

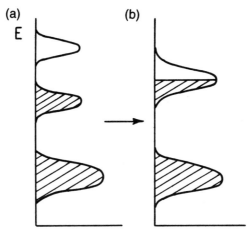

Figure 5.21 (a) Schematic electronic description of an antiferromagnetic insulating oxide with large κ. (b) Comparable description for a typical metallic oxide where κ is small. The oxygen p band (filled) is at lowest energy.

4.4, and metallic behavior from this source is particularly important for the low oxidation-state metals at the left-hand side of the Periodic Table. When the bandwidth does not arise through metal–metal interactions, but through metal–nonmetal ones, then metallic conduction is via a pathway linking metal and nonmetal rather than one connecting metal atoms. Sometimes if the nonmetal is an electronegative one, such as oxygen, then this description is at first sight surprising.

A variant of the localization picture of Fig. 5.18(a) is one due to Mott (Ref. 252), where the effective bandwidth W is controlled by the density of the material or by the density of that subset of the electrons that may lead to conduction. Thus studies of cesium vapor show a drop in conductivity as the density decreases. U remains roughly constant, but W decreases as the average distance between the atoms increases (and thus their overlap integral decreases). This is a case where the total number of electrons remains constant. An example of a system where the number of electrons changes is solutions of sodium in liquid ammonia. These are blue at low sodium concentrations, and contain solvated electrons, but show metallic behavior above a critical concentration, where the solution turns bronze. In both cases whether localized or delocalized behavior is seen is determined by the ratio $|U/W|$. Nonconducting compounds where such localization effects are the key to understanding electrical properties are frequently called Mott–Hubbard insulators.

Another facet of this idea is due to Anderson (Ref. 15), who considered the effects of disorder on the electronic structure problem. The general idea is as follows. In the disordered solid, the effective values of the H_{ii} parameters will be different depending upon the local environment in which the atom finds itself. There will thus be a distribution of H_{ii} values [Fig. 5.22(a)] centered around some point. Atoms lying in the middle of this distribution will have many neighbors with similar H_{ii} values, and will overlap to form an energy band. Atoms in the wings of this distribution will have far fewer atoms of similar energy, and their orbital description can be a localized one. Thus the set of levels of Figure 5.22(b) will consist of a delocalized region sandwiched between sets of localized orbitals. A mobility edge separates the two. The

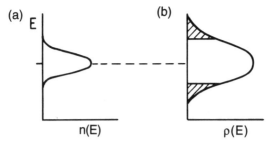

Figure 5.22 The origin of Anderson localization. (a) The distribution of "effective" H_{ii} values in the disordered solid. (b) The localized states (shaded) at the bottom and top of an energy band.

model suggests that at a critical doping level, metal–insulator transitions may be induced by small changes in temperature, pressure, or dopant level.

5.5 Transition metal and rare earth oxides

Perhaps one of the most striking observations concerning the oxides of the first transition metal series is that, on moving from the left to the right of the transition metal series, metallic behavior is gradually replaced by insulating character, as mentioned briefly in the previous section. Thus at the left-hand end TiO is a metal and VO is semimetallic, but at the right-hand end NiO and CoO are antiferromagnetic insulators. How this occurs is readily understandable using the discussion of the previous section. Figure 4.4(a) shows the density of states expected for a metal oxide with the NaCl structure from the viewpoint of band theory. A gap between e_g and t_{2g} bands may be present if the e_g–t_{2g} splitting is large compared with the bandwidth. However, for the elements of both transitional series, the electronic behavior often lies close to the localized description, and frequently changes in stoichiometry or pressure can dramatically influence properties.

One way to visualize the energy difference (U) between the electronic states of H_2 at infinite separation, discussed in Sections 1.2 and 5.4, is to consider it as equal to the energy difference between the two processes $s^1 \rightarrow s^0$ (ionization potential) and $s^2 \rightarrow s^1$ (electron affinity). In an exactly analogous way (Ref. 144, 145, 250, 269, 354) the localized description of the transition metal–containing system employs the energetic separation between the processes $d^n \rightarrow d^{n-1}$ and $d^{n+1} \rightarrow d^n$. An analogous pair of processes but for the f^n configuration would be appropriate for the lanthanides. Just as in Figure 5.18 these states are broadened by overlap interactions such that the description of a localized transition metal oxide may be visualized as in Figure 5.21(a). Eventually [Fig. 5.21(b)] with increasing overlap between metal and oxygen levels, the band model of Figure 4.4(a) results. The picture of Figure 5.21(a), although qualitatively correct, has some severe quantitative problems. The values of U derived from gas-phase data are usually in the region of 15 eV, but for the solid, values of 3–5 eV are commonly found. The difference is difficult to pinpoint exactly but is usually ascribed to polarization effects in the solid. In the orbital model this is the change in metal-neighbor interactions induced by the change in oxidation state.

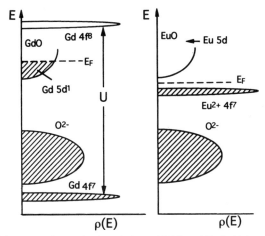

Figure 5.23 Schematic electronic description of GdO and EuO. (Adapted from Ref. 145.)

In rare earth oxides such as GdO and EuO, although the metal atoms contain partially occupied $4f$ orbitals, the Coulombic repulsion U is so high compared to the metal f-oxygen interaction integrals β, that localized behavior is always seen for these electrons. Although the overlap between the tightly bound, almost core-like, f orbitals, and the oxygen orbitals is small, the $5d$ orbitals of these metals have large enough interactions with the oxygen orbitals so that a delocalized band picture is often appropriate. The pictures (Fig. 5.23) for these two systems are therefore composite ones (Ref. 145) containing both localized and delocalized contributions. The oxygen bands are filled with two electrons donated by the rare earth atom, and thus there are several possible choices for the electronic configuration of the metal in GdO. The configurations $4f^8$, $4f^7 5d^1$, $4f^6 5d^2$, etc., are obtained by distributing electrons in the two atomic orbitals $4f$ and $5d$, which are close in energy. As Figure 5.23(a) shows, the lowest-energy arrangement is $4f^7 5d^1$ and contains a localized $4f^7$ state and a delocalized $5d$ band, which contains one electron. The energy gap between $4f^7$ and $4f^8$ localized states is large in this case because of the special exchange-stabilized half-filled f orbital configuration. The energy difference between the two states is labeled U on the diagram, since the energy difference involved is that of placing two electrons in the same orbital. GdO is thus a metal from this picture. EuO is different in two ways. It has one electron less, and the $4f^7$ state lies much higher in energy, as shown in Figure 5.23(b). Whereas the $4f^7$ state lies below the oxygen band in GdO, in EuO it lies between the oxygen band and the $5d$ band. EuO is thus an antiferromagnetic insulator, its electrons localized in the $4f$ levels. This electronic picture leads to a ready way to understand the electrical properties of nonstoichiometric EuO. Generation of $EuO_{1-\delta}$ leads to the addition of electrons to metal-located levels. From Figure 5.23(b) these go into the $5d$ band and give rise to metallic conduction. The situation in $Eu_{1-\delta}O$ is rather different. Here electrons are removed from the localized $4f^7$ state. What happens is the creation of a small polaron (Ref. 3), a region associated with a local geometrical distortion around the site where one Eu atom has lost an electron. (These are discussed in more detail in Section 5.6.) There is now a mixture of metal sites in the crystal, some Eu(II) and

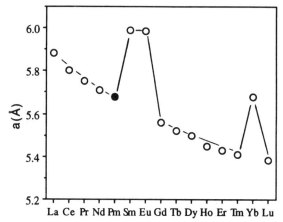

Figure 5.24 Cell parameters for the rocksalt sulfides of the rare earths. The solid point is an interpolation for PrS.

some Eu(III). Conductivity is now activated and has strong similarities to electron transfer reactions in solution. The result is semiconducting behaviour.

The behavior of some rare earth compounds is interesting in that their electronic description may be completely altered by pressure. Figure 5.24 shows the variation across the series in the length of the side of the cubic unit cell of the rare earth sulfides. Notice that the points corresponding to SmS and EuS are displaced from the rest. Whereas most of the rare earth sulfides have the electronic configuration $4f^{n-1}d^1$, these two sulfides at ambient pressures have (Ref. 391) a localized description $4f^n$. Thus SmS is an insulator but under pressure becomes a metal with a cell side that now is in line with the rest of the compounds. Simply, pressure has led to a change of electronic state from a localized to a delocalized one. Figure 5.25 shows schematically, potential energy curves for the two electronic situations and the two different equilibrium Sm–S distances.

The valence d orbitals of the transition metal interact much more strongly with the oxide orbitals than in the case of the $4f$ orbitals of the rare earth oxide. This means that the formation of energy bands is a possibility if the on-site Coulomb repulsion U is not high. The $3d$ orbitals at the left-hand side of the series are more

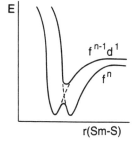

Figure 5.25 Schematic showing the functional form of the energies of the two lowest-lying states of SmS. The delocalized metallic $f^{n-1}d^1$ state and the localized insulating f^n state have minima at different values of the Sm–S distance and hence cell parameter. The two states are of the same symmetry and will interact with each other as shown.

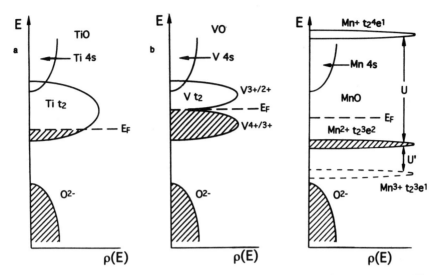

Figure 5.26 A comparison of the electronic situations in metallic TiO, semimetallic VO, and antiferromagnetically insulating MnO. (Adapted from Ref. 145.)

diffuse than those at the right-hand side, which implies an increasing U on moving from left to right. This is reflected in practice by the observation of metallic oxides at the left-hand side of the series [Fig. 5.21(b)] and insulating oxides [Fig. 5.21(a)] in the middle and right-hand side. Figure 5.26 shows (Ref. 145) in a qualitative way the cases of TiO, VO, and MnO. (TiO has a defect NaCl structure that is described in Section 4.4. No special statement concerning it is made here.) for TiO, U/W is sufficiently small that the two d electrons occupy a partially filled d band. For MnO this is not the case, and the electrons are localized in the $t_2^3 e^2$ state of Mn(II). To higher energy, separated approximately by U is the $t_2^4 e^1$ state, where two electrons are paired in the t_2 level. VO is a particularly interesting case since it lies between the two extremes; the upper and lower Hubbard bands overlap to give rise to a situation where there is a reduced density of states at the Fermi level and a semimetal results.

These ideas may be quantified and made more general to view (Ref. 344) the electronic properties of a wide range of transition metal oxides. Three parameters are important. The electron–electron repulsion parameter U and the bandwidth W (proportional to β) are the obvious pair, but Δ, the separation between oxygen and metal levels, is also important, as shown in the Zaanen–Sawatzky–Allen classification (Ref. 397) of metals and insulators of Figure 5.27 for oxides. Here two types of electrons are considered, those that lie in metal d orbitals and those that fill the oxygen $2p$ bands. Five different possibilities are shown that differ in the way the metal levels, localized in four of the five cases, are energetically situated relative to the oxygen ones. Two of the five cases give rise to metallic behavior. Notice the difference between the Mott–Hubbard insulator and the charge-transfer insulator. Both arise via a large value of the "effective value" of U (labeled U') the difference between the two lying in the ratio of Δ/U'. As described, there is a simple way to quantify U from the various atomic properties of the metals concerned (Ref. 16). This is just the sum of the ionization potential, I_{v+1}, and the electron affinity, $-I_v$, but in addition

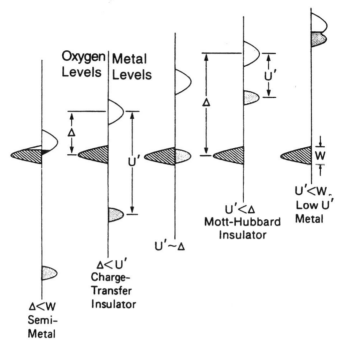

Figure 5.27 The Zaanen–Sawatzky–Allen classification of metals and insulators. (Adapted from Ref. 397.)

for the solid the corresponding electrostatic energy of the pair created $(-e^2/d_{MM})$ needs to be added. This term is absent for the hydrogen molecule at infinite separation. A similar process allows computation of Δ. This is the energy difference involved in moving an electron from O to M, and is thus the sum of the difference in Madelung energy between the two sites (ΔV_M), the electron affinity of M $(-I_v)$, the ionization potential of O^{2-} (I_O), and the corresponding electrostatic energy of the pair created $(-e^2/d_{MO})$. All of these terms may be simply calculated from the details of the crystal structure (the electrostatic parts) or come from atomic data. Figure 5.28 shows a plot (Ref. 344) of Δ versus U for several transition metal–containing oxides and is effective in separating most of the examples into regions where only either metals or insulators are found. (See Ref. 344 for the labeling scheme.) Systems that show metal–insulator transitions lie close to the boundaries. As noted earlier, the absolute size of U or U' is far too large, but the qualitative picture is very useful. **5.14** shows the form of the plot expected from the considerations between Figure 5.27 and enable specific reasons to be given for metallic or insulating behavior in each instance. For example, $LaCuO_3$ (the point labeled Cu3) is metallic because of metal d-oxygen $2p$ overlap. Holes now appear in the oxygen $2p$ levels. LaO (La2) is a metal because it has a low U. Neither the bandwidth W nor the structure appears in this treatment, suggesting that the broad picture is controlled by gross *atomic* properties. Structure is often important, though, as the preceding discussion concerning the perovskite and cadmium halide structures indicates.

This discussion suggests an alternative explanation for the metal–insulator behavior of the $RENiO_3$ systems (Ref. 215) of Section 5.4. As expected from the implications

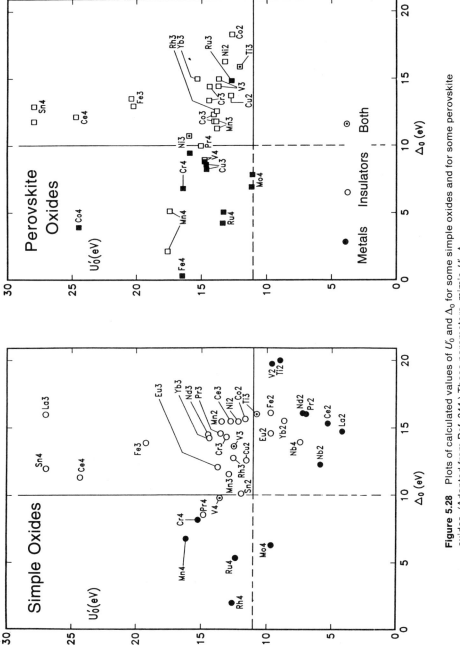

Figure 5.28 Plots of calculated values of U'_0 and Δ_0 for some simple oxides and for some perovskite oxides. (Adapted from Ref. 344.) These parameters mimic U', Δ.

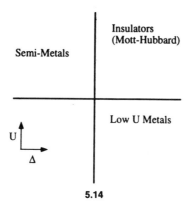

5.14

of **5.11**, the wide band of the RE = La case leads to a semimetal, but the narrower band for the smaller REs leads to a charge-transfer insulator. Which explanation is correct has yet to be established.

5.6   Effect of doping

It is particularly interesting to examine the physical behavior expected on the addition of electrons or holes to the half-filled level situation shown in Figure 5.18. There are two electronic extremes. When U/W is small and a delocalized band picture is appropriate, then addition or removal of electrons just changes the occupancy of the relevant energy band. When U/W is large and the electrons are localized, then the extra electron or hole may lead to a local distortion at the site where it is trapped, a state of affairs described as a small polaron (Ref. 3). A clear picture of how the details of the electronic picture change with doping level at intermediate values of U/W, and especially close to the metal–insulator transition, although very important, is unfortunately not available. The state of affairs is complicated by the presence of the dopant itself. Thus disorder, bandwidth, and on-site repulsion are all general ingredients of the problem. A common feature of many systems is that doping an oxide insulator (for example, $La_{2-x}Sr_xMO_4$, where M is a first row transition metal (Ref. 309) often leads to a metal–insulator transition at a critical value of x. This is sometimes described in terms of an effective U that decreases on doping. Alternatively this Mott transition may be regarded in terms of a screened Coloumb potential between the conduction electrons and the nuclei. There is a critical screening constant (Refs. 210, 252) determined by the electron density, at which the electrons condense around the nuclei to give an insulator.

There are some interesting connections between this behavior and that associated with the various types of mixed valence compounds (Refs. 109, 110, 316). In these systems if there are ions of different formal charge from consideration of the chemical formula, is the best electronic description $M^{n+} + M^{(n+1)+}$ or as $2M^{(n+1/2)+}$? Robin and Day (Ref. 316) devised a useful classification scheme for these materials by defining three classes of compounds.

Class I mixed valence compounds are those materials where there is a large difference in atomic environment between the two ions such that they are immediately geometrically distinguishable. In this case, KCr_2O_8, containing Cr(III) and Cr(IV), the two separate valences, are trapped, and there is virtually no interaction between the ions. Another example is $RECu_2O_4$, which contains Cu(II) and Cu(III) ions both in square-planar coordination but with very different Cu–O distances.

Class II mixed valence systems contain different environments for the two metal atoms, but here there is enough interaction between the two that electron transfer is possible. It is often difficult to set a firm border between this class and Class I. Thus in the platinum mixed valence compound (Ref. 233) shown in 2.30 there is significant interaction between the two sites.

Class III mixed valence systems are materials where all of the metal atoms lie in equivalent sites so that the electronic description is a delocalized one. This class is subdivided into two. Class IIIA compounds contain small delocalized clusters ($Nb_6Cl_{15}\cdot8H_2O$ is an example), whereas in Class IIIB compounds the electrons are delocalized throughout the entire solid. There are many examples of this type; the heavily doped superconductor $La_{2-x}Sr_xCuO_4$ ($x > 0.2$) is one.

The transition from Class II to Class III behavior is perhaps best seen using the picture of Figure 5.29 (Ref. 109), which shows energy as a function of some coordinate. In this case it is r, the M–X bond length associated with the local coordination of the two metals. Here the two diabatic curves represent the electronic stages appropriate for the two localized states, M^{n+} and $M^{(n+1)+}$. The two classes differ only in the magnitude of the interaction (V) between the two states. If the interaction is small, then Class II behavior is found. There is a thermally activated barrier for electron transfer (E_{th}) between the two states and the possibility of an electronic excitation, the "intervalence transfer" band, which takes an electron from one state to the other. If the two diabatic curves are assumed to be parabolae, then it is easy to show (Ref. 238) that hv (intervalence) $= 4E_{th}$. In systems of this type conduction is via activated hopping from one site to another. Sometimes it is not possible to distinguish one class from another crystallographically. Eu_2S_3, for example (Ref. 103), has Eu atoms that are indistinguishable from x-ray crystallography, but that give rise to two Mössbauer peaks at low temperature, indicative of Eu^{2+} and Eu^{3+} ions. As the temperature is increased, then the Mössbauer peaks coalesce due to rapid hopping between the two. Such behavior is readily confirmed since the activation energy is the same as that obtained from the temperature dependence of the conductivity.

Class III systems are described by the state of affairs shown in Figure 5.29(b). Here

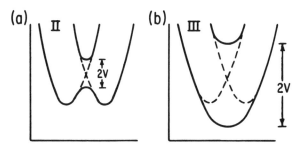

Figure 5.29 The Robin–Day classification scheme for mixed valence compounds. (a) Class II, where the interaction, V, between the two states corresponding to localized electrons is small. (b) Class III, where the interaction between the two states corresponding to localized electrons is large.

the interaction between the two curves is large enough to create a single minimum in the distortion coordinate, the wave function is delocalized over both atoms, and if this occurs throughout the solid, a metal results. The crucial parameter in Figure 5.29 is the interaction energy V, which will lead eventually to the creation of a metallic band and is a measure of the electronic coupling via overlap of the two states. Thus the picture described here is of an insulator to metal transition that takes place as described for the Mott transition. As the doping level increases, the density of small polarons increases, and in this way the picture is similar to that for solutions of sodium in liquid ammonia.

The type of behavior expected on doping a system with extra electrons or holes thus depends on the magnitude of the interaction between adjacent sites. Where this is large, as in TiO_2, for example, the extra electrons will occupy an energy band, and a metal will be produced. In FeO, where the bandwidth is smaller, the effect of nonstoichiometry is the generation of local distortions that trap the extra electrons or holes. These "Koch clusters" involve more than one metal atom in the case of FeO. Thus wide bands generally lead to metals on doping, narrow bands to the formation of local distortions such that the electrons are trapped.

Localization of holes also occur when polyacetylene is doped (Ref. 93). The band model leads to the expectation that the underlying skeleton remains basically unchanged for small doping levels, and that the electron density remains fairly uniformly distributed over the atoms of the solid. In fact, the solid-state analog of a carbocation or carbanion is generated. The carbanion is shown in **5.15**, around which

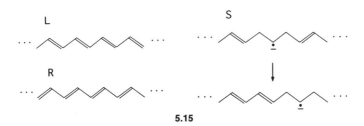

5.15

the single- and double-bond alternation is now out of phase. The carbanion site is frequently called a domain wall, since it separates these two parts of the chain (domains) with bond alternation patterns that are mirror images of each other (labeled L and R in **5.15**). The term *soliton* (labeled S in **5.15**) is also used to describe this state of affairs. There is evidence that the distortion associated with the soliton is not as tightly localized as shown, but extends over perhaps 15 or so atoms of the chain (Refs. 93, 341). Charge transport is associated with the movement of this localized unit along the chain as also shown in **5.15**. (In practice the conductivity is limited by interchain contacts in this polymer rather than the intrinsic properties of an individual chain.)

The platinum mixed valence compounds of **2.29** and **2.30** and the behavior of their nickel analogs illustrate the interesting competition between the several factors involved in determining whether a metallic or insulating state is found. The nickel compound (Ref. 343) has a structure where all of the Ni–X distances are equal (**2.29**). It is then described as Ni(III) and is an antiferromagetic insulator [right-hand side of Fig. 5.18(a)]. The platinum compound, best written as Pt(II)/Pt(IV), is a class II mixed valence compound with unequal distances (**2.30**). The electronic picture for the platinum compound is shown in Figure 2.32, and shows a Peierls distortion of the half-filled band [left-hand side of Fig. 5.18(a)]. The comparison between the two solids may be linked immediately to the discussion concerning the high- and low-spin nickel- and platinum-containing molecules of Section 1.6. In both cases, molecules and solids, the two-electron terms in the energy are larger for the first row metal, and the one-electron terms in the energy are larger for the second and third row metals. Thus $|U/W|$ is smaller for the heavier metal, and so it is to be expected that $|\Delta(U/W)|$ is smaller, too, where ΔW is the change in W during the geometrical distortion. Note that in this analogy the high-spin $Ni(H_2O)_6^{2+}$ species really corresponds to the ferromagnetic rather than the antiferromagnetic insulator.

To complete this section, it is interesting to distinguish between the localization leading to soliton and polaron. Consider an undoped platinum chain with the structure shown in **2.30**. If it is doped with an electron acceptor, the chain becomes partially oxidized (Ref. 263). If the undoped state of the chain is described as $\cdots 2424242424\cdots$, emphasizing the oxidation states of the two types of platinum atoms, then the soliton is written as $\cdots 242432424\cdots$ and the polaron as $\cdots 242434242\cdots$. The label 3 then refers to the site of local oxidation.

5.7 Superconductivity in the C$_{60}$ series

One very interesting property of many solids is that of superconductivity or the presence of zero resistance towards the passage of an electrical current. The connection between structure and superconductivity is at present still not very well defined, especially since the physical mechanisms for the phenomenon is unresolved for many systems, and particularly for the high-temperature superconducting cuprates. However, superconductivity in many systems appears to be described by the BCS theory (Ref. 322), named for its discoverers, Bardeen, Cooper, and Schrieffer. It relies on electron–phonon coupling as a route to generate a Cooper pair of electrons. Within the theory the dependence of the critical superconducting temperature (T_c), above which the superconductivity disappears, is given by

$$kT_c = 1.14\hbar\omega \exp[-1/\rho(E_F)V] \qquad (5.3)$$

where $\rho(E_F)$ is the density of states at the Fermi level, V is the attractive electron–electron potential at the Fermi level (the product of the two being the electron–phonon coupling constant), and ω is the average phonon frequency responsible for the coupling. All three parameters will be determined by the chemical composition and crystal structure. In general, however, light atoms will lead to higher T_c through higher values of ω. [A recent theoretical prediction (Ref. 19) of T_c for a form of metallic hydrogen led to a value of 140–170 K, with a large contribution from the small mass of hydrogen.]

The importance of structure is usefully shown in the superconducting properties of molecular metals (Ref. 318), which include the alkali metal—doped C$_{60}$ solids (Refs. 128, 153), of stoichiometry A_3C_{60}. The C$_{60}$ molecules themselves, the fullerenes

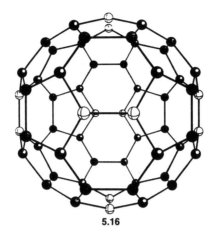

5.16

(**5.16**), are "spherical polyenes" and contain, in addition to many six-membered rings, six five-membered rings. The A_3C_{60} solid shows a cubic eutactic array with the A atoms in all the octahedral and tetrahedral holes. Recalling the stability of $C_5H_5^-$, high electron affinities for the attachment of electrons to C_{60} might be expected from the presence of the five-membered rings. Indeed, from the Hückel level structure of the system (Ref. 153) shown in Figure 5.30, six electrons are readily accommodated to give a semiconductor (filled t_{1u} band). With three electrons a half-filled band is found for A_3C_{60} and leads not only to a metal but to a superconductor with a T_c as high as ~ 35 K for $A_3 = Rb_2Cs$.

Figure 5.31 shows the very interesting dependence (Ref. 128) of T_c upon cell

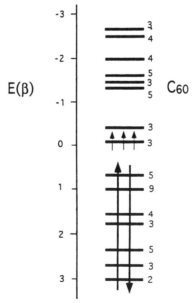

Figure 5.30 The Hückel orbitals of C_{60}, the degeneracy of each level indicated by the figure at the far right-hand side. The filled levels in the neutral species are shown by a pair of large arrows. The three extra electrons in M_3C_{60} are shown by smaller arrows and occupy an orbital of t_{1u} symmetry.

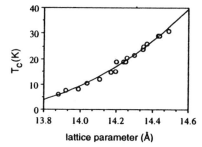

Figure 5.31   Critical superconducting transition temperature (T_c) versus lattice parameter (which varies with applied pressure). (Adapted from Ref. 128.)

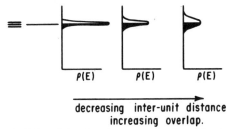

decreasing inter-unit distance
increasing overlap.

Figure 5.32   Change in bandwidth and density of states at the Fermi level, $\rho(E_F)$ for the half-filled t_{1u} band of doped C_{60}.

parameter in a variety of systems containing mixtures of alkali metal atoms as fullerene dopants. This behaviour may be traced back to a simple structural dependence of the band structure. As the metal size increases, so the C_{60} units are pushed farther apart with a corresponding decrease in the overlap integral between adjacent molecules. As this overlap integral determines the bandwidth, the variation with distance shown in Figure 5.32 has a straightforward structural interpretation. Since the total integrated density of states is constant, as the bandwidth decreases the density of states at the Fermi level must increase. Assuming that ω and V of Eq. (5.3) remain invariant to metal atom substitution (i.e., that the vibrations determining ω are C_{60} located), then T_c varies simply with $\rho(E_F)$. Figure 5.33 shows the correlation

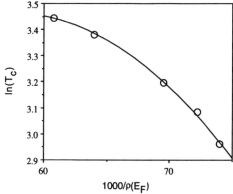

Figure 5.33   Plot of $\ln(T_c)$ against $\rho(E_F)^{-1}$ from an extended Hückel band structure calculation for some doped fullerene superconductors. (Adapted from Ref. 128.)

found, readily understood on the model described. Since the bandwidth decreases as the distance between C_{60} units increases, there will be some limiting separation where the delocalized Mulliken–Hund model becomes inappropriate and the electrons localize as in Figure 5.18 at large κ. Such an electronic instability will not only destroy superconductivity but also normal metallic behavior. The C_{60} superconductors have higher T_cs than most others (excepting the cuprates), almost certainly because of the relatively high value of ω associated with the vibrations around 1400 cm^{-1} of the C–C bonds of the C_{60} shell.

In more general terms, superconductivity, usually associated with a high density of electronic states at the Fermi level, $\rho(E_F)$, often competes with other electronically driven processes in solids. The possibility of electron localization in A_3C_{60} as the interunit separation becomes large has been noted, but such high values of $\rho(E_F)$ often also signal geometrical distortions of the CDW type and ferromagnetic behavior from the Stoner criterion of Section 4.6. However, the dependence of superconducting properties on structure seems to be particularly clear in this case.

5.8 High-temperature superconductors

Onnes discovered superconductivity in elemental mercury in 1911, and in 1986 Bednorz and Müller discovered (Refs. 23, 259) superconductivity in the ceramic cuprates, materials previously regarded by many as unlikely candidates for metallic behavior, let alone superconductivity. What sets these systems apart from other superconductors is that their critical superconducting transition temperatures are an order of magnitude larger than those known before (Refs. 87, 311, 330, 353, 388, 261, 262, 311). The highest T_cs achieved to date (1994) are (Ref. 353) around 125 K in the complex materials $Tl_2Ba_2Ca_2Cu_3O_{10}$, $TlBa_2Ca_3Cu_4O_{11}$, and $(Tl, Pb)Sr_2Ca_2Cu_3O_9$ and 153 K under pressure in the system $HgBa_2Ca_2Cu_3O_{8+\delta}$. In this section possible superconducting mechanisms will not be discussed, but how the geometrical structure of these materials controls their electronic structure and ultimately therefore their superconducting properties.

All of the superconducting cuprates contain (Ref. 47) sheets of stoichiometry CuO_2 [Fig. 5.35(b)] shown in Figure 5.34 by the structure of $La_{2-x}Sr_xCuO_4$, the first "high-temperature" superconductor with a T_c of 35 K for $x \sim 0.15$. In accord with its stoichiometry, this has been called the 2–1–4 compound. The structure is basically quite simple with distorted CuO_6 octahedra vertex fused in two dimensions. The analogous structure where vertex fusion occurs in all three directions is the perovskite structure of Figure 5.35(a), which leads to the frequently used description of this family of compounds as "perovskite superconductors." The two long axial and four short equatorial Cu–O distances are typical of Jahn–Teller distortions of Cu(II). La (and Sr) atoms lie between slabs of these units.

The second "high-temperature" superconductor to be discovered was the first with a T_c above liquid nitrogen temperature (77 K), the 1–2–3 compound (Ref. 181) $YBa_2Cu_3O_7$, with a T_c of $\sim$95 K. Figure 5.36 shows the structure of the compound and highlights an important property, namely the range of oxygen stoichiometry possible in the interplanar region leading to the formulation $YBa_2Cu_3O_{7-\delta}$, $0 < \delta < 1$. The dependence of T_c on δ is shown (Refs. 88, 220) in **5.17**, and shows a

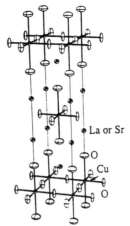

Figure 5.34   The structure of $La_{2-x}Sr_xCuO_4$. (In fact, the geometry is frequently distorted to the orthorhombic structure shown in Fig. 6.14.)

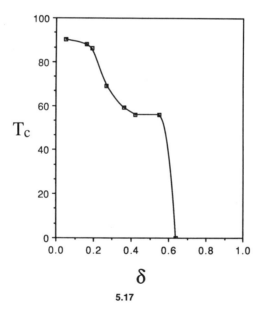

5.17

precipitous drop at around $\delta = 0.65$, where the compound becomes an antiferromagnetic insulator. Here is a very striking dependence of superconducting properties on stoichiometry. In geometrical terms $YBa_2Cu_3O_7$ (Fig. 5.36) contains two types of copper atom: square pyramidal Cu(1) and square planar Cu(2). Since the fifth Cu(1)–O distance is quite large, a good description of the structure is of chains of copper [Cu(2)] in square-planar coordination sandwiched between planes of copper [Cu(1)]. Oxygen atoms are lost from the chains as δ increases, and copper atoms in linear two coordination are generated. In $YBa_2Cu_3O_6$ all of the interplanar copper atoms are two coordinate. Some 50 other superconducting cuprates are now known, and all may be described geometrically as in Figure 5.37. Sheets of CuO_2 stoichiometry where superconduction takes place alternate with sheets or slabs of insulating

(a)

(b)

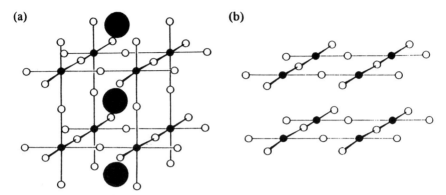

Figure 5.35 (a) The perovskite structure, ABO_3. (b) A CuO_2 sheet emphasizing its derivation from (a).

material. The latter will be described as "reservoir" material for reasons that will become apparent in the following, but its description in terms of a spatially distinct region separated from the CuO_2 sheets is validated by the long axial Cu–O distances in the octahedra The reservoir material in the 2–1–4 compound, $(La_{2-x}Sr_xO_2)$ (CuO_2), is a slab of rocksalt-like $La_{2-x}Sr_xO_2$. In the 1–2–3 compound, $(Y)(Ba_2CuO_3)$

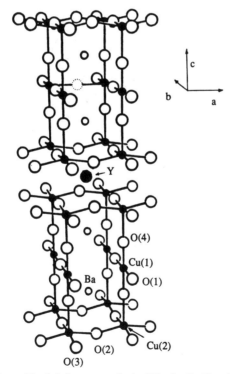

Figure 5.36 The structure of the 1–2–3 superconductor $YBa_2Cu_3O_7$. The structure consists of puckered CuO_2 sheets [Cu(2), O(2) O(3)] and CuO_3 chains [Cu(1), O(1), O(4)]. In $YBa_2Cu_3O_6$ all of the O(1) sites are empty. For oxygen stoichiometries between 6 and 7, both the O(1) site and the O(5) site (shown dotted in the upper half of the picture) may be occupied.

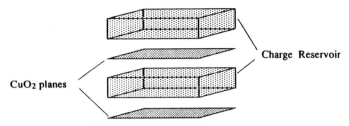

Charge Reservoir

CuO₂ planes

Figure 5.37 Schematic showing the structure of all high-T_c superconducting cuprates. CuO_2 planes are separated by reservoir material that may just be composed of atoms [Ca, Sr in the infinite layer compound (Ca, Sr)CuO_2] or more complex (the CuO_3 chains in $YBa_2Cu_3O_7$).

$(CuO_2)_2$, there are two types of reservoir material (Fig. 5.36), namely layers of Y atoms and the CuO_3 chains of square-planar Cu(2) linked by Ba atoms. In superconductors containing bismuth and thallium, Bi_2O_2 and Tl_2O_2 rocksalt layers are found.

The dependence of the properties of many of these compounds on composition are described by the generic diagram (Ref. 379, 380) of Figure 5.38. As the copper oxidation state changes, then the material may be converted from an antiferromagnetic insulator (probably a charge-transfer insulator using the scheme of Fig. 5.27) to a superconducting metal and then to a normal metal. Specifically for the 2–1–4 compound T_c reaches a maximum at $x \sim 0.15$. This is a hole-doped system. A smaller number of superconductors are formally electron doped (Refs. 195, 340). The change in copper oxidation state is controlled by the composition of the reservoir material. Anion (e.g., F for O) or cation (e.g., Sr for La) substitution, or the presence of defects (nonstoichiometric amounts of oxygen or bismuth, for example), is effective in this regard. For the region where the molecular orbital, delocalized approach is valid, the band structure of the CuO_2 sheet is readily derived. Figure 5.39 shows the d orbital level splitting pattern expected for the planar CuO_4 fragment and how a band is generated in the extended array. The electronic description is just the two-dimensional analog of the z^2 band in the undistorted platinum complex of **2.29**. The $x^2 - y^2$ orbital of the metal is involved in σ interactions with the ligands and lies at highest energy of the d orbital set. For a d^9 configuration it is the highest occupied orbital of the fragment. The z^2 orbital lies at low energy, stabilized in this geometry

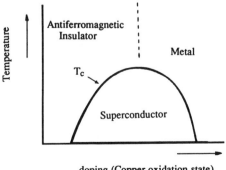

doping (Copper oxidation state)

Figure 5.38 Typical behavior of a cuprate superconductor on doping. The break between metallic and insulating behavior above T_c is often ill-defined.

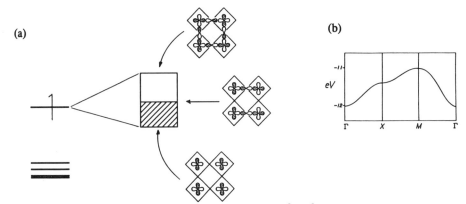

Figure 5.39 (a) The simple band structure derived from the $x^2 - y^2$ orbital on copper in the flat CuO_2 sheet. The orbitals shown are oxygen $2p$ and copper $3d$. (b) The energetic dispersion from a calculation.

by d–s mixing (see Section 4.4). The atomic energies of the copper $3d$ and oxygen $2p$ orbitals are very similar, and so strong mixing between them is to be expected. (The physics community calls this *hybridization*, although chemists reserve this term for orbital mixing on the same atom.) The description of the "metal $x^2 - y^2$" orbital is one, therefore, where there is heavy mixture of the two. The exact electronic description of the highest occupied levels of the solid, whether they are largely copper or oxygen located, may be strongly influenced by the copper oxidation state. One key to the superconductivity in these systems may in fact be the degeneracy of copper and oxygen levels achieved by varying the doping level.

Whatever its exact description, the level occupied by the single unpaired electron of Cu(II) is antibonding between copper and oxygen (Refs. 48, 60), a result that shows up experimentally as a strong dependence of Cu–O distance on doping (Ref. 360). Since the band is metal–oxygen antibonding, addition of electrons (as in the n-doped systems $Nd_{2-x}Ce_xCuO_4$) should lengthen, and removal of electrons as in the p-doped systems ($La_{2-x}Sr_xCuO_4$) should shorten, the Cu–O distance. As Figure 5.40 shows (Ref. 360), although this effect is seen, the demands of the reservoir material also control these distances. In Figure 5.39(a), the effect of substitution of the small La^{3+} by the larger Sr^{2+} in the $La_{2-x}Sr_xCuO_4$ series is an increase in the c axis

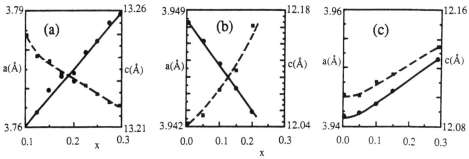

Figure 5.40 The variation with x of the crystallographic parameters in (a) $La_{2-x}Sr_xCuO_4$, (b) $Nd_{2-x}Ce_xCuO_4$, (c) $Nd_{2-x}La_xCuO_4$. (Adapted from Ref. 360.)

(perpendicular to the sheets), indicative of an increase in the "size" of the rocksalt layers. However, substitution of Sr for La leads to oxidation of the CuO_2 sheet with the electronically predicted decrease in the a axis (which is set by the Cu–O distance in the sheets). Figure 5.40(b) shows the opposite behavior for the electron-doped (Ref. 340) series $Nd_{2-x}Ce_xCuO_4$. When the smaller Ce^{4+} ion replaces the larger Nd^{3+} ion, the c axis contracts, but since one more electron is provided by the reservoir on substitution of Ce for Nd, the Cu–O bonds become longer and the a axis increases. Figure 5.40(c) shows the result where there is no change in copper oxidation state but just one of reservoir size, namely substitution of Nd^{3+} by the larger La^{3+}. The electron density in the $x^2 - y^2$ band of the sheets remains unchanged on substitution and *both* a and c increase. The overall picture therefore is one where the electronic demands of the sheets are the dominant effect in setting the Cu–O distance, but the effect of the reservoir material should not be ignored.

The half-filled band situation for the La_2CuO_4 material favors the antiferromagnetic insulator (large κ) using arguments similar to those used in Section 5.4 for the nickel analog of the platinum mixed valence compounds **2.29** and **2.30**. The on-site Coulomb repulsion for this first row metal (Cu) is expected to be large. Addition of holes suppresses the CDW instability expected for the half-filled metallic band, and on doping a metal is generated. Similar arguments for the second row silver analog (not known) would lead to a Peierls distorted solid for the undoped material.

Straightforward electron counting gives a simple description of the 2–1–4 compound in terms of copper oxidation state and the picture of Figure 5.38, but the 1–2–3 compound, $YBa_2Cu_3O_{7-\delta}$, is certainly more complex (Ref. 60). How does the oxidation state of the planar copper atoms depend on oxygen stoichiometry? Since square-planar coordination is found for both Cu(II) and Cu(III), but square-pyramidal geometries, always with long apical bonds, only for Cu(II), the formula $YBa_2(Cu(II)O_2)_2(Cu(III)O_3)$ should be a chemically satisfying way to describe the $\delta = 0$ compound. Analogously, $YBa_2(Cu(II)O_2)_2(Cu(I)O_2)$ is a chemically satisfying way to describe the $\delta = 1$ compound, linear two-coordination being widespread for Cu(I). But why is $YBa_2Cu_3O_7$ a metal (and a superconductor) but $YBa_2Cu_3O_6$ an antiferromagnetic insulator? To relate the behavior of the 1–2–3 compound to the generic picture of Figure 5.38, it is imperative to examine the form of the band structure of the material.

Figure 5.41 shows a schematic of the band structure of the 1–2–3 material. The $x^2 - y^2$ band associated with the square planes is constructed in the same fashion as shown in Figure 5.39. The z^2 orbital lies deep in energy for the same reasons as noted before. Since the $x^2 - y^2$ band is of δ symmetry with respect to the interplanar material, it is essentially electronically decoupled from the energy bands of the chains since there are no oxygen orbitals of that symmetry. For the square-planar copper atoms of the chain, an analogous $x^2 - y^2$ band may be constructed. This needs to be labeled $z^2 - y^2$ since these chains run in the yz plane. The actual electronic description of this compound depends crucially on the relative locations of these two "valence" bands, namely the $x^2 - y^2$ band of the square pyramids and the $z^2 - y^2$ band of the chains. Writing the $\delta = 1$ system as $YBa_2(Cu(II)O_2)_2(Cu(III)O_3)$ forces the description of Figure 5.41(a), where the chain $z^2 - y^2$ band is empty [Cu(III)] and the plane $x^2 - y^2$ band is half-full [Cu(II)]. However, the true situation is subtly different. These two bands overlap [Fig. 5.40(b)] in such a way that electron transfer

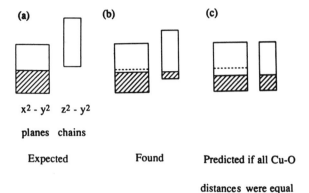

$x^2 - y^2$ $z^2 - y^2$

planes   chains

Expected   Found   Predicted if all Cu-O

distances were equal

Figure 5.41 Some possibilities for the relative location of the $x^2 - y^2$ and $z^2 - y^2$ bands in YBa$_2$Cu$_3$O$_7$. (a) Ensuring that the $z^2 - y^2$ band is empty [thus leading to the formal description YBa$_2$(Cu(II)O$_2$)$_2$(Cu(III)O$_3$)] by making the Cu–O distances in the chain very short. (b) The actual state of affairs where the $z^2 - y^2$ band plays a doping role analogous to the introduction of Sr in the 2–1–4 compound. (c) The situation where both bands are of the same width.

occurs from plane to chain. So the interplanar CuO$_3$ unit plays the same role as substitution of Sr for La in the La$_2$CuO$_4$ system, the overall result being removal of electron density from the CuO$_2$ planes.

The geometrical details of the copper coordination are vital in controlling this charge transfer. If the square-planar bond lengths and the in-plane square-pyramidal bond lengths are set equal, the $x^2 - y^2$ and $z^2 - y^2$ bands, while being of different widths, will be very similarly energetically located as in Figure 5.41(c). This leads to 0.33 holes on each planar copper atom. Because the bond lengths in the planes [1.927 Å (twice) and 1.961 Å (twice) with an average of 1.944 Å] are substantially longer than those in the chains [1.833 Å (twice) and 1.942 Å (twice) with an average of 1.888 Å], although the result is destabilization of the $z^2 - y^2$ band relative to $x^2 - y^2$, there is still overlap between them [Fig. 5.41(b)] and thus charge transfer between the two. Since in YBa$_2$Cu$_3$O$_6$ there is no $z^2 - y^2$ band with which to overlap, the planar copper oxidation state is that given by the formula YBa$_2$(Cu(II)O$_2$)$_2$-(Cu(I)O$_2$), and the electronic state of affairs is similar to that of undoped 2–1–4 in Figure 5.38. **5.18** shows the two distinct ways to remove electron density from the

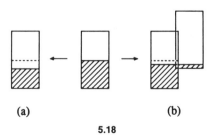

(a)   (b)

5.18

planar $x^2 - y^2$ band, by doping and by band overlap (Ref. 48).

Use of this model shows immediately why the YBa$_2$Cu$_3$O$_6$ and YBa$_2$Cu$_3$O$_7$ compounds behave differently. The latter, with its square planes between the sheets, allows plane–chain charge transfer. In the former there are no chains, and so this

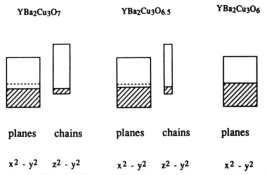

$YBa_2Cu_3O_7$ $YBa_2Cu_3O_{6.5}$ $YBa_2Cu_3O_6$

planes   chains   planes   chains   planes

$x^2 - y^2$ $z^2 - y^2$ $x^2 - y^2$ $z^2 - y^2$ $x^2 - y^2$

Figure 5.42 The variation with stoichiometry of the extent of depletion of the $x^2 - y^2$ band in $YBa_2Cu_3O_{7-\delta}$.

mechanism for generating holes is missing (Fig. 5.42). Somewhere between the $YBa_2Cu_3O_{6.5}$ and $YBa_2Cu_3O_6$ stoichiometries a metal–insulator transition is therefore expected. This shows up in the T_c dependence plot of **5.17** and in structural changes in the material (Ref. 88, 200, 3, 15).

The importance of the structural details when bands overlap is found in other superconductors. $Tl_2Ba_2Ca_{n-1}Cu_nO_{2n+4}$ contains distorted Tl_2O_2 double rocksalt layers between the CuO_2 sheets, whereas $TlBa_2Ca_{n-1}Cu_nO_{2n+3}$ contains single TlO layers. In the former there is band overlap (Ref. 201) between the bands of the CuO_2 sheets and those of the reservoir material, but in the latter there is no such overlap, as shown in Figure 5.43. There is a simple electronic explanation for the broader band for the double-layer case. Just as the "bandwidth" of the polyene of Figure 1.6 increases as the number of atoms increases, so the contribution to the bandwidth of the Tl_nO_n unit increases with the number of atoms or units (n) in the direction perpendicular to the slab.

All of the currently known high-T_c superconductors may be described in the general way shown in Figure 5.37. The electronic picture from the band model for Cu^{2+} leads to a half-filled $x^2 - y^2$ band, a configuration susceptible to localization or to a Peierls distortion as described in Figure 5.18, two routes that lead to the creation of an insulator. To generate a metal, electron density needs to be added or removed from this band. Most superconductors are hole doped, where electrons are removed from the half-full band, a process effected either by overlap with an empty band or a change in stoichiometry (**5.18**). The latter may be a simple variation in cation substitution, the generation of cation vacancies, or the presence of extra oxygen. The

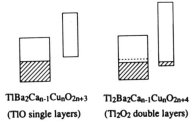

$TlBa_2Ca_{n-1}Cu_nO_{2n+3}$ $Tl_2Ba_2Ca_{n-1}Cu_nO_{2n+4}$

(TlO single layers)   (Tl_2O_2 double layers)

Figure 5.43 The effect of the reservoir thickness on band overlap in $Tl_2Ba_2Ca_{n-1}Cu_nO_{2n+4}$ (Tl_2O_2 double layers) and $TlBa_2Ca_{n-1}Cu_nO_{2n+3}$ (TlO single layers).

second mechanism is easily understandable by consideration of the chemical formula, but the first requires a more detailed understanding of the electronic structure of each part of the solid. Many of these cuprates are very complex, and certainly there are examples of bismuth and thallium superconductors where both mechanisms are at work.

How metallic and eventually superconducting behavior, is produced on doping is not yet well established. The small-polaron model for small doping levels, which eventually becomes metallic, along the lines of the Class II/Class III mixed valence problem of Section 5.4, is a reasonable one. The balance between localized and itinerant electrons is a difficult one to get right from calculation (Ref. 351). However, not yet established is the nature of the superconducting state that lies between normal metal and doped semiconductor.

6 □ The structures of solids and Pauling's rules

On many occasions, and almost always when a new solid has been synthesized and structurally characterized, a question frequently asked is why it has the structure it does. The question goes back to the earliest days of crystal structure determination, where it was addressed in terms of atom "size" and classical sphere packings (Refs. 127, 365, 366). The next three chapters are devoted to structural problems. Almost any discussion of the structure of solids should begin with the rules enunciated by Pauling (Ref. 275) concerning the factors that control the structures of that class of materials described by him as being "ionic" in nature. However, he did recognize later (see the footnote on p. 547 of Ref. 276) that many of the ideas should be applicable to "covalent" compounds, too. The approach in this chapter will be to use the rules as a skeleton on which to build a more general approach to structure. Some of the old ideas will have to be discarded and some carefully qualified.

6.1 General description of ion packings

The long preoccupation with the idea of atomic size was certainly suggested initially by the pleasing geometrical picture that results in terms of the structures that are adopted by arranging a collection of spheres such that the packing density is highest (Refs. 31, 275). Thus much of inorganic crystal chemistry is described in terms of so-called "close-packed" arrangements (usually of anions, but sometimes in terms of cations) in which ions of the opposite type occupy holes in the structure. The usual description of many compounds is of small cations located within the interstices of a close-packed array of larger anions. But is this correct, or even a good model? One test (Ref. 264) of this idea is to see how the volume per anion varies from one MO oxide to another. If the dimensions of the structure are determined by close-packed oxide ions with small metal ions sitting in the holes, then the volume per anion should be constant from system to system. In fact, this parameter varies wildly, being 13.8 Å^3 in BeO and 42.1 Å^3 in BaO. This observation certainly argues against the idea of close packing. The origin of this result lies in the sizes of the radii themselves. From the compilation of radii by Shannon and Prewitt (Ref. 324), it is found for many systems, such as NaF or CaO, that the cations (Na$^+$: 1.16 Å; Ca^{2+}: 1.14 Å) are of similar size to the anions (F$^-$: 1.19 Å; O^{2-}: 1.26 Å).

There are also problems with the concept of close packing itself. A collection of like-charged spheres is hardly likely to pack together at all, because of their mutual electrostatic repulsion. The same result comes from the orbital model of Figure 4.28. Here the chlorine atoms repel one another because of overlap forces. The top of the chlorine band is pushed up in energy more than the bottom is stabilized. What then are the forces holding the solid together? On the orbital model these are the stabilizing interactions between the Na and Cl levels. By analogy with the way that molecular geometries are viewed (such as that of the CCl$_4$ molecule), the structure of the solid is set by minimizing the anion–anion (Cl–Cl contacts in both solid and molecule) for a given anion–cation interaction (Na–Cl and C–Cl, respectively). Thus the "close-packed" anion array is best regarded as that arrangement that *minimizes* such repulsions (Ref. 264). From this viewpoint the most stable structure is expected to be the one of *maximum* anion volume. In this sense the array is not close packed in the traditional sense, the state of affairs being the exact opposite. One way of circumventing this language problem is to use the concept of eutaxy (from ἐντακτός, Greek for well ordered). Here the atoms are envisaged located at the positions appropriate for the close-packed structure, but no claims are made as to whether the density is minimized such that the structure really qualifies to be called "close packed". Thus the NaCl structure (Fig. 4.1) is described as being assembled from a cubic eutactic arrangement of chloride ions with sodium ions in all of the octahedral holes, and the ZnS (wurtzite) structure (Fig. 6.1) as a hexagonal eutactic array of sulfide ions with zinc ions in one-half of the tetrahedral holes. Since both of these

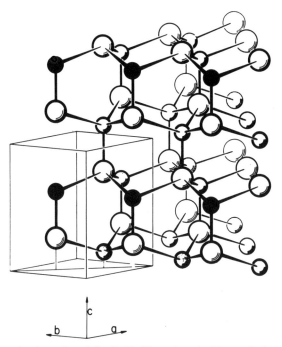

Figure 6.1 The structure of wurtzite (ZnS). (Reproduced with permission from Ref. 280.)

structures are their own antistructure, the eutatic array is equally well decribed as being composed of cations.

The idea of these structures being determined by the minimum repulsion of the anions subject to a fixed anion–cation distance and, therefore, the arrangement that maximizes the anion volume may be simply used to understand structural details. Take first of all the wurtzite structure. This is hexagonal ($P6_3mc$) with the cations in the sites 2(a): $0,0,u_1$; $\frac{1}{3},\frac{2}{3},\frac{1}{2} + u_1$, and the anions in 2(a): $0,0,u_2$; $\frac{1}{3},\frac{2}{3},\frac{1}{2} + u_2$. There are therefore two parameters that are needed to determine the structure, $\gamma = c/a$ and $u = u_1 - u_2$. The four close anion–cation distances (l) are equal for $\gamma^2 = 4/(12u - 3)$, which gives the volume as

$$V = \sqrt{27}l^3(u - \tfrac{1}{4})/2u^3 \tag{6.1}$$

This is maximized for $u = \frac{3}{8}$, which leads to $\gamma^2 = \frac{8}{3}$ or $\gamma = 1.633$, with both ions in hexagonal eutaxy. In BeO the observed values are $u = 0.378$ and $\gamma = 1.602$. Perhaps a bigger challenge is the determination of the details of the rutile and $CaCl_2$ structures. The $CaCl_2$ structure is orthorhombic (Pnnm) with the cations in 2(a): $0,0,0$; $\frac{1}{2},\frac{1}{2},\frac{1}{2}$ and the anions in 4(g): $\pm(u,v,0)$; $\pm(u + \frac{1}{2},\frac{1}{2} - v,\frac{1}{2})$. If $a = b$ and $u = v$, the tetragonal rutile structure results. The parameters of this structure are thus u, v, $\beta = b/a$, and $\gamma = c/a$. For six equal anion–cation distances then

$$\gamma^2 = 4u + 4v\beta^2 - 1 - \beta^2 \tag{6.2}$$

and the volume with this restriction becomes

$$V^2 = l^6\beta^2(4u + 4\beta^2v - 1 - \beta^2)/(u^2 + \beta^2v^2)^3 \tag{6.3}$$

V is maximized for $u = v = 0.300$, $\beta = 1$, $\gamma = \sqrt{2/5} = 0.632$. Rutile itself has $u = 0.305$, $\gamma = 0.634$, and, of course, since it is tetragonal, $\beta = 1$. This structure is geometrically a long way away from the eutactic array, which corresponds to $u = 0.289$, $v = 0.25$, $\beta = 0.816$, $\gamma = 0.6$. These results suggest that although the literature is replete with descriptions of structures in terms of close packings of anions one should be careful both in geometrical interpretations of these terms and in the nature of the electronic models such language suggests.

The eutactic array is a highly symmetric one. The planar two-dimensional array contains atoms that have six equally spaced neighbors. These highly symmetric arrangements may be shown to be the ones that are to be expected from such repulsive forces by study of the one-dimensional chain of **5.8**. Keeping the average nearest-neighbor interionic distance (a) constant leads to an energetic destabilization on distortion (Δa), which is proportional to $(\Delta a/a)^2/a$. So just as for the Hooke's Law situation of Eq. (5.1), these symmetric structures ($\Delta a = 0$) are the ones that are most stable.

6.2   The first rule

"A coordinated polyhedron of anions is formed about each cation, the cation–anion distance being determined by the radius sum and the coordination number of the cation by the radius ratio" (Ref. 275).

There are two parts to this rule. The first is the focus on cation-centered polyhedra. While being of unquestioned utility as a purely geometrical description [witness the large number of structures described in this way (Refs. 94, 331)], it does concentrate electronic attention at the cation, whereas the electrons are located more on the anions. This problem comes to the fore in subsequent discussions of the second rule. The second part of the first rule introduces the idea of an ionic radius.

There are two observations that enable comment as to the "size" of atoms and ions, and especially their transferability. The first is the quite good additivity relationships in terms of interatomic separation for many compounds, especially oxides and fluorides. Such additivity indeed means that interatomic distances in solids can be predicted, sometimes very well indeed. Since $K–X$ distances are larger than $Na–X$ distances, then from these results K is larger than Na. Also, when discussing the structures of both molecules and solids, the concept of a "typical" $A–B$ distance is often used. For bonds to the heavier main group elements the additivity is not quite so good. Some quantitative data, however, show that all is not completely clear even for the light elements. Table 6.1 shows some cell volumes (Ref. 325) for some pairs of isotypic compounds. Can it be concluded from these data that Mg^{2+} is larger than Ni^{2+}, or that Ge^{4+} is larger than Si^{4+}? There are, therefore, some obvious caveats concerning the general philosophy of the extraction of a set of "radii" from crystal data. The second observation comes from the study of electron distributions in solids via x-ray diffraction. If atoms and ions do have a well-defined "size," then in a solid such as sodium chloride a minimum in the electron density between the two nuclei should represent the point at which the two ions "touch." These two considerations are behind the generation of collections of ionic radii, of which those of Shannon and Prewitt are best established (Ref. 324, 325). In addition, earlier ideas of ionic radii had their origin in considerations of free atoms. In crystals, anions, often regarded as being large (vide supra), are, crudely speaking, compressed by the presence of the surrounding cations, and vice versa (Ref. 86).

The third part of the rule introduces the idea of radius ratio as a controlling factor in determining the structures of solids. Perhaps more than any other of the "rules" that have made their way into elementary chemistry books, this one has less justification in terms of being factually correct. Figure 6.2 shows a plot (Ref. 73) of the pairs of anion and cation radii for 98 known octet solids. Pauling's initial suggestion from purely geometrical arguments is that there are critical ratios, 0.732 and 0.414, that separate eight- from six-, and six- from fourfold coordination. Lines with these slopes are show solid in Figure 6.2. Of the 98 points, 38 fall in the wrong

Table 6.1 Cell volumes for some pairs of isotypic compounds ($Å^3$)

NiO	72.4	NiF_2	33.4	NiI_2	88.3
MgO	74.7	MgF_2	32.6	MgI_2	102.1
Fe_2SiO_4	307.9	Fe_2GeS_4	530.7		
Mg_2SiO_4	290.0	Mg_2GeS_4	570.3		

Source: Refs. 46 and 325.

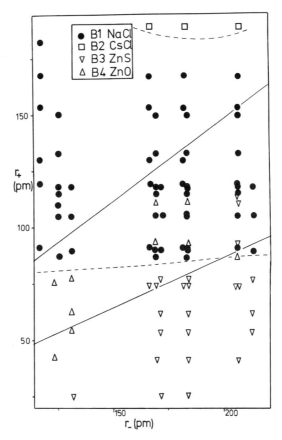

Figure 6.2 Structure map for *AB* octets using the Shannon and Prewitt radii. Solid lines show the Pauling predictions for the boundaries between 4 and 6 and between 6 and 8 coordination. The dashed lines use a "Mendelevian" approach.

place—hardly encouraging for a basic rule of crystal structure. The radii used in this figure are those of Shannon and Prewitt, but irrespective of the choice of set of radii, it is not possible to obtain agreement for even the small subset of the alkali halides, for example, some of the most "ionic" examples. There is also a question when constructing these plots whether the radii appropriate for 4-, 6-, or 8-coordination should be used. Certainly by appropriate choice of radius the "agreement" is better than if one set were used. However, there are real problems with such simple and appealing ideas (Refs. 300, 301). The dashed lines in Figure 6.2 show the separation using a "Mendelevian" approach; namely, draw the best lines to sort the structures.

On the Pauling model critical values of the radius ratios come from simple geometrical concepts, but may be approached in a more rigorous way by looking at the energetic dependence of structure using the ionic potential of Eq. (6.25). The ratio of the electrostatic energies of two structures (i,j) may be shown (Ref. 265) to be equal to

$$E_i/E_j = (M_i/M_j)^{n/(n-1)}(Q_j/Q_i)^{1/(n-1)} \qquad (6.4)$$

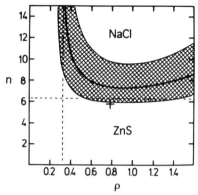

Figure 6.3   The computed stability of the rocksalt and sphalerite structures using the ionic model. The single point corresponds to that for LiF. The shaded region allows for a 2% error in the energy difference. (Adapted from Ref. 265.)

where

$$Q_i = m_i + (1.25m_{+i}/\gamma^n_{+i})[2\rho/(1 + \rho)]^{n-1}$$
$$+ (0.75m_{-i}/\gamma^n_{-i})[2/(1 + \rho)]^{n-1} \tag{6.5}$$

The model includes a Madelung constant (M) for each structure and repulsive terms between nearest-neighbor like, and unlike, pairs. If n, the Born exponent, is very large, then the ions approximate hard spheres, but as n becomes smaller, then the ions become more squashy. There are m_+ (m_-) cation–cation (anion–anion) nearest-neighbor distances at a distance γ_+r (γ_-r), where r is the anion–cation distance. ρ is the radius ratio, and m the coordination number of anions by cations and vice versa. Figure 6.3 shows how (Ref. 265) from Eq. (6.4) the more stable crystal structure depends on the radius ratio ρ, and n for the sphalerite and rocksalt structures. The special Pauling value of 0.414, which is claimed to separate four- and six-coordinate structures has no special significance on this plot. Indeed for hard spheres a limiting value of 0.325 is found, the vertical dashed line on the plot. Not shown in Figure 6.3 is the analogous plot separating six- and eight-coordination. The values of n required for eight-coordination are so high so as to be chemically unreasonable. The point labeled with a cross is that for LiF that adopts the rocksalt structure. Notice that irrespective of the choice of radius ratio it will never fall into the correct region of the diagram. Thus the ionic model is remarkably poor in getting structure correct.

Clearly though the idea of "size" in general is a very useful concept. In qualitative terms, the idea that "small" cations generally lead to low coordination numbers, and that "large" cations often have larger coordination numbers, is well established. The real question at stake here is how to define the "size" of the atom, or whether this has any meaning at all. There are two approaches that have been used, and both have their merits. The first is to continue to use the Shannon and Prewitt radii, but bearing in mind the preceding comments concerning quantitative aspects of this question. The second is to search for a less ambiguous way of approaching atomic size (Refs. 300, 301). The first route is well explored. There are many structure maps available (Refs. 258) of the type shown in Figure 6.4, for a variety of stoichiometries. Here a plot of the radii of A and B leads to regions where the majority of structures

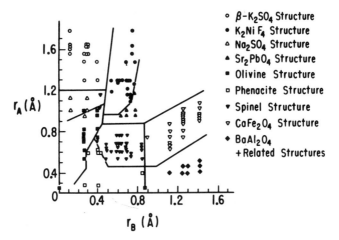

- ○ β-K_2SO_4 Structure
- ● K_2NiF_4 Structure
- △ Na_2SO_4 Structure
- ▲ Sr_2PbO_4 Structure
- ■ Olivine Structure
- ▢ Phenacite Structure
- ▼ Spinel Structure
- ▽ $CoFe_2O_4$ Structure
- ◆ $BaAl_2O_4$
- + Related Structures

Figure 6.4   Structural sorting of A_2BO_4 systems using ionic radii as indices. (Adapted from Ref. 258.)

of a given type fall. The second route leads to the idea of a pseudopotential radius.

The pseudopotential (Refs. 25, 96, 167, 398) is a way of describing the potential felt by an outer, or valence, electron in an atom or solid. The core electrons are assumed to be frozen, and because of the Pauli principle there is a repulsion between the core and valence electrons that nearly cancels the Coulombic attraction of the electron with the nucleus. (Figure 6.5 shows the contributions to the pseudopotential

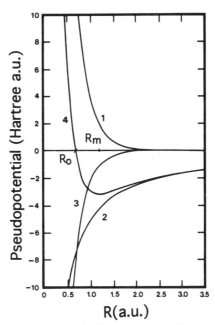

Figure 6.5   Ingredients of the pseudopotential. (1) the Pauli potential, a repulsion, (2) the Coulomb attraction, and (3) the effect of core screening, (4) the net pseudopotential. R_0 is the crossing point of the psuedopotential and R_m the position of the minimum. (Adapted from Ref. 398.) (1 Hartree = 27.2 eV)

for Sb.) So the pseudopotential is rather weak when compared both to the raw Coulombic attraction and that screened by the core. Numerically, pseudopotentials have been very useful in reproducing the energy levels of the valence electrons in atoms and in solids. In Section 8.4, pseudopotential theory will be useful in studying structural preferences using a nearly-free-electron model. In terms of the discussion here the important point about the pseudopotential is that it allows a unique definition of a radius. This is the point where there is a radial maximum in the pseudo-wave function. In classical terms this occurs at the point where the repulsive and attractive forces exactly balance. These radii are "nonlocal," meaning that they depend upon the nature of the orbital under consideration, namely s, p, or d. Even these radii differ from one author to another, but not by much.

Table 6.2 lists one set of radii that will be used here. It is important to note that these radii are atomic parameters, and as such are independent of the chemical environment. Figure 6.6(a) shows a structure map for the AB octets using these pseudopotential radii (Ref. 73). The parameter r_σ is used as a measure of size. This is just the sum of the valence s and p pseudopotential orbital radii. Lines are drawn to separate the four-, six-, and eight-coordinate structures. A "Mendelevian" approach has been used here just as in Figure 6.2. The lines are not set by some preconceived theoretical notion but are drawn to best separate structure types. A comparison of Figures 6.2 and 6.6 suggests that these radii are somewhat better in

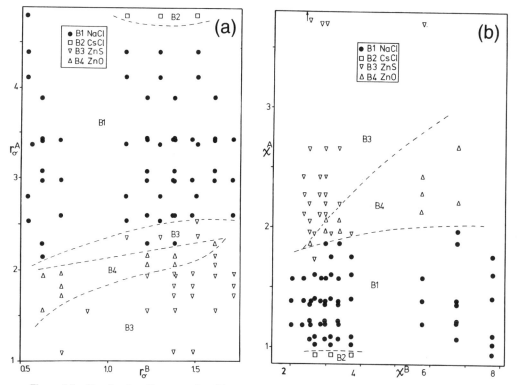

Figure 6.6   Structural sorting maps for AB octets using pseudopotential radii as indices. (a) using r_σ^A and r_σ^B, (b) using χ^A and χ^B, where χ is the electronegativity as defined as in the text.

Table 6.2   Pseudopotential radii

	r_s	r_p		r_s	r_p
Li	0.934	1.879	C	0.384	0.472
Na	1.020	2.365	Si	0.657	0.888
K	1.363	2.757	Ge	0.651	0.919
Rb	1.453	2.947	Sn	0.761	1.068
Cs	1.624	3.176	Pb	0.751	1.086
Be	0.630	0.932	N	0.322	0.381
Mg	0.863	1.433	P	0.589	0.759
Ca	1.221	1.869	As	0.620	0.834
Sr	1.356	2.088	Sb	0.722	0.966
Ba	1.553	2.341	Bi	0.730	1.025
Sc	1.097	1.504	O	0.277	0.319
Y	1.252	1.732	S	0.534	0.665
La	1.452	1.982	Se	0.591	0.764
			Te	0.695	0.888
Ti	0.996	1.286	Po	0.710	0.980
Zr	1.179	1.546			
Hf	1.370	1.800	F	0.243	0.274
			Cl	0.488	0.592
Cu	0.761	1.607	Br	0.565	0.705
Ag	0.877	1.663	I	0.666	0.822
Au	0.820	1.396	At	0.690	0.80
Zn	0.722	1.220			
Cd	0.835	1.322			
Hg	0.796	1.260			
B	0.477	0.625			
Al	0.743	1.084			
Ga	0.685	1.037			
In	0.795	1.170			
Tl	0.773	1.165			

Source: Ref. 25.

terms of highlighting the importance of size than the traditional crystal ones. The pseudopotential radii, since they are directly related to atomic properties, have some other advantages that will become apparent in Chapter 7.

Added insight into this problem comes through a different manipulation of the pseudopotential radii. The function $(1/r_l)$ scales with energy, and indeed there is a good correlation between the reciprocal of the pseudopotential radius and the ionization energy from a given orbital. Recalling Mulliken's definition of electronegativity, or the related ones of Allen or Martynov and Batsanov (Ref. 234), one estimate of χ might be simply as the sum $(1/r_s) + (1/r_p)$. A plot of the same data in Figure 6.6(a) is shown (Ref. 73) in Figure 6.6(b), but this time using $\chi^{A,B}$ as indices. The separation into structural types is very good. The results shown in these two figures underline a philosophical problem in identifying the factors that determine structure type. Figure 6.6(a) indicates that "size" is important, Figure 6.6(b) suggests that the relative orbital locations are important. The interdependence of such parameters, of importance in different models, will make it very difficult to pinpoint unambiguously a reason for the adoption of one co-ordination number over another.

6.3 The second rule

"In a stable co-ordinated structure the total strength of the valency bonds which reach an anion from all the neighboring cations is equal to the charge of the anion" (Ref. 275).

In the earliest use of this idea, Pauling defined the electrostatic bond strength (EBS) as equal to the cation charge divided by the coordination number. Thus in NaCl each cation is attached to six anions with an EBS of $\frac{1}{6}$. The sum of the EBS of the bonds to the cation, $\sum_{\text{EBS}} = 6 \times \frac{1}{6} = 1$, and thus the rule is obeyed. There is of course a strong topological and stoichiometric background to the role, associated with the fact that the total anion and cation charges must balance, and that the coordination numbers and stoichiometry are also related. Imagine the simplest case, that of a structure containing anions and cations, there being $N_{a,c}$ anions and cations with charges $C_{a,c}$, and coordination numbers $\Gamma_{a,c}$. The EBS of one limb of the cation coordination sphere is C_c/Γ_c and so

$$\sum_{\text{EBS}} = \frac{C_c}{\Gamma_c} \Gamma_a \tag{6.6}$$

Because of electrical neutrality, $N_a C_a = N_c C_c$, and, since each anion is connected to a set of cations and vice versa, $N_a \Gamma_a = N_c \Gamma_a$. Thus

$$\sum_{\text{EBS}} = C_c \frac{\Gamma_a}{\Gamma_c} = C_c \frac{N_c}{N_a} = C_c \frac{C_a}{C_c} = C_a \tag{6.7}$$

There are some more general arguments that generalize statements such as these (Ref. 125). Clearly though stoichiometry and connectivity ensure that the rule works in many examples.

The rule does not hold in many cases, however. It does not work, for example, for the unsymmetrical structure of the mineral angellelite (Ref. 246), and it is interesting to ask why. This material, $Fe_4^{3+}As_2^{5+}O_{11}$, is composed of linked $Fe^{3+}O_6$ octahedra and $As^{5+}O_4$ tetrahedra. Each of these polyhedra lie in quite unsymmetrical environments. Simple arguments, similar to those just discussed, allow insight into the situation here. The electrostatic bond strength associated with one limb of the iron octahedra is $\frac{3}{6}$, and that for the arsenic tetrahedron is $\frac{5}{4}$. Thus for an oxygen atom situated at a location where it is at the same time a vertex of x octahedra and y tetrahedra, the electrostatic bond strength sum is just $x(\frac{1}{2}) + y(\frac{5}{4})$. From Pauling's version of the second rule then this sum should equal 2 such that x and y are positive integral solutions of

$$2x + 5y = 8 \tag{6.8}$$

The properties of this simple equation are interesting. As $2x$ and 8 are even, $5y$ must be even, too. Since 5 is odd, y must be even, and so the only solutions with positive integers are $x = 4$, $y = 0$. Thus the conclusion may immediately be drawn that there is no structure containing linked iron octahedra and arsenic tetrahedra with the stoichiometry of angellelite and where Pauling's second rule is satisfied at all sites. The only solution contains isolated arsenic cations, un-coordinated ($y = 0$) to oxygen. The result is a very simple one to derive, but of considerable importance. It shows that purely on topological grounds, a "simple," "high-symmetry" structure for this material, given the cation coordination geometries, is not possible.

These electrostatic bond strength ideas have been developed in a more general way to study the factors that control bond lengths in solids (Refs. 33–37). In modern usage the electrostatic bond strength is replaced by the bond valence, defined as

$$s = \exp[(r - r_0)/B] \tag{6.9}$$

or as

$$s = (r/r_0)^{-N} \tag{6.10}$$

The term r_0 is a reference bond length for a given pair of atoms, as are the parameters B and N. The first definition is somewhat more general then the second in that there apears to be a universal value of $B = 0.37$. The reference bond length sometimes is allowed to depend upon oxidation state. Brown has extended Pauling's original ideas and has formulated two rules that are very useful in understanding how solids are assembled (Ref. 35). Brown's first rule is effectively the same as Pauling's second, but uses the modern definitions of s. At each atom $\sum s = v$, the valency of the atom. The valency of the atom is introduced in practice by the determination of the parameters r_0 and N for a large structural data set that is fitted to $\sum s = v$. Brown's second rule is that at each atom the values of s are as equal as possible. Table 6.3 gives some values of the parameters, used, and Table 6.4 shows how well the first rule works in practice. This is a particularly interesting case (Ref. 235), where the Mo–O distances can be used to evaluate the different "valences" of the four different types of Mo atom in the structure of $Ca_{5.45}Mo_{18}O_{32}$. Along with the oxygen stoichiometry, this gives a value for the total charge on the $Mo_{18}O_{32}$ unit. Now a prediction may be made as to the value of x in $Ca_x Mo_{18}O_{32}$, which is needed for charge balance. The result, $x = 10.7/2 = 5.35$, is very close to the actual value of 5.45.

Figure 6.7 shows an example of a frequently observed structural result (Ref. 75) understandable in terms of the bond valence sum rule. It shows the two I–I distances from a series of crystallographical studies of the I_3^- ion. Depending on the counterion the molecule can be found as a symmetric species ($I–I–I^-$) at one extreme or as a I^- ion weakly coordinated to an I_2 molecule as $I–I \cdots I^-$ at the other. Notice that as r_1 becomes shorter, r_2 becomes longer. Using the expression for s in Eq. (6.10)

Table 6.3 Some values of the parameters used in bond valence sum calculations[a]

			r_0				
Ag	1	O	−2	1.842	Mo	6	O	−2	1.907
Al	3	O	−2	1.651	N	3	O	−2	1.361
As	3	O	−2	1.789	N	5	O	−2	1.432
As	5	O	−2	1.767	Na	1	O	−2	1.803
B	3	O	−2	1.371	Nb	5	O	−2	1.911
Ba	2	O	−2	2.285	Ni	2	O	−2	1.654
Be	2	O	−2	1.381	P	5	O	−2	1.617
Bi	3	O	−2	2.094	Pb	2	O	−2	2.112
C	4	N	−3	1.442	Pb	4	O	−2	2.042
C	4	O	−2	1.390	Pt	4	O	−2	1.879
Ca	2	O	−2	1.967	Rb	1	O	−2	2.263
Cd	2	O	−2	1.904	S	4	O	−2	1.644
Co	2	O	−2	1.692	S	6	O	−2	1.624
Cr	3	O	−2	1.724	Sb	3	O	−2	1.973
Cr	6	O	−2	1.794	Sb	5	O	−2	1.942
Cs	1	O	−2	2.417	Sc	3	O	−2	1.849
Cu	2	O	−2	1.679	Si	4	O	−2	1.624
Fe	3	O	−2	1.759	Sn	4	O	−2	1.905
Ga	3	O	−2	1.730	Sr	2	O	−2	2.118
Ge	4	O	−2	1.748	Ta	5	O	−2	1.920
H	1	N	−3	0.885	Te	4	O	−2	1.977
H	1	O	−2	0.882	Te	6	O	−2	1.917
Hg	2	O	−2	1.972	Ti	4	O	−2	1.815
In	3	O	−2	1.902	U	6	O	−2	2.075
K	1	O	−2	2.132	V	3	O	−2	1.743
La	3	O	−2	2.172	V	4	O	−2	1.784
Li	1	O	−2	1.466	V	5	O	−2	1.803
Mg	2	O	−2	1.693	W	6	O	−2	1.917
Mn	2	O	−2	1.790	Y	3	O	−2	2.019
Mn	3	O	−2	1.760	Zn	2	O	−2	1.704
Mn	4	O	−2	1.753	Zr	4	O	−2	1.928

Source: Adapted from Ref. 36.

[a] B is set equal to 0.37. The valencies of "anion" and "cation" are also shown.

leads to a correlation between the two distances of the form

$$r_1^{-N} + r_2^{-N} = \text{constant} \tag{6.11}$$

a relationship in accord with the experimental plot.

Some rather straightforward theoretical arguments (Ref. 55) allow access to the origin of Brown's rules. The orbital picture of Figure 1.3 for the two-atom (*AB*)

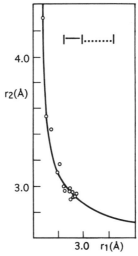

Figure 6.7 I–I distances in a series of solids containing the I_3^- ion. Notice that as one I–I distance (r_1) is shortened, the other (r_2) lengthens.

problem led to the interaction energy between the two orbitals as

$$\Delta E \sim \beta^2/(\alpha_1 - \alpha_2) - \beta^4/(\alpha_1 - \alpha_2)^3 \qquad (6.12)$$

The corresponding picture for the AB_2 system shown in Figure 6.8 arises via solution of the secular determinant

$$\begin{vmatrix} \alpha_1 - E & \beta & \beta \\ \beta & \alpha_2 - E & 0 \\ \beta & 0 & \alpha_2 - E \end{vmatrix} = 0 \qquad (6.13)$$

One of the resulting orbitals is nonbonding ($E = \alpha_2$), and the stabilization energy of the bonding orbital is given by

$$\Delta E \sim 2\beta^2/(\alpha_1 - \alpha_2) - 4\beta^4/(\alpha_1 - \alpha_2)^3 \qquad (6.14)$$

Table 6.4   The use of the bond valence sum rules in Ref. 235. $Ca_{5.45}Mo_{18}O_{32}$

Repeat unit	No. of units	Valence of Mo	Anion charge per repeat
MoO_3	2	3.75 (5)	2.26
$Mo_2O_{3.5}$	4	3.42 (2)	0.22
		3.36 (3)	
Mo_4O_6	2	2.20 (1)	2.65
		2.47 (4)	
Total charge			10.7

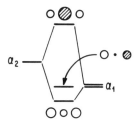

Figure 6.8 Orbital diagram for the AB_2 system, where B carries a single deep-lying orbital.

Importantly, although the quadratic term is linear in the coordination number in Eqs. (6.12) and (6.14), the quartic term is not. This means that the stabilization energy per linkage, $\Delta\varepsilon$, is coordination number (Γ) dependent. ($\Delta\varepsilon$ is obtained just by multiplying the stabilization energies of the orbitals by the number of electrons in that orbital and dividing by the number of ligands). In general, it may be shown quite simply from this treatment that

$$\Delta\varepsilon \sim 2\beta^2/(\alpha_1 - \alpha_2) - 2\Gamma\beta^4/(\alpha_1 - \alpha_2)^3 \qquad (6.15)$$

The stabilization energy clearly decreases with an increase in the coordination number, a result in accord with the general observation of longer bonds the higher the coordination number. In order to quantify this result, however, the model needs to be developed further.

The one-electron model used here does not allow an accurate determination of the equilibrium internuclear separation. Consider, for example, the case of the H_2 molecule. The maximum stabilization is found when the overlap integral is unity, that is, at $r(\text{H–H}) = 0$. However, changes in interatomic distances found when structures are distorted, and when the coordination number changes, are often well described by using such a model but with the constraint that the second moment of the energy density of states is kept constant. This model was used in Section 1.5 to evaluate the energetics of the isomers of H_3^+ and in Chapter 7 is used to look at a wide spectrum of solids. This restriction means here that

$$\sum_j H_{ij}^2 = \Gamma\beta^2 = k = \text{constant} \qquad (6.16)$$

Since in general the overlap integral, and thus β, varies in the region of chemical interest as some function such as Ar^{-m}, Eq. (6.16) may be rewritten using the equilibrium bond length r_e as

$$\Gamma A^2 r_e^{-2m} = k \qquad (6.17)$$

or

$$r_e^{-2m} = (1/\Gamma)(k/A^2) \qquad (6.18)$$

The overlap dependence on distance in the region of chemical interest for pairs of orbitals has been studied for many systems. That between two main group elements varies as r^{-2}, and that between a main group element and a transition metal as $r^{-7/2}$. [Ref. 160] The quantitative prediction of Eq. (6.18) may be tested for one of the most important bonds in the mineral world, Si–O. Given that a typical value of $r(\text{Si–O})$ is 1.63 Å for four-coordinate silicon, then a typical value for six-coordinate silicon will be 1.80 Å. This is indeed the case.

The extension of this result to Brown's first rule (the modern analog of the electrostatic sum rule) is quite simple (Ref. 56). Eq. (6.18) is first rewritten as

$$(r_e/r_0)^{-2m} = (1/\Gamma)(k/A^2)(1/r_0^{-2m}) \tag{6.19}$$

The term $(k/A^2)(1/r_0^{-2m})$ is set equal to a new parameter, v (whose meaning will be clear in the following) such that

$$(r_e/r_0)^{-2m} = (1/\Gamma)v \tag{6.20}$$

and the sum over all Γ linkages leads to

$$\sum (r_e/r_0)^{-2m} = v \tag{6.21}$$

Equation (6.21) is just the bond valence sum rule with the "valency" v, being set by the parameter $(k/A^2)(1/r_0^{-2m})$ and $N = 2m$. In practice the values of the parameters of the model are chosen such that v represents the traditional valency of the atom in question. Of course in real systems there will be a sum over all the pairs of orbital interactions involved, rather than the single one of this bare-bones model. The rule works because AB bond lengths generally shrink as the oxidation state of the cation increases.

A similar picture may be extracted from an ionic model. Using the Born–Landé potential of Eq. (6.28) the energy of an infinite AB solid with charges on the anion and cation of $q_{i,j}$ may be written

$$E(r_{ij}) = A_M q_i q_j/r_{ij} + \Gamma B/r_{ij}^n \tag{6.22}$$

where A_M is the Madelung constant and the repulsive part of the energy written as dependent directly on the coordination number Γ. Minimization of the energy leads to

$$r_e^{1-n} = (1/\Gamma)(A_M q_i q_j/Bn) \tag{6.23}$$

and thus

$$(r_e/r_0)^{1-n} = (1/\Gamma)(A_M q_i q_j/Bn)(1/r_0)^{1-n} \tag{6.24}$$

There is now a similar dependence of bond length on coordination number as found in the orbital model. The sum over all Γ linkages leads to

$$\sum (r_e/r_0)^{1-n} = v \tag{6.25}$$

where v, is now $(A_M q_i q_j/Bn)(1/r_0)^{1-n}$ and $N = 1 - n$.

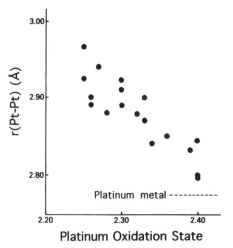

Fig ire 6.9 Variation in the nearest-neighbor Pt–Pt distance with oxidation state in Pt-chain com-
pc nds. (Adapted from Ref. 110.)

Although the second rule and the idea of bond valence originated in the area of oxide chemistry, bond length changes with oxidation state are well established in other areas of the solid state. Figure 6.9 shows the variation (Ref. 110) in the Pt–Pt distance in these mixed valence compounds with the average Pt oxidation state.

The nature of the orbitals under discussion above were not vital to understanding Brown's first rule. However, discussion of Brown's second rule is much more system-specific. The rule states that the bond valences of the linkages to a given center are as nearly equal as possible. This result comes from the simplest mechanical model of eq. (5.1), but is certaintly not true for many solids with Fermi surface or Jahn–Teller instabilities. Thus as noted in Chapter 1, the bond lengths in cyclobutadiene are clearly not equal. Brown's second rule seems to be obeyed, however, in sp octet systems, or for the σ manifold of systems containing both σ and π manifolds. Imagine the orbital connectivity of the π manifold of cyclobutadiene as being described by **6.1**, where adjacent sites are linked by a π-interaction integral of β. The σ framework is then characterized as in **6.2** by an sp bonded orbital manifold

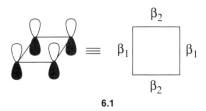

6.1

which we can model by on-site (γ) and inter-site (β) interactions between hybrids. The moments approach is particularly useful here. The π case treated in Section 1.7. There the largest fourth moment is found for the undistorted situation where $\beta_1 = \beta_2$. Since the electronic situation of interest is the one where the fractional orbital filling is 0.5, one where half of the available orbitals are filled with electrons, the more stable

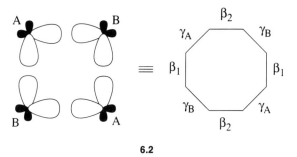

6.2

arrangement is the distorted structure. This is no more than a restatement of the Jahn–Teller or Peierls ideas. The situation is the opposite for the σ framework. Here the moments are readily evaluated from **6.2** and for simplicity the homoatomic case is used with $\gamma_A = \gamma_B = \gamma$. The contributions to the fourth moment from walks originating in a pair of adjacent orbitals are readily found to be $\beta_1^4 + \gamma^4 + 2\beta_1^2\gamma^2 + \beta_1^2\gamma^2 + \beta_2^2\gamma^2$ and $\beta_2^4 + \gamma^4 + 2\beta_2^2\gamma^2 + \beta_2^2\gamma^2 + \beta_1^2\gamma^2$, which when summed leads to the expression $(\beta_1^2 + \beta_2^2)^2 - 2\beta_1^2\beta_2^2 + 2\gamma^4 + 4\gamma^2(\beta_1^2 + \beta_2^2)$. Thus, in contrast to the π case the largest fourth moment is now found for the distorted case where $\beta_1 \neq \beta_2$. This implies, also in contrast to the π case, that at the half-filled point the symmetric structure should be of lower energy.

This important result means that for this electron count there is an electronic stabilization of the state of affairs where adjacent bonds are equal in length. All nearest neighbor distances will be equal if this is topologically possible for the octet system. Thus Brown's second rule, which requires the bond valences (bond lengths if the coordinating atoms are identical) at a given center to be as equal as possible, is simply a restatement of the fourth moment result. Notice that the predictions of the π and σ discussions are in opposition. So although a half-filled band in a solid should lead to a Peierls distortion and a (first or second) Jahn–Teller instability in a molecule to a symmetry lowering, in opposite to this is the tendency of the underlying σ manifold to favor the undistorted arrangement. In cyclobutadiene, the first order Jahn–Teller distortion dominates. In substituted cyclobutadienes, however, a symmetrical structure is found. Here the nonbonding pair of π orbitals of the homoatomic molecule has split apart in energy with an increase in the HOMO-LUMO gap. Thus Jahn–Teller activity has been converted from a first order into a second order one. A regular structure for the push-pull cyclobutadienes is now found since the structural preferences of the σ manifold now outweigh the second order Jahn–Teller driving for distortion. A similar effect is found for benzene, and the structure of graphite contains regular hexagons for the same reason. One might then be tempted to ascribe the origin of the "elastic forces" of Section 5.3 to this source.

6.4 The third rule

"The existence of edges and particularly of faces, common to two anion poly-hedra in a coordinated structure decreases its stability; this effect is large for cations with high valency and small coordination number, and is especially large when the radius ratio approaches the lower limit of stability of the polyhedra" (Ref. 275).

This rule has a clear electrostatic origin. As polyhedra share vertices, edges, and faces, there is a destabilization arising from cation–cation Coulombic repulsions (**6.3**). The cation–cation distance is shortest for face sharing, and here the repulsion

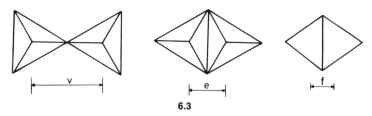

6.3

is most severe. There are many examples of structures that violate this rule, β-BeO (Ref. 331) and the isostructural compounds $BeCl_2$, SiS_2, and SiO_2-w being some of them. In dumortierite there are examples of sharing of all three types. However, the rule is satisfied often enough to remain useful.

One way (Refs. 67, 68) to examine the electronic reasons behind this rule is to compare the observed wurtzite-type structure of α-BeO, for example, with other alternative tetrahedral structures that have the same oxygen framework but where different sets of tetrahedral holes are occupied. In wurtzite itself all of the upward-pointing tetrahedra are occupied, but if some upward- and some downward-pointing holes are occupied, then edge sharing of tetrahedra occurs. Wurtzite itself is shown in Figure 6.1, and, as may be seen in the view of Figure 6.10(a), all the tetrahedra point up, and no edges are shared. In the alternative structure of Figure 6.10(b), some tetrahedra point up, others point down, and edge sharing occurs. Since each Be atom lies in a site of tetrahedral coordination, both possibilities satisfy Pauling's second rule exactly, and they should differ in energy only because they differ in the extent of edge sharing.

Figure 6.11 shows the energies computed (Refs. 67, 68) for a set of 22 BeO structures with small unit cells as a function of the number of shared edges. Structurally they have been generated using the ideas of Figure 6.10. The plot shows the energies

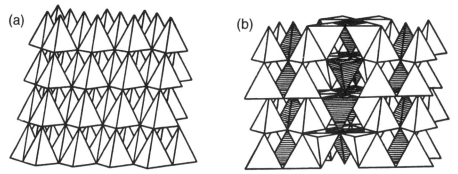

Figure 6.10 (a) The wurtzite structure showing occupation of upward-pointing tetrahedra only and so sharing of edges. (b) An alternative structure where some downward-pointing tetrahedra are occupied and edges are shared (the structure shown is in fact that of cubanite, $CuFe_2S_3$).

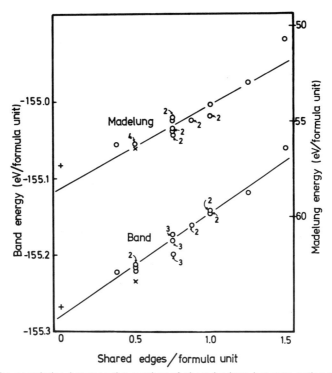

Figure 6.11 The correlation between the number of shared edges between cation tetrahedra and the computed values of the Madelung energy and the energy from an extended Hückel tight-binding calculation on a set of 22 structures based on filling half the tetrahedral holes of a eutactic oxide array, with the stoichiometry BeO.

calculated from the band model and also that using the simplest ionic model, namely a purely point charge Madelung sum, $E = \frac{1}{2}\sum q_i q_j (r_{ij})^{-1}$, where q_i and q_j are the charges on atoms i and j separated by a distance r_{ij}. Interestingly, either a purely electrostatic or a purely orbital model predicts that the number of shared edges controls the total energy. The origin of the electrostatic result is easy to see, but what about the orbital one? It is important to remember that in this structure the majority of the electron density resides on the anions. It encourages a study, not of the cation geometry, stressed in the drawing of Figure 6.10(a), but that of the anion. There are four anion coordination geometries found in these structures, which are shown in **6.4**. They are labeled 0, 1, 2, and 3, the figures indicating the number of shared edges

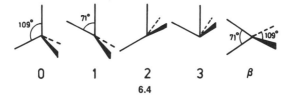

6.4

in which anion in that geometry participates. (Also shown is the local geometry in β-BeO.) The energetic differences between these geometries can be assessed by computing the energies of eight-electron OBe_4^{6+} molecules. The energies of these units

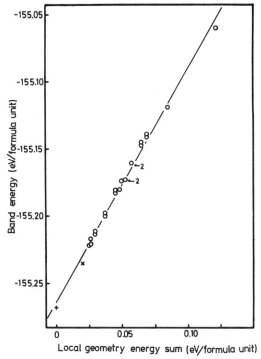

Figure 6.12   The relation of the computed energy from an extended Hückel tight-binding calculation, and the local geometry energy sum (α-BeO = +, β-BeO = x).

increase in the order $0 < 1 < 2 < 3$ in accord with the simplest ideas of chemical bonding. The energy goes up when the unit is distorted away from tetrahedral. Approximate energies for the 22 crystal structures can be computed by simply evaluating $\sum n_i E_i$, where n_i is the number of anions per unit cell having coordination geometry i, and E_i is the energy of this unit. Figure 6.12 shows the very good correlation between the energy calculated via these molecular fragments and the crystal energy from the full band calculation.

The orbital explanation for Pauling's third rule, therefore, is directly associated with the energetic penalty of the anion when it is forced to adopt a distorted local arrangement as a result of edge sharing of the tetrahedra.

The Fourth Rule states that: "In a crystal containing different cations, those of high valency and small coordination number tend not to share polyhedra elements with each other" (Ref. 275). This rule is a simple corollary to the third.

6.5   The fifth rule

"The number of essentially different kinds of constituent in a crystal tends to be small" (Ref. 275).

This rule of parsimony is an interesting one, and implies that stable structures should be simple. It has several aspects. Imagine the rocksalt structure distorted as shown in projection in Figure 6.13 so that there are two different types of sodium

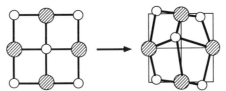

Figure 6.13   A distorted rocksalt structure.

coordinated by four long linkages. The mechanical model described by Eq. (5.1) suggests that this distorted structure should be less stable than the undistorted one. In electronic terms Brown's second rule, as described earlier leads to the same conclusion. It is the symmetric, undistorted structure, where the bond valences around the chlorine atoms are equal and where the bond valences around the sodium atoms are equal, which is more stable. Because of the connectivity of the structure, the different sodium environments are reflected in the unequal distances around chlorine and vice versa. Brown's second rule (Ref. 35) and the electronic arguments behind it go a long way to explain the small unit cells and structural simplicity of many structures. As mentioned for angellelite, simple structures are impossible for some systems on purely topological grounds given the stoichiometry and local coordination geometry. Distortions that are found away from such symmetrical arrangements demanded by the rule of parsimony are often accessible by study of the Fermi surface of the undistorted parent as described in Sections 3.2 and 4.8. However, it should be pointed out that such distortions are rarely found in the sp-bonded mineral world, which provided the database for the rules.

6.6   The description of solids in terms of pair potentials

One of the obvious problems associated with the description of solids in terms of the delocalized orbital ideas behind band theory is the frequent complexity of the results. A search, therefore, for an alternative viewpoint in terms of the energetics associated with small atomic groupings is certainly an advantageous one. In Chapter 8 the structural ideas associated with the method of moments will be developed, but here it is appropriate to examine ways to describe the energetics of solids in terms of many-atom potentials of various types. Most use the idea of a pair potential, namely one between a pair of atoms. Such approaches are well developed for both "ionic" and "covalent" materials (Refs. 54, 83, 85, 148, 169). In the molecular area, the techniques of molecular mechanics have been well used to correlate energetics with structure. Here the potential is written as if the molecule were held together by balls and springs, in terms of the displacements (δx_i) away from an equilibrium position.

$$2V = \sum_{i,j} f_{ij}\delta x_i \delta x_j \tag{6.26}$$

There are energetic contributions from bond stretching ($i = j$ = stretch), bond angle bending ($i = j$ = bend), and stretch–stretch and stretch–bend interactions. In the solid state such an approach has been used to calculate (Ref. 140) a priori mechanical and structural properties of solids such as the various polymorphs of SiO_2, by taking calculated values of the force constants from an ab initio calculation on a molecular analog, disiloxane $H_3Si–O–SiH_3$, for example (Ref. 141). The assumption here is that the dominant electronic interactions are of sufficiently short range that a few local parameters will well describe the structure. This view is one that is encouraged by the observation that many of these octet solids are just giant molecules.

Whatever the exact nature of the chemical bonding in a molecule or solid, the actual structure that is adopted is a balance between attractive and repulsive forces. This does not appear in Eq. (6.26), since the energy is written in terms of displacements away from some chosen equilibrium geometry. Probably the most basic potential is that of Lennard–Jones, often used to describe the interatomic forces between atoms from Group 18.

$$\Phi(r_{ij}) = -A_{ij}/r_{ij}^6 + B_{ij}/r_{ij}^{12} \tag{6.27}$$

Here $A, B > 0$ and depend on the identity of the atoms. The potential consists of a van der Waals attractive part, of longer range than the harder repulsive potential that sets in at shorter distances. Of more general use in materials regarded as "ionic" is a potential that includes an electrostatic term. The simplest form is of the Born–Landé type (Ref. 28)

$$\Phi(r_{ij}) = q_i q_j/r_{ij} + B_{ij}/r_{ij}^n \tag{6.28}$$

but a more frequently used expression is of the Born–Mayer type (Ref. 29) with the addition of a van der Waals attraction.

$$\Phi(r_{ij}) = q_i q_j/r_{ij} + B_{ij}\exp(-r_{ij}/\rho_{ij}) - C_{ij}/r_{ij}^6 \tag{6.29}$$

The first term in both equations is attractive for unlike ions and repulsive for like ions. The B_{ij}, ρ_{ij}, and C_{ij} are adjustable parameters (Ref. 85) and are determined by fitting to an experimentally observed structure. The van der Waals constant is set by the relationship $C_{ii} = 3I_i\alpha_i^2/4$, where I_i and α_i are the ionization potential and polarizability, respectively, for the atom i. Often the values B_{ij}, ρ_{ij}, and C_{ij}, where $i \neq j$ are estimated from those of the like atoms as

$$B_{ij} = \sqrt{B_{ii}B_{jj}}, \qquad C_{ij} = \sqrt{C_{ii}C_{jj}}, \qquad \rho_{ij}^{-1} = (\rho_{ii}^{-1} + \rho_{jj}^{-1})/2 \tag{6.30}$$

The adjustable parameters are usually obtained from simple structures where comparable values of r_{ij} are found and then transferred to a case of chemical interest. An interesting structural transformation found for one of the high-temperature superconducting cuprates will be used as an example here (Ref. 126).

The basic structure of the 2–1–4 superconductor $La_{2-x}Sr_xCuO_4$ is shown in Figure 5.33. Above 533 K the parent compound La_2CuO_4 is tetragonal, a result implying that the CuO_2 layers are flat. Below this temperature the structure distorts to an orthorhombic arrangement where the layers are buckled. A view is shown in

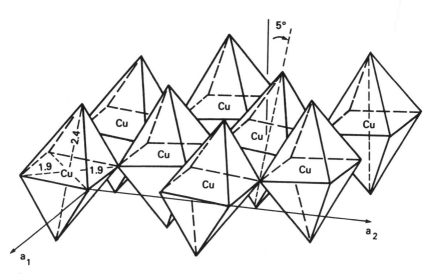

Figure 6.14 The relationship between orthorhombic and tetragonal 2–1–4 structures. In the ortho-rhombic structure the octahedra are tilted off their axes as shown. In the tetragonal structure there is no such tilting, and the CuO_2 planes are flat. Some approximate Cu–O distances are also shown.

Figure 6.14. As doping with Sr (or Ba) increases, the temperature of the tetragonal to orthorhombic transition decreases, and eventually the tetragonal structure is the one stable at all temperatures. There is an electronic explanation for the distortion, but the use of the potential of Eq. (6.24) shows the importance of the "size" of the doping atoms. First, the values of B, ρ, and C can be determined for $Cu^{2+}\cdots Cu^{2+}$, $La^{3+}\cdots La^{3+}$, and $O^{2-}\cdots O^{2-}$ interaction by choosing values that reproduce the cell and positional parameters in the binary oxides CuO and La_2O_3. The values are given in Table 6.5. Using these values the structure of the superconductor may be studied, both in tetragonal and orthorhombic forms. The energy difference between the two comes to 1.85 kcal/mol formula unit, the orthorhombic structure being predicted to be more stable, a result qualitatively in accord with experiment. The energy difference is also of the right order of magnitude. On doping with Sr^{2+} or Ba^{2+}, two of the electronic parameters change. The copper atoms are oxidized and thus become "smaller" (the average Cu–O distance decreases) and the larger 2+ cation leads to larger nonbonded repulsions than the 3+ cation. These two changes may be simulated by repeating the calculation but by increasing the value

Table 6.5 Ionic parameters for modeling the 2–1–4 superconductor

Pair	B(eV)	ρ(Å)	C(eV Å^6)
$O^{2-}\cdots O^{2-}$	1387.7	0.375	63.31
$Cu^{2+}\cdots Cu^{2+}$	269.10	0.264	0.5586
$La^{3+}\cdots La^{3+}$	28 855	0.250	325.0

Source: Ref. 126.

of B for $La^{3+} \cdots La^{3+}$ interactions and decreasing it for the $Cu^{2+} \cdots Cu^{2+}$ inter-actions. The result of such a calculation is a gradual decrease in the tegragonal/orthorhombic energy difference as the extent of doping increases. Since Ba^{2+} is larger in size than Sr^{2+}, the model predicts a lower transition temperature for the barium system than for its calcium analog, a result in agreement with experiment. Although this result is in accord with the experimental trend, there is a simple orbital explanation of the orthorhombic/tetragonal transition in terms of the variation in the stability of the two structure with electron count (Refs. 59, 60).

Although such studies using the potential of Eq. (6.26) imply an ionic model, it should be realized that over the range of internuclear distances sampled in many structural problems, other potentials may be equally as good, as long as they adequately represent the attractive and repulsive parts of the potential. Thus just because an "ionic" model is used, this does not demand an "ionic" description of the chemical bonding. Note the parallel formulations of the second and third rules in terms of the "ionic" and orbital models. The formulation of Eq. (6.26), though, often leads to excellent values of the cohesive energies of solids. The values calculated using either this equation or the Born–Landé variant, Eq. (6.25), where the repulsive potential depends on r^{-n}, lead to values of the cohesive energy that are in close agreement (Ref. 345) with those from the Born–Haber cycles. In terms of mimicking the structural part of the energy, this approach is usually not at all good, as described in Section 6.2. The reasons for this are not too hard to see. The ionic approach models well the monopole of the electronic charge distribution, but very poorly higher terms in a multipole expansion, those that are needed to describe the "polarizability" of the atoms, for example. The latter are important in structural problems (Ref. 266). Use of Eq. (6.26) does lead to a good correlation with some of the dynamical properties of solids, SiO_2, for example (Ref. 84). In addition, the use of MAPLE calculations (Madelung part of the lattice energy) have been very effective in the right hands (Ref. 180) in guiding synthetic strategy, as are general ionic arguments concerning a variety of chemical problems (Ref. 198). Pair potentials between atoms also come from pseudopotential theory, a topic taken up in Chapter 8.

The results described in Section 6.4 concerning the energies of the BeO structural alternatives wrote the total energy as $E_T = \sum n_i E_i$, where n_i is the number of anions per unit cell having coordination geometry i, and E_i is the energy of this unit. There are other structural problems that may be tackled in the same way by using a many-atom potential or energetic contributions from a set of units of different types. For example, the stability of the different ordering patterns of the oxygen atoms between the planes of the 1–2–3 superconductor may be studied by writing the total energy in this way (Ref. 305). A similar approach allows access to the observed structures of arsenic and phosphorus (Refs. 55, 65). Figure 3.11 shows the breakup of the simple cubic structure to give the arrangements found for these elements. It was noted in that section that there are a total of 36 different structures that may be generated using this bond-breaking algorithm and a unit cell of this size. Why are the other possibilities not observed? **6.5** shows one possibility, the P_8 cube, a species that is unknown. All of the possibilities contain trigonal pyramidal atoms, and so the features that energetically distinguish between them must involve second- and higher nearest-neighbor interactions. Four types of interactions appear

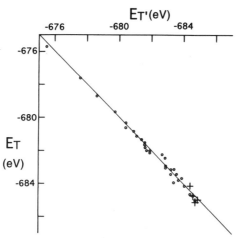

Figure 6.15 Fitting of the calculated tight-binding energies of the 36 possible arsenic-like structures to a simple local fragment energy sum. The known structures, indicated by crosses, are located around the bottom of the plot.

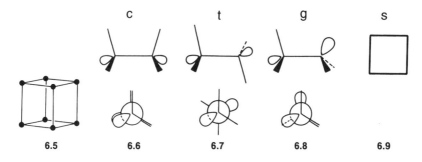

to be important and are shown in **6.6–6.9**. They are the presence of syn, gauche, and anti interactions between pairs of atoms and the presence of four-membered rings. The values of the E_i are determined by a least-squares fit of the 36 calculated electronic energies E_T' to that expected from the algorithm $E_T = \sum n_i E_i$. The results are shown in Table 6.6 and the fit, E_T' versus E_T, plotted in Figure 6.15. The agreement is excellent. Also shown in Table 6.6 are the values of the E_i determined from calculations on $P_n H_m$ molecules, where the ends of the molecules are tied off with hydrogen atoms to mimic the behavior in the solid. (This is similar to the philosophy noted for SiO_2.) While the numbers are not the same, the general picture is quite similar. Observed structures are indicated by a cross in the plot of Figure 6.15. The black phosphorous arrangement is the most stable one (satisfyingly, the calculations used phosphorous parameters), and the arsenic structure is the seventh lowest. Two other known structures are also indicated on the plot. Thus four known structures with nets derived in this way occur in the seven structures calculated to lie at lowest energy. So writing the energy in terms of contributions from small units captures much of the essence of this structural problem. From the figures of Table 6.4, the reason the structure **6.5** is not found is because of the large number of squares. The lowest-energy structures are those where there are no squares and the number of destabilizing syn and anti interactions is minimized.

Table 6.6   Parameters used to fit the energies of the 36 "phosphorous" structures[a]

	1	2	3
syn	− 56.787	0.284	0.486
gauche	− 57.071	0	0
anti	− 57.036	0.035	0.407
4-ring	0.994	0.994	0.90

Source: Ref. 55.

[a] Column 1 lists the values of E_s, E_g, E_a, and E_q derived from least-squares fit to the band structure total energies. Column 2 shows the same numbers with the first three rows relative to $E_g = 0$. Column 3 is the values expected for column 2 from the calculations on molecules. All energies are in eV.

6.7   More about the orbital description of silicates

Related to the ideas of this chapter, a particularly useful approach to the study of silicates has been through the use of molecular orbital calculations of various types on molecular fragments torn from the extended solid. In order to satisfy the dangling bonds generated in such a process, hydrogen atoms are used to "tie off" these loose ends (Refs. 141, 346, 347). One molecule used to study solids is the disiloxane molecule of **6.10**. Many of the structural observations of the extended array are neatly

6.10

reproduced by studies on this small molecule. In quantitative terms the bending force constants calculated for $H_6Si_2O_7$ are in rough agreement with those derived for α-quartz from lattice dynamical studies (Ref. 140). Figure 6.16 shows how the equilibrium Si–O–Si angle varies (Ref. 141) as a function of Si–O distance (the three distances 1.65, 1.62, 1.59 Å were chosen). Quite a range of equilibrium bond angles are found in practice for a relatively small variation in the Si–O distance. Particularly interesting is the rather soft bending force constant for this system, especially at the shorter distance. Figure 6.17 shows how the Si–O bond length and Si–O–Si bond angle are related in the silica polymorph coesite. The upper curve reports the experimental data; the lower curve comes from the results of three calculations on disiloxane where the Si–O–Si angle is fixed and the energy minimized with respect to the Si–O distance. In qualitative terms the agreement is very good; certainly the

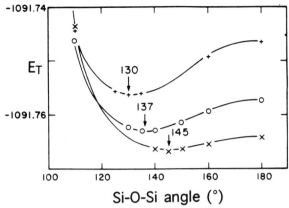

Figure 6.16   The total energy (atomic units) for $H_6Si_2O_7$ as a function of the Si–O–Si angle for three different values of the Si–O distance. The minimum in each curve is indicated. (Adapted from Ref. 141.)

general behavior of the two curves is quite similar. Of considerable importance is the observation that there is quite a spread in the observed values of the Si–O–Si angle in these solids, and that the energy differences between the different geometries are small. Such considerations go quite a way towards understanding why there are such an enormous variety of solid silicates. Whereas in organic chemistry there is a considerable strain energy associated with deformations away from tetrahedral carbon, which sets limits on some of the possible structures, variations in Si–O–Si angle of over 60° do not appear to cost much energy. So the angles in these compounds may be readily adjusted to take into account local bonding and packing requirements without a severe energetic penalty. Several theoretical explanations have been put forward to rationalize the existence of this soft potential, ranging from electronic arguments (Ref. 46) to those involving nonbonded repulsions between silicon atoms (Refs. 267, 268). However, it is worthy of note that XYZ systems, where Y is a first row atom and X and Z are π-acceptor ligands, invariably have low bending

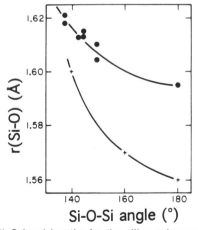

Figure 6.17   Experimental Si–O bond lengths for the silica polymorph coesite (upper curve) and theoretical plot (from the results of Fig. 6.16) as a function of Si–O–Si angle. (Adapted from Ref. 141.)

force constants. Examples include C_3O_2 (OC–O–CO) and its isoelectronic ylid analog R_3P–C–CO as well as the series of siloxanes themselves. An electronic model that is particularly satisfying (Ref. 46), and one that connects the structures of these molecules to those of the solid silicates, is that of stabilization of the linear geometry by higher-energy orbitals on silicon. This is just the two-coordinate analog of the square-planar oxygen atom in NbO discussed in Section 4.4 and the planar three-coordinate oxygen atom in rutile (Ref. 44) and trisilyamine of Section 7.8.

7 □ The structures of some covalent solids

Atoms are not characterless spheres like billiard balls. If they were, then such objects with a uniform diameter and an incompressible exterior would be expected to pack together in space according to the principles enunciated by Laves. He argued (Ref. 218) that such systems would lead to crystal structures with the highest symmetry, highest coordination numbers, and densest packing of atoms. Such structures are in fact found for many of the metallic systems of the following chapter (although there are electronic reasons for such arrangements, too). This chapter will describe the structures of some materials where from the ball-and-stick models of their geometrical structures it is clear that some type of "directional bonding" is at work. Present theoretical methods (Refs. 95, 97, 148, 396) are able to distinguish energetically between simple structures. This chapter continues the theoretrical philosophy of exploring the chemistry behind the numbers provided by a calculation.

7.1 Electron counting

As pointed out in Section 5.1, the electronic structures of " tetrahedral" solids present a special case for their theoretical description. Even though their orbitals are delocalized, they may be described in terms of two-center two-electron bonds between pairs of atoms and lone pairs where needed. **5.1** shows a description in terms of "bonding" "nonbonding," and "antibonding" bands. The rules governing the assembly of these systems are particularly simple (Refs. 186–188, 247, 274), given this one-to-one correspondence between the localized and delocalized pictures. In fact the connectivity of the atoms that make up the solid is simply determined by the stoichiometry and electron count. The rules are only just a little more complex than those for octet molecules, the infinite structure of the material playing a role now. If per formula unit, n_a is the number of valence electrons on the anions, n_c is the number of valence electrons on the cations (less any unshared electrons), b_a is the number of electrons forming anion–anion bonds, b_c is the number of electrons forming cation–cation bonds, then if there are N_a anions

$$\frac{n_a + n_c + b_a - b_c}{N_a} = 8 \tag{7.1}$$

So the tetragonal structure of ZnP_2 (Fig. 7.1) is valence compound because

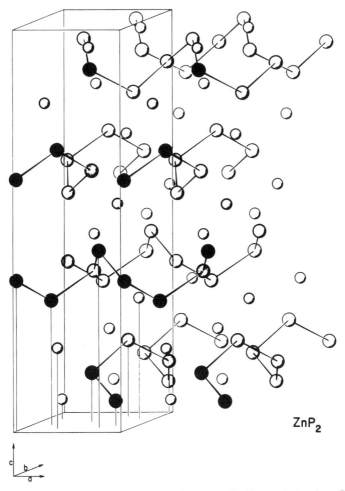

ZnP₂

Figure 7.1 The red, tetragonal form of ZnP_2. (Reproduced with permission from Ref. 280.)

$n_a = 2 \times 5 = 10$, $n_c = 2$, $b_a = 4$, $b_c = 0$, $N_a = 2$. In SnSe, a derivative structure of Figure 7.2, the tin atom has two unshared electrons in a lone pair and so $n_a = 6$, $n_c = 2$, $b_a = 0$, $b_c = 0$, and $N_a = 1$. The application of ideas such as these to tetrahedral structures is a simple and powerful one (Ref. 274).

The number of anion–anion bonds formed by each anion is b_a/N_a, and the number of cation–cation bonds formed by each cation is b_c/N_c. If the coordination numbers of the anions and cations are Γ_a and Γ_c, then the anions form $N_a(\Gamma_a - b_a/N_a)$ bonds with the cations that must equal the $N_c(\Gamma_c - b_c/N_c)$ bonds formed by the cations with the anions. So $N_a(\Gamma_a - b_a/N_a) = N_c(\Gamma_c - b_c/N_c)$. Therefore, $b_a - b_c = N_a\Gamma_a - N_c\Gamma_c$, leading to

$$\frac{n_a + n_c + N_a\Gamma_a - N_c\Gamma_c}{N_a} = 8 \tag{7.2}$$

For the case of elemental solids then $n_c = N_c = 0$ and the coordination number

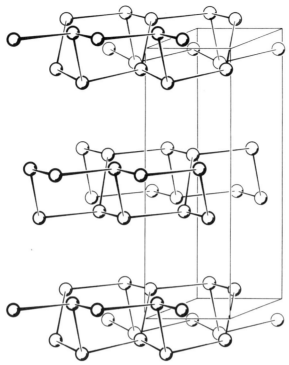

Figure 7.2   The structure of black phosphorus. (Reproduced with permission from Ref. 280.)

around each atom ($N_a = 1$) is given by $8 - n_a$, which is just one of Hume–Rothery's rules. So black phosphorous (Fig. 7.2) contains $8 - 5 = 3$-coordinate atoms, elemental selenium (Fig. 5.5) $8 - 6 = 2$-coordinate atoms. Sometimes the structures of solids with quite complex formulae are simple to understand. The solids that comprise the series $A_{14}MX_{11}$ (A = Ca, Sr, Ba; M = Al, Ga, Mn; X = As, Sb, Bi), for example contain (Ref. 132) fourteen A^{2+} cations, one linear X_3^{7-} unit (isoelectronic with I_3^-), one MX_4^{9-} tetrahedron (isoelectronic with $SiCl_4$), and four X^{3-} ions (isoelectronic with I^-). There are many novel species known of similar type. Thus both GaN_3^{6-} (Ref. 99) and $InAs_3^{6-}$ (Ref. 24) are isoelectronic with carbonate ion.

Equation (7.1) is very useful as a predictor of semiconductivity (Refs. 280, 281). As shown in **5.1**, if there are sufficient gaps between the bonding, nonbonding, and antibonding bands involved, then the solid will be nonmetallic. However, it is not always obvious how many bonds, especially anion–anion or cation–cation, should be included. PbO and InBi have the same structure type, but whereas PbO is a semiconductor, InBi is metallic. At first sight, $b_a = b_c = 0$ for both systems, but on closer inspection the Bi atoms are found to be quite close (3.68 Å) in InBi. This distance is shorter than the van der Waals separation of ~ 4.4 Å. So clearly $b_a \neq 0$ and the system is not a valence compound, does not obey Eq. (7.1) and so will not be a semiconductor (Ref. 280). Interestingly InSb has the sphalerite structure, and is a semiconductor. Here, though, the Sb–Sb distance is much larger at 4.58 Å ($b_a = 0$). Similar considerations were behind the discussion of Section 5.1. Recall that intersheet interactions in arsenic-like structures are responsible for a small amount of band overlap and the generation of a semimetal.

7.2 Change of structure with electron count

The model shown in **5.1**, which does not rely on a detailed knowledge of the electronic band structure of the solid, can be used to provide a qualitative understanding of the structures of many solids composed of main group elements. The approach is readily appreciated by consideration of the series of molecules of **7.1**. As the number

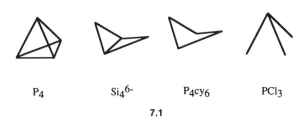

| P_4 | Si_4^{6-} | P_4cy_6 | PCl_3 |

7.1

of electrons increases, notice the structural changes that result (Ref. 245). The tetrahedral P_4 structure with 20 electrons is an electron precise molecule, one where all bonding and nonbonding orbitals are occupied. Six P–P bonds result. Addition of extra electrons has to lead to occupation of orbitals that are antibonding between the phosphorous atoms if this tetrahedral structure is maintained. As a result, one of the bonds of the tetrahedron is broken to give the geometry found for Si_4^{6-} (Ref. 124) or B_4H_{10}, and lone pairs of electrons are generated at the ends of the severed bonds. By the time the PCl_3 molecule is generated with 26 electrons, three linkages of the parent have been broken. Similar results are found in solids (Refs. 40, 41) Figure 6.1 shows the framework structure of wurtzite (found for 8-electron systems such as ZnS). Figure 5.3 shows the slab structure of GaS (found for 9-electron systems), and Figure 5.4 the layer structure of SnS (found for 10-electron systems). Elemental arsenic, antimony, and bismuth adopt the degenerate SnS structure. The important observation is that as the electron count increases, so the wurtzite structure becomes more broken up. This shows up electronically in the computed band structures for these materials.

Figure 7.3 shows how (Ref. 64) the band structure of a single sheet of the graphite structure changes as each site distorts from the planar to pyramidal geometry. The resulting structure is, of course, that of a single sheet of the arsenic structure. This is a different but related problem to the one just discussed. Here no bonds are broken, but the angular geometry changes. The stabilization of the highest occupied orbital of the graphite structure at the arsenic electron count is clear to see. In this sense changing the number of electrons results in small perturbations of large parts of the structure.

An alternative way to break up the wurtzite structure is to cleave the structure vertically rather than horizontally (which gave the GaS structure). The result is the

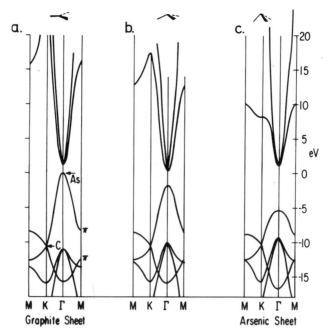

Figure 7.3   The evolution of the band structure of the planar graphite sheet as it distorts toward the puckered arsenic sheet. The calculations are two-dimensional ones for a single sheet.

structure of wolfsbergite ($CuSbS_2$), another 9-electron (per two atoms) system, shown in Figure 7.4. The structure of GaS and GaSe is a little different from that expected by the simple fission of the wurtzite structure. Ga–Ga linkages have been formed and the chalcogenide has been displaced to the outside of the slab. The reason for this is simple to appreciate. The emphasis has been on the generation of lone pairs of electrons as a major contributor to the energetics driving the structural change. This is shown for the solid in **7.2**. Extending ideas well known for molecules

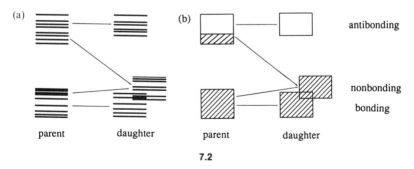

7.2

(Refs. 41, 92), the most stable product will then be the one where the lone pairs lie on the most electronegative atom, in this case selenium. Invariably the most electronegative elements of the structure lie at the sites of smaller coordination, where energetically stable lone pairs can be formed. Such an argument is readily extended to the structures of realgar, AsS (As_4S_4) and S_4N_4 (**7.3**). Notice the different sulfur

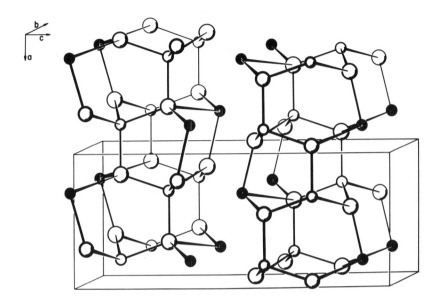

Figure 7.4 The structure of wolfsbergite ($CuSbS_2$). (Reproduced with permission from Ref. 280.)

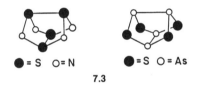

7.3

coordination numbers in the structure. These ideas are related to those of topological charge stabilization (Ref. 142) of use in predicting the identity of stable isomers of molecules.

There is a variation of the foregoing route to generate a stable solid by the ejection of an atom. This is just the process of cleaving all the linkages to one particular atom. Figure 7.5 shows the structure of $CdIn_2Se_4$. It is geometrically related to that of sphalerite. In fact, the selenide is simply an ordered defect derivative of sphalerite. Now four lone pairs point into the empty site. The generalized Grimm–Sommerfeld

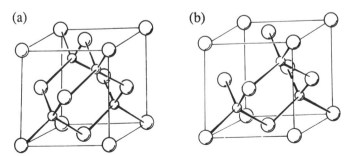

Figure 7.5 The structure of (a) sphalerite and (b) $CdIn_2Se_4$, a defect sphalerite structure. (Reproduced with permission from Ref. 280.)

valence rule (Ref. 274) is useful in counting electrons here. The number of electrons per site should be equal to 4 for a stable octet compound based on four-connected networks. This is true for ZnS, $(2 + 6)/2 = 4$, but not true at first sight for $CdIn_2Se_4$, $(2 + 6 + 24)/7 \sim 4.7$. However, if the material is written as $\square CdIn_2Se_4$, where $\square$ represents a vacant site, then the rule holds exactly, $(0 + 2 + 6 + 24)/8 = 4$. Similar rules also hold in molecules. **7.4** shows the structures of the

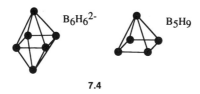

7.4

closo borane $B_6H_6^{2-}$ (only the boron atoms are shown for simplicity) and the *nido* borane B_5H_9. In terms of the number of skeletal electrons, they both have the same number, namely, 14. The number of skeletal electrons per site is only $14/6 = 2.33$ in both species but only if the vacant site of the *nido* octahedron is included.

7.3 Structures of some AX_2 solids

A similar approach gives insights (Ref. 66) into the structures of some AX_2 solids. As Figure 7.6 shows, the CaC_2 structure is a tetragonal derivative of rocksalt, with C_2 units lying parallel to c. Locally a linear Ca–C–C–Ca geometry results. FeS_2 has four more electrons, and the structures of marcasite and pyrite result, which contain skewed Fe–S–S–Fe units (Fig. 7.7). With two more electrons, the X–C linkage is broken, and the rutile structure generated from marcasite (Fig. 7.8). Figure 7.9 shows

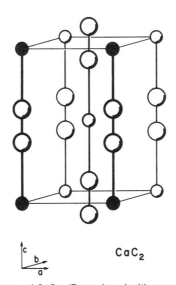

CaC_2

Figure 7.6 The structure of CaC_2. (Reproduced with permission from Ref. 280.)

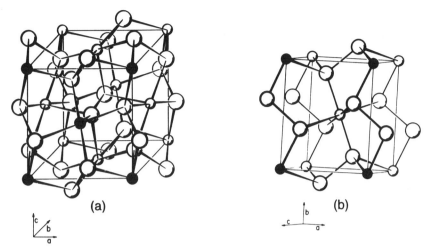

Figure 7.7 (a) The structure of pyrite (FeS_2). (b) The structure of marcasite (FeS_2). (Reproduced with permission from Ref. 280.)

in a different pictorial fashion how an analogous distortion of pyrite leads to the structure of fluorite. With only one extra electron ($IrSe_2$) only half of the S–S linkages are broken (Fig. 7.10). With yet more electrons the *molecular* crystal structure of XeF_2 may be regarded as being derived from that of fluorite (Fig. 7.11) by moving the A atoms from the centers of X_8 cubes to their edges. These geometrical changes found in the solid are readily related to analogous changes found in molecular A_2X_2 systems as the electron count increases and are shown in **7.5**. Notice that not always does the addition of electrons give rise to the breaking apart of the structure as in Section 7.2. Lone pair orbitals on the anion are generated simply by changing the local geometry as in $CaC_2 \rightarrow FeS_2$. Figure 7.12 shows this structural-electronic process in a more general way. There are nine possible ways (Ref. 66) to distort the CaC_2 cell of Figure 7.6 to a pyrite-like structure. In principle all may respond

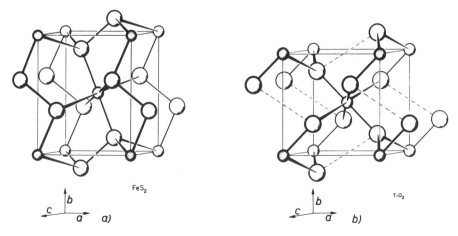

Figure 7.8 Relationship between the structures of marcasite (a) and rutile (b). (Reproduced with permission from Ref. 280.)

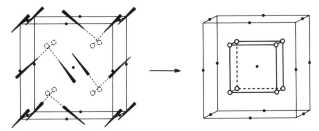

Figure 7.9 Generation of the fluorite structure from that of pyrite by bond breaking.

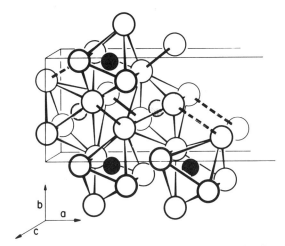

Figure 7.10 The structure of IrSe$_2$. (Reproduced with permission from Ref. 280.)

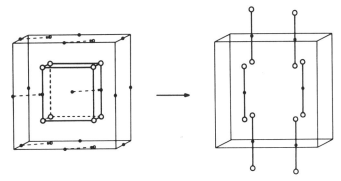

Figure 7.11 Generation of the structure of solid XeF$_2$ from that of fluorite by bond breaking.

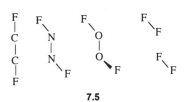

7.5

227

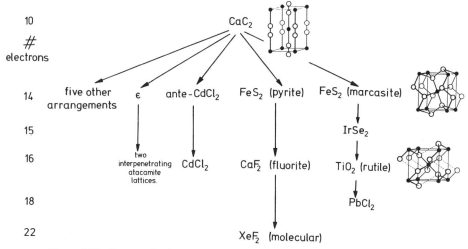

Figure 7.12 Structural scheme linking some AX_2 structures with electron count.

geometrically to changes in electron count. The "ante-$CdCl_2$" structure consists of sheets of the $CdCl_2$ structure linked together, just as in the structure of delafossite shown in Figure 7.13, which contains a spacer between the sheets.

As a final example Figure 7.14 shows how (Ref. 51) the structures of PdP_2, PdPS, and PdS_2 are interrelated in the same way to those of ZnS, GaS, and GeTe. The building block in the ZnS series is the puckered 6^3 sheet, but is the $5^4 + 5^3(2:1)$ sheet in the PdP_2 series.

7.4 Structures derived from simple cubic or rocksalt

The description in Chapter 3 of the structures of elemental arsenic, antimony, and bismuth (s^2p^3) in terms of a Peierls distortion of the parent simple cubic structure (Refs. 61, 134) is an interesting approach that may be extended to other s^2p^n configurations. If the p^3 configuration is unstable to bond length alternation in all three directions, then the p^2 configuration should be unstable to bond length alternation in two, and the p^1 configuration unstable to bond length alternation in just one, direction. This idea leads to some interesting results.

Figure 7.15 shows how a specific bond length alternation in one direction only for a 4^4 sheet leads to the generation of the 6^3 sheet. This is just the geometrical arrangement found in the structure of elemental gallium of Figure 7.16. The way these sheets are connected to each other is more complex than expected from the bond-breaking algorithm shown in Figure 7.15. From Figure 7.16 it may be seen that the sheets are connected to each other by sliding one layer relative to the next such that triangles of atoms are created. The hexagons are more difficult to see (they are vertical) in the diagram of Figure 8.3, but these triangles are clearly visible. The simplest structure expected from bond breaking in one direction is that of the cadmium net of the structure of $CeCd_2$ (Fig. 8.10). This is a structure usually described as the cadmium iodide type, but the octahedra containing the Ce atoms are quite

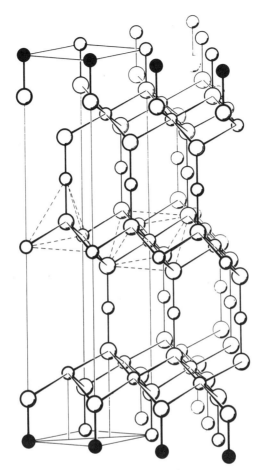

Figure 7.13 The delafossite structure (CuFeO$_2$). (Reproduced with permission from Ref. 280.)

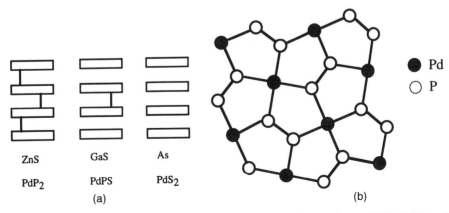

Figure 7.14 (a) The electronic–geometrical relationships between the structures of PdP$_2$, PdPS, and PdS$_2$ compared to those between ZnS, GaS, and GeTe. (b) The sheet involved in the Pd sequence.

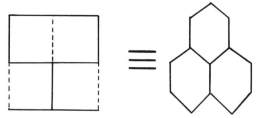

Figure 7.15 Breakup of the square lattice (4^4) to give the graphite net (6^3).

distorted, and making connections between the atoms of adjacent sheets leads to the network shown in the figure. This structure is found not only for $CeCd_2$ itself but also for YCd_2.

For solids with the s^2p^2 configuration, structures containing the 6^3 net of Figure 7.15 are expected, coupled together in a manner derived by bond alternation along the direction perpencidular to the sheets. First pucker the 6^3 sheet, then make one connection at each node, either up or down. Figure 7.17(a) shows how a specific bond length alternation pattern in this third direction leads to generation of the structure of cubic diamond. Another pattern [Fig. 7.17(b)] leads to the generation of the hexagonal diamond structure. Other possibilities exist. If the bond alternation in the third direction is such that squares are created linking the 6^3 sheets, then the "tetragonal carbon" structure is found [Fig. 7.17(c)]. This is the arrangement found for the boron atoms in the structure of CrB_4, for example, and as a derivative structure in β-BeO (Ref. 331), but is unknown for carbon itself. It is worth noting that the structure of diamond is close to that of white tin. These results lead

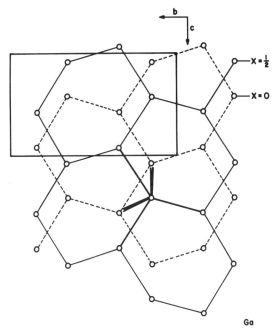

Figure 7.16 The structure of gallium. (Reproduced with permission from Ref. 280.)

Cubic Diamond   Hexagonal Diamond   Tetragonal Diamond
ZnS sphalerite   ZnS wurtzite   β–BeO
(Mg)AgAs   (Ca)In2   (Cr)B4

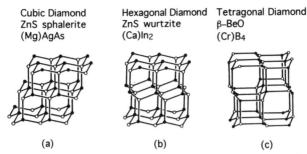

(a)   (b)   (c)

Figure 7.17 The structures of (a) cubic diamond, (b) hexagonal diamond (lonsdaleite), and (c) tetragonal diamond (unknown for carbon itself).

to interesting insights concerning the factors influencing the structures of diamond (or its derivative structure, zincblende) and rocksalt, which are described in the next section.

Another group of solids may be readily derived from the simple cubic structure, or its AB derivative by extending these ideas. They invariably contain elements from the right-hand side of the periodic table. Bi_2Te_3 of Figure 7.18 is an example. Notice that the bismuth and a third of the tellurium atoms are six-coordinate, with the remainder of the Te atoms being three-coordinate. It appears to be a piece of the rocksalt structure but one that is just five atoms thick along the [111] direction. SnS_2 has the cadmium halide structure (Figs. 7.19 and 7.20), which contains six-coordinate tin but three-coordinate sulfur. This appears to be a piece of the rocksalt structure three atoms thick along the [111] direction. The structures of gallium, diamond, arsenic, and black phosphorus could be regarded as arising via a Peierls instabilities in one, two, or three dimensions driven by a half-filled collection of p bands at the simple cubic structure. Bi_2Te_3 and SnS_2 may be regarded as arising via bands of different fillings. As discussed in Section 5.3, the shortest-wavelength distortion is the one of lowest energy, and thus for an n/N fractional p band occupancy an N-merization is to be expected. Applying this rule to Bi_2Te_3, there are a total of $10 + 18 = 28$ valence electrons, of which $5 \times 2 = 10$ are in s orbitals. This leaves a fractional p band occupancy of $(28 - 10)/(5 \times 6) = 3/5$. A pentamerization is thus expected for this system, indeed found for Bi_2Te_3 (Fig. 7.18), where a triple-layer Te structure, with Bi in the octahedral holes, is found. Three perpendicular, five-atom Te–Bi–Te–Bi–Te strings are easily identified. One of the largest slab structures of this type that is accessible using this approach is that of Bi_8Se_9, which, with a (10/17)-filled band, gives a 17-merization to form a very large slab of the NaCl structure. Table 7.1 shows (Ref. 134) a collection of structures of binary and ternary systems composed of elements from Groups 14–16, where such ideas work well.

Of course, small structural modifications of the products of the N-merization are to be expected in terms of the nonequivalence of bond lengths and angles now not constrained to be 90°, etc. Also, the three-dimensional packing of the slabs or fragments produced might be different from that predicted by simple bond fission. The $CdCl_2/CdI_2$ pair of structures and the many polytypes in between are a case in point. The $CdCl_2$ structure is one based on trimerization of rocksalt, and leads to the generation of slabs separated by a van der Waals gap. The anions are arranged

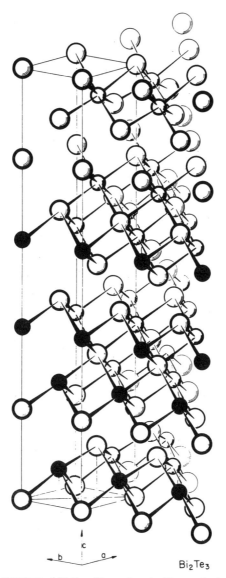

Figure 7.18 The structure of Bi_2Te_3. (Reproduced with permission from Ref. 280.)

close to cubic eutaxy. The CdI_2 structure results simply from a shift of one such slab relative to another so that now hexagonal eutaxy is found for the anions. The two arrangements are close in energy. Conceptually, however, the generation of the $CdCl_2$ variant provides the simpler picture.

As the electronegativity difference between the atoms concerned increases, then the number of electrons occupying the s orbital manifold may not be obvious. Consider, for example, the cadmium halide structure found for SnS_2. If the tin $5s$ orbitals lie above the sulfur $3s,3p$ orbitals, then they will be unoccupied. If this is the case, then only four electrons from the total of 16 s and p electrons are located in sulfur $3s$

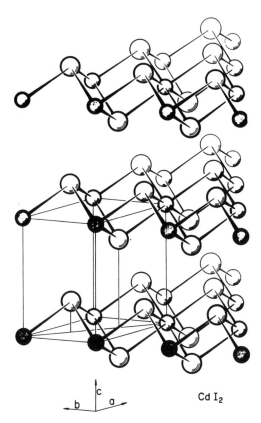

Cd I$_2$

Figure 7.19   The structure of CdI$_2$ [sometimes called the brucite, Mg(OH)$_2$ structure). (Reproduced with permission from Ref. 280.)

orbitals, leaving 12 electrons for the set of tin and sulfur p orbitals. The fractional occupancy of the p orbital set is thus $(16 - 4)/18 = \frac{2}{3}$, and a trimerization is expected in all three directions. This is just what is found in the structure composed of three perpendicular S–Sn–S chains centered at each metal site. The trimerized cadmium halide structure would not be predicted for this species if the Sn s orbitals were occupied. That these structures are controlled by atomic identity in addition to electron count is shown by the observation of the same structure for In$_3$Te$_4$, SnSb$_2$Te$_4$, P$_3$Sn$_4$, As$_3$Sn$_4$, and Bi$_3$Se$_4$.

The structure of elemental Se or Te presents a different problem. Here the p band is $\frac{2}{3}$ full, but a trimerization is not found. Instead (Fig. 7.21) the orbital occupancy shown is appropriate for a distortion, which leads at each atom to complete bond breaking in one direction (completely filled band) and alternate breaking (half-filled bands) in the other two. The result is an interesting spiral structure (Fig. 5.5) with two-coordinate atoms. The picture of Figure 7.21 is a general one for many systems. Although it is often a straightforward matter to understand the origin of a particular pattern, such as Se or SnS$_2$, rationalization of the stability of one possibility over another is frequently not easy. In the case of polonium it is clear that the driving force for the Peierls distortion is not large, since the undistorted simple cubic structure is the one that is found.

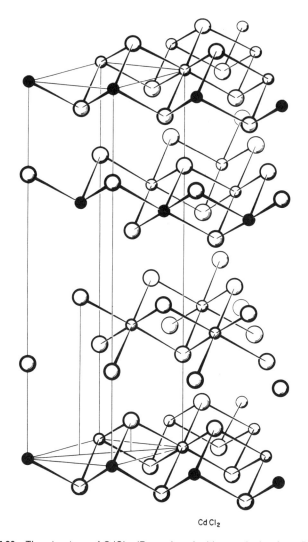

CdCl₂

Figure 7.20 The structure of CdCl$_2$. (Reproduced with permission from Ref. 280.)

7.5 The stability of the rocksalt and zincblende structures

The discussion of the previous section raises the question of whether the simple cubic structure be stabilized against a Peierls distortion by substitution, in an exactly analogous way to that described for molecules in Section 1.4 and some simple solids in Section 2.4. The answer in yes (Ref. 43). Using the same pictorial style as in Figure 3.10, Figure 7.22 shows how the energy bands of the constituents vary as their electronegativity difference increases. If the atoms of the simple cubic structure are colored in two colors (A, B), the result is the AB rocksalt structure. Notice though that the electron configuration has changed on moving from left to right. Whereas the four-electron element has the $2(s^2p^2)$ configuration, the AB compound at the far

Table 7.1 The structures of some solids from the viewpoint of partially filled bands at the rocksalt structure

Compound	f_i	N-merization
P (black)	$\frac{1}{2} = 0.500$	2
As		
Sb		
Bi		
$As_2Ge_8Te_{11}$	$\frac{11}{21} = 0.524$	21
$As_2Ge_5Te_8$	$\frac{8}{15} = 0.533$	15
$As_2Ge_4Te_7$	$\frac{7}{13} = 0.538$	13
$As_2Ge_3Te_6$	$\frac{6}{11} = 0.545$	11
$Ge_3Bi_2Te_6$		
$Bi_{14}Te_6$	$\frac{11}{20} = 0.550$	20
Sb_2Te	$\frac{5}{9} = 0.556$	9
$As_2Ge_2Te_5$		
Bi_4Se_3	$\frac{4}{7} = 0.571$	7
Bi_4Te_3		
As_2GeTe_4		
Bi_6Se_5	$\frac{19}{33} = 0.576$	33
Bi_8Se_7	$\frac{26}{45} = 0.578$	45
$SbTe$	$\frac{7}{12} = 0.583$	12
$BiSe$		
$BiTe$		
As_4GeTe_7		
Bi_8Se_9	$\frac{10}{17} = 0.588$	17
$\beta\text{-}As_2Te_3$	$\frac{3}{5} = 0.600$	5
Sb_2Te_3		
Bi_2Se_3		
Bi_2Te_3		

Source: Ref. 134.

right-hand side has the configuration $s^0p^0 + s^2p^6$. Electrons have, therefore, been formally transferred from s to p orbitals. At intermediate electronegativity the electron configuration is of course $s^2p^0 + s^2p^4$.

Clearly from Figure 7.22, the electronegativity difference between the anion and cation is an important consideration in switching off the Peierls distortion of the simple cubic structure. The rocksalt structure is thus favored for the highly ionic situation. The structures of these AB octets, therefore, should be very dependent on the energetic location of the valence s and p orbitals of the atomic constituents. This is true to quite a remarkable degree (Ref. 71). With reference to Figure 7.23, without exception all the rocksalt and CsCl examples are of type 1. All of the wurtzite

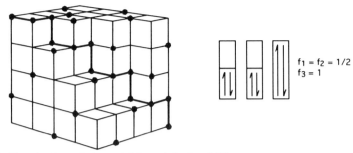

Figure 7.21 The structure of elemental Se and the band filling at the simple cubic structure needed to reach it.

examples are of type 1 with the exception of MgTe, which is also a borderline case in several of the plots to be described. Most sphalerite examples are of types 2, 3 (BeTe and GaSb), or 4 (the Group 14 elements). CuBr, AgI, ZnS, and HgSe (sphalerite structure) are type $1\frac{1}{2}$. Only CuCl and SiC do not fall into this simple classification scheme. They have the sphalerite structure but are both of type 1. The presence of a band gap also depends upon the widths of both conduction and valence bands. Even if the electronic situation is described by case 1 of Figure 7.23, only if the bandwidths are small will a gap exist. Section 5.1 described a simple model that suggested that, for the tetrahedral solids, the bandwidth depends upon the difference

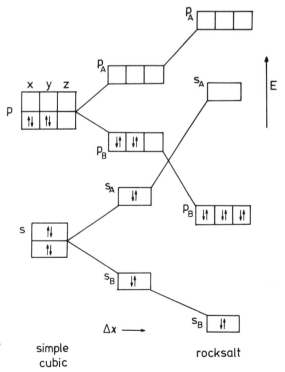

Figure 7.22 Schematic of the variation in the band structure of the simple cubic structure as the electronegativity of the atoms is varied and the rocksalt arrangement generated.

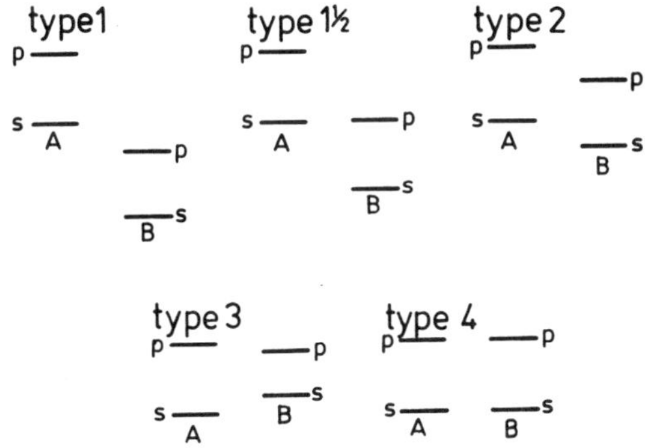

Figure 7.23   Types of orbital energy relationships between the atoms A and B of the rocksalt structure.

of atomic s and p levels, $e_s - e_p$ on the same atom. This becomes smaller as the element becomes heavier, the largest change from the pseudopotential radii of Table 6.2 being between first and second row elements.

Mooser and Pearson constructed a structure map (Refs. 248, 280) that enabled a good separation of not only examples of structures with different coordination number, but of the two four-coordinate structures, zincblende (cubic) and wurtzite (hexagonal). Two octet diagrams are shown in Figure 7.24. The indices are simple ones, just the difference in Pauling electronegativity between A and B, and the average value of the principal quantum numbers of the valence shell. A perfect separation for the octets was achieved by Phillips and Van Vechten (Refs. 300, 301, 303) (Fig. 7.25) using the division of the band gap into "covalent" (E_c) and "ionic" (E_i) parts suggested by the result derived from Figure 1.3, namely, $\Delta^2 = (\Delta\alpha)^2 + 4\beta^2$. For the solid, Δ should perhaps represent some average excitation from the bonding to antibonding band rather than the band gap itself. This is extracted from spectral measurements. The covalent part is all that is present for the homonuclear systems $(\Delta = 2\beta)$. Using the measured values of Δ for the elements and assuming that the covalent contribution and internuclear separation d are related by $E_c \propto d^{-2.5}$, values of E_c and E_i may be calculated for all AB octets. The plot of these pairs is shown in Figure 7.25, and is remarkable in that it cleanly sorts four- and six-coordinate structures. The two plots are similar; the electronegativity difference and E_i are obviously closely related. Yet a third important way to study this problem uses a structure map that employs pseudopotential radii as indices (Refs. 25, 398).

The pseudopotential radius, as described in Section 6.2, measures the distance from the nucleus where the radial maximum in the probability distribution is found. Earlier, the advantages of using such an atomic measure of "size" was emphasized. The ratio $1/r_i$ has units of energy and thus should scale with orbital ionization potential (Refs. 25, 334). Indeed there are good correlations between calculated values of the inverse radii and the multiplet-averaged ionization energies of atoms. The average of the ionization potential and electron affinity, of course, is just Mulliken's definition of electronegativity. These relationships between size, electronegativity, and orbital ionization energies show that models based on one of the three are immediately transferable to another that uses a different language. The indices that are used for the third map (Fig. 7.26) are two simple combinations of the pairs of pseudopotential radii for the atoms that make up the AB solid. If $r_\sigma = r_s + r_p$ and

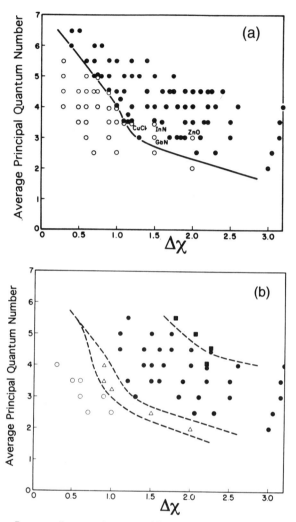

Figure 7.24 Mooser–Pearson diagrams for some *AB* octets. (a) A diagram for non-transition-metal containing examples. Here the sorting is between octahedral coordination (NaCl structure, filled circles) and tetrahedral coordination (sphalerite or wurtzite structures, open circles). This diagram has been modified compared to the original by the removal of the point for CuF, which does not exist. (b) A similar map but with metals from groups 11 and 12 removed. Here the sorting is between the CsCl structure (solid squares), the rocksalt structure (filled circles), the wurtzite structure (open triangles), and the sphalerite structure (open circles). (Adapted from Ref. 280.)

$r_\pi = |r_s - r_p|$, then the structural indices used are just

$$R_\sigma = |r_\sigma^A - r_\sigma^B|, \qquad R_\pi = |r_\pi^A + r_\pi^B| \qquad (7.3)$$

R_σ is a measure, not only of the difference in "size" of the two atoms, but also of their electronegativity difference.

From the similarities in the three plots, it appears that E_c, $\tilde{n}$, and R_π measure a similar property of the *AB* pair. The model of Figure 7.22 indicates that this is the

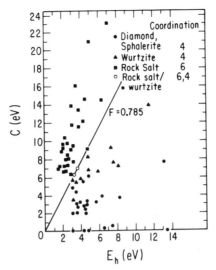

Figure 7.25 Structural sorting of octet *AB* systems using the Phillips–van Vechten approach. The critical slope that separates the two is shown (*F*). (Adapted from Ref. 303.)

sum of the widths of valence and conduction bands. E_i, χ, and R_σ measure the electronegativity difference. Their values put given AB solids into one of the classes of Figure 7.23, and the plots of Figures 7.24–7.26 all show ways to differentiate between these electronic situations. The rocksalt structure is the one found when the Peierls instability has been suppressed. Electronic arguments such as these are really only appropriate when there are two structures with different coordination numbers related electronically as described here. In many, perhaps most, cases where coordination number questions arise, the structures are not so related. However, the same indices, involving the "sizes" of atoms, are appropriate. As described in Section 6.2, arguments concerning the relative stability of structures involving sizes of atoms are related to those involving electronegativities and hence "electronic" arguments. The use of structure maps to sort structure and properties of solids is widespread (Refs. 308, 356–358). Their interpretation, however, is usually not simple. Finally, it should be stated that there is at present no theoretical model that allows absolute prediction of structure type (other than by brute force calculation), especially when structures of different coordination number are involved.

7.6   The structures of the spinels

The question of the structures that are adopted by the spinels, systems having the formula AB_2O_4 (Fig. 7.27), is an old one (Ref. 122). The spinel structure is based on a cubic eutactic array of oxide ions in which the metal atoms occupy some of the octahedral and tetrahedral holes. In the "normal" spinel structure the A ions occupy one-eighth of the tetrahedral holes and the B atoms one-half of the octahedral holes. In the "inverse" structure all of the A ions and one-half of the B ions have changed places. Most spinels are of the normal type, but there are several with the inverse

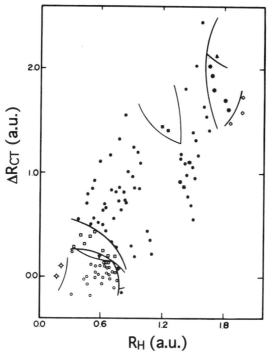

Figure 7.26 Structural sorting map for AB octets using combinations of pseudopotential orbital radii [Eq. (7.3)] as indices. (Adapted from Ref. 25.) $\bar{R}_H$ is related to R_π and ΔR_{CT} to R_σ.

structure, and some others where the best description is of a disordered variant somewhere between the two. Table 7.2 gives some examples of each type. The challenge is to find a model that leads to an explanation of this result. The traditional approach to the problem focuses on the d orbitals of the ions involved and uses the crystal-field model (Ref. 122). Calculation of the crystal-field stabilization energy for each possibility enables an estimate to be made of the value of the difference in CFSE, Δ(CFSE), which favors the inverse structure $[A(\text{oct})B(\text{tet})B(\text{oct})]$ over the normal one $[A(\text{tet})B_2(\text{oct})]$. This approach is limited to systems containing asymmetric d orbital occupancy (d^0, d^5, and d^{10} ions are excluded) since only these ions will have a nonzero Δ(CFSE). Figure 7.28 shows the crystal-field splitting of the five d orbitals in an octahedral field and Figure 7.29(a) the corresponding variation in CFSE with d electron configuration. Figure 7.29(a) shows the analogous picture also using an orbital model with the parameters of the angular overlap approach (Refs. 41, 47). Irrespective of the approach used, the characteristic double-humped curve of CFSE or MOSE (molecular orbital stabilization energy) as a function of d count results. By comparison of Figure 7.29(b) with numerical data it may be shown that the energetics of the problem are dominated in the orbital model by interaction with the valence $(n + 1)s$ and $(n + 1)p$ orbitals of the metal, the nd orbitals playing a much smaller energetic role. The nd orbitals are energetically more important in the orbital than electrostatic model.

This picture is reinforced when a structure map using the s and p pseudopotential orbital radii of the A and B metals is constructed (Refs. 74, 306) (Fig. 7.30). Notice

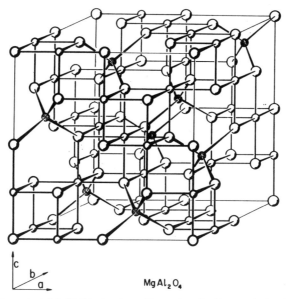

Figure 7.27 The spinel ($MgAl_2O_4$) structure. (Reproduced with permission from Ref. 280.)

that the normal and inverse structures are, by and large, well separated by the use of these indices. There are four "normal" errors in the "inverse" region, of which two correspond to poorly characterized systems and two are real errors. (Such a clean separation between the structure types is not achieved if Shannon and Prewitt crystal radii are used.) At the boundary between the two regions is found a mixture of normal and inverse structures. In every case the structure that is found here is the one predicted to be more stable from consideration of the value of Δ (CFSE). So the

Table 7.2 Some examples of materials with the spinel structure[a]

A						B_2			
	Al	Ga	In	Ti	V	Cr	Mn	Fe	Co
Mg	N	I	I	N	N	N	N	I	
	0.1	0.7–0.8	1.0	0.0	0.0	0.0	0–0.4	0.7	
Cu		N	I			N	N	I	
		0.4	1.0			0	0.2–0.4	0.7–1	
Zn		N	N		N	N	N	N	N
		1.0	1.0		1.0	1.0	1.0	1.0	1.0
Co	N	I			N	N	N	I	
	0.2	0.9			0	0	0–0.2	1.0	
Ni	I	I				N	I	I	I
	0.75	1.0				0	0.74	1.0	1.0

[a] x, the parameter that defines the ordering of B cations over the octahedral sites, is shown. This is equal to 1 and 0 for the inverse and normal structures, respectively.

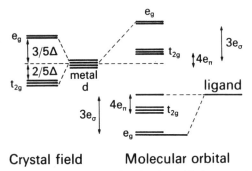

Figure 7.28 A comparison of the crystal field and molecular orbital approaches to the "stabilization energy" of a given d orbital configuration.

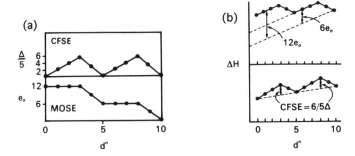

Figure 7.29 (a) The crystal field stabilization energy (CFSE) and the molecular orbital stabilization energy (MOSE) as a function of d count. (b) How both may be correlated with the sawtooth behavior of the experimental values of the heat of formation of some MX and MX_2 systems.

model that emerges is the following. The dominant effect controlling the coordination number problem here is, as in main group octet compounds, associated with the electronic demands of the valence s and p orbitals. At the boundary between the two regions where the energetic driving force for one arrangement over the other is not large, then the smaller d orbital forces become important. Statistically the following figures are of some interest. There are about 187 known spinels. They are mainly oxides and chalcogenides, but some halides and cyanides adopt this structure. Of these there are only 80 that contain asymmetric d orbital configurations for which the d-orbital model can make a prediction. The success rate from this model in 67/80, there being 13 errors. The structure map gives a total of 4 errors out of 187 examples.

7.7 Distortions of the cadmium halide structure: Jahn–Teller considerations

Figure 7.31 shows a structure map for some 16 electron AB_2 solids (the double octets). Not all examples with this composition have been included. Left off the diagram are examples of structures that contain B–B dimers such as those of the 14 electron systems found in marcasite and löllingite. The diagram is a useful one. It tells us that most compounds with oxygen or fluorine have the rutile structure, although the

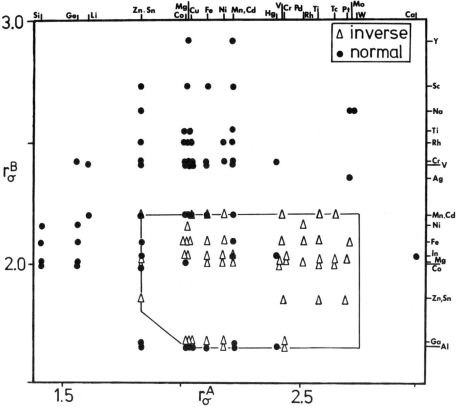

Figure 7.30 A structure map for the spinels using the pseudopotential radii combinations r_σ^A and r_σ^B as indices.

fluorite structure is an alternative for large cations. There is a large region where either the cadmium chloride or iodide structure is found. These structures differ only in the type of stacking associated with the anions, cubic for the chloride and hexagonal for the iodide. The cadmium iodide structure type is often referred to as the brucite [Mg(OH)$_2$] structure. These are just the two simplest polymorphs of structures arising from completely filling the octahedral holes between alternate pairs of eutactic anion layers. There are three islands of structures found in the middle of the cadmium halide region, the structures of the copper (II) halides (d^9), those of the mercuric halides (d^{10}), and those of the chalcogenides of some of the early transition metals ($d^{1,2}$). Bearing the result of the spinel problem in mind concerning the relative importance of valence nd and the $(n + 1)(s + p)$ orbitals, the d-orbital manifold is probably the place to look for an electronic explanation of the structural features of these islands. The discussion begins with the examples from the right-hand end of the transition metal series.

The structures of both the d^9 and d^{10} examples are significantly distorted away from the "ideal" cadmium halide structure (Fig. 7.32). Mercuric chloride and bromide show a shortening of two *trans* linkages, and in HgCl$_2$ this is so pronounced that the structure consists of linear HgCl$_2$ units packed together and the structure lies far

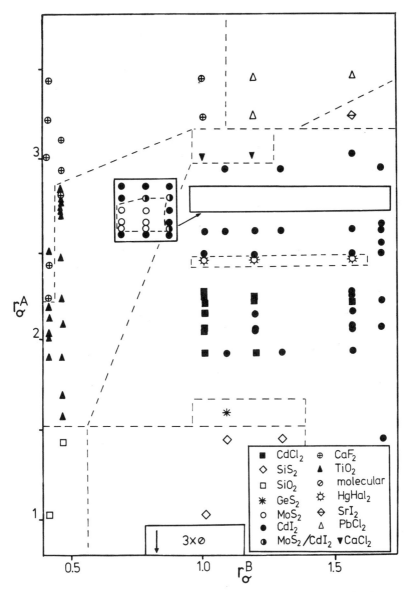

Figure 7.31 A structure map for the AB_2 double octets using the pseudopotential radii combinations r_σ^A and r_σ^B as indices.

from that of cadmium halide. In $HgBr_2$ the distortion is smaller and the resemblance to the parent is easy to see with two sets of distances [2.48 Å(2) and 3.23 Å(4)]. In the copper halides the opposite distortion is found, and two *trans* linkages are lengthened. The structure is now well described as being made up of one-dimensional chains of edge-sharing square planes (Fig. 7.32). There are many examples of distorted Cu(II) structures. The bond lengths in $CuBr_2$, for example, are 3.18 Å(2) and 2.40 Å(4). There are analogous structural observations found in AB systems. The structure of cinnabar, HgS (Fig. 7.33), contains spirals containing linear S–Hg–S

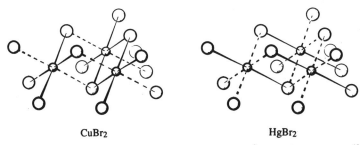

CuBr₂ HgBr₂

Figure 7.32 Distortions of the local octahedral geometry in d^9 (e.g., CuBr$_2$) and d^{10} (e.g., HgBr$_2$) systems.

units $[r(\text{Hg–S}) = 2.39 \text{ Å}(2)]$ rather like the structure of selenium (Fig. 5.5) but with mercury atom spacers between each pair of nonmetal atoms. However, there are four longer Hg–S distances $[3.10 \text{ Å}(2), 3.30 \text{ Å}(2)]$ that complete a distorted octahedron around the metal. AuCl and AuBr show zigzag chains containing linear X–Au–X units of a similar type. Tetrahedrally coordinated d^{10} systems show a similar effect. The pathways connecting different coordination geometries are low-energy ones, Especially interesting is the observation that the distortion coordinate leading from the tetrahedron to trigonal planar coordination (**7.6**) places the metal ion in the face of the anion tetrahedron and so to a route from one interstice to

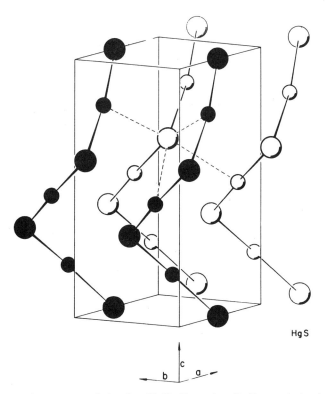

HgS

c
b a

Figure 7.33 The structure of cinnabar (HgS). (Reproduced with permission from Ref. 280.)

7.6

another of a eutactic anion array. Such low-energy pathways lead to ionic conductors (Ref. 184) found for many Cu(I) and Ag(I) systems.

Jahn–Teller arguments have long been used to describe the distorted structures often found in Cu(II) chemistry (Refs. 41, 120, 121, 194, 270–272). Hybridization between nd and $(n + 1)s$ orbitals is the theoretical tool used to understand the mercury d^{10} case. The actual electronic picture is a much richer one. Equation (7.4) gives the expansion of the electronic energy, $E_0(q)$, for the electronic ground state $|0\rangle$ as a function of a distortion coordinate q in terms of perturbation theory (Refs. 22, 41, 277). Within the spirit of the Jahn–Teller approach the stability of some (usually high-symmetry) structure is tested, initially in first order and then in second order if necessary.

$$E_0(q) = E_0^0(0) + E_0^1(q) + E_0^2(q) + \cdots$$
$$= E_0^0(0) + \langle 0|\partial\mathcal{H}/\partial q|0\rangle q$$
$$+ (\tfrac{1}{2})(\langle 0|\partial^2\mathcal{H}/\partial q^2|0\rangle - 2\sum_j |\langle 0|\partial\mathcal{H}/\partial q|j\rangle|^2/[E_j(0) - E_0(0)])q^2 + \cdots \quad (7.4)$$

Symmetry considerations allow conclusions to be drawn quite rapidly concerning the properties of the various terms. Γ_q, the symmetry species of q, must be contained within the symmetric direct product of Γ_0 for a nonzero value of the first-order term, $\langle 0|\partial\mathcal{H}/\partial q|0\rangle$. This symmetry result leads to the first-order Jahn–Teller theorem, namely, that orbitally degenerate electronic states of high-symmetry molecules will distort so as to remove the degeneracy. For the present case of an octehedral d^9 arrangement the electronic ground state of this localized e_g^2 system is ^2E_g. This leads to the symmetry prediction of a distortion coordinate of e_g symmetry via this first-order effect. The d^{10} system is electronically a closed-shell system, and its $^1A_{1g}$ symmetry means that the first-order term is zero here and so octahedral Hg(II) is first-order Jahn–Teller inactive.

There are an infinite number of choices available for the degenerate, e_g, Jahn–Teller distortion coordinate. One pair is shown in **7.7**, but any linear combination of the two is also a valid choice. Notice that one of them is of the correct form to take the

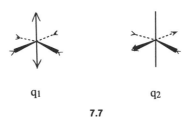

q_1 $\qquad\qquad\qquad$ q_2

7.7

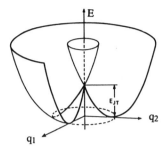

Figure 7.34 The "Mexican hat" potential arising from consideration of the first-order Jahn–Teller term for an E state.

octahedral structure with six equal distances to the geometry with four short and two long distances and found for virtually all d^9 systems with six identical ligands. However, the *symmetry* considerations of the first-order Jahn–Teller stabilization give no reason to expect the observation of a distortion of this type over one of the many other possibilities. This may be viewed as in Figure 7.34, where the "Mexican hat" potential is drawn as a function of the two coordinates q_1 and q_2 of **7.7**. The surface is equienergetic for a given $|q|$ for all ϕ if the distortion coordinate is generalized as $q = q_1 \cos \phi + q_2 \sin \phi$. Figure 7.35 shows the d orbitals of the octahedron and how they change in energy during the distortion for the two cases $\pm q_1$. There is no energetic distinction between the two, the only difference being the identity of the upper and lower orbitals. This problem may be "solved" by addition of a judiciously chosen "crystal potential" to the Mexican hat, which leads to minima appearing along $q = -q_1$ and the two other equivalent distortions set by the octahedral symmetry. However, the best solution is to see how the terms in q^2 of Eq. (7.4) influence the picture. This leads to consideration of the second-order Jahn–Teller effect (Refs. 42, 137).

From Eq. (7.4) the second-order term, $E_0^2(q)$, consists of two parts. $\langle 0|\partial^2 \mathcal{H}/\partial q^2|0 \rangle$ is the classical force constant, which describes motion of the nuclei in the electronic

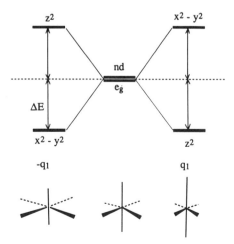

Figure 7.35 Behavior of the "d levels" on distortion of the octahedron, using a d-orbital-only model.

state $|0\rangle$, namely, the frozen electronic charge distribution of the undistorted, $q = 0$, structure. The second term, $-\sum'_j |\langle 0|\partial\mathcal{H}/\partial q|j\rangle|^2/[E_j(0) - E_0(0)]$, is always negative (i.e., stabilizing). The sign of $E_0^2(q)$ is set by the competition between these two terms. The energy gap expression $E_j(0) - E_0(0)$ appears in the denominator, implying that it will be most important for close-lying electronic states. If one such state, $|j\rangle$, is of the correct symmetry ($\Gamma_q = \Gamma_0 \otimes \Gamma_j$), then the lead term in this expansion can become important and $E_0^2(q)$ may become negative. The system will now distort away from the symmetrical structure. Another way of viewing the distortion is that it is driven by the mixing between HOMO and LUMO induced by the distortion q. The HOMO is stabilized as the distortion proceeds. Because the symmetry species of the states $|0\rangle$ and $|j\rangle$ are determined by the symmetry of the orbitals that are occupied, the problem often reduces to an orbital rather than a state picture. Thus if the HOMO and LUMO are nondegenerate, then the symmetry species of q is given by $\Gamma_q = \Gamma_{HOMO} \otimes \Gamma_{LUMO}$.

At the right-hand end of the transition metals the energy separation between nd and $(n + 1)s$ orbitals is small and the interaction between $(n + 1)s$ and nz^2 orbitals is allowed by symmetry during the distortion from octahedral to D_{4h}. Importantly, the $x^2 - y^2$ orbital is of the wrong symmetry to mix with $(n + 1)s$. The $(n + 1)s/nz^2$ orbital interactions of this type depicted in Figure 7.36 show immediately why the two long/four short distortion (q_1 of **7.7**) is found for the d^9 configuration but the two short/four long distortion ($-q_1$) for d^{10}. If the gap $E_j(0) - E_0(0)$ of Eq. (7.4) is set equal to the $(n + 1)s/z^2$ separation, then it is clear that the stabilization of z^2 will be larger in the two short/four long case than for the converse ($|\Delta E_1| < |\Delta E_2|$). For

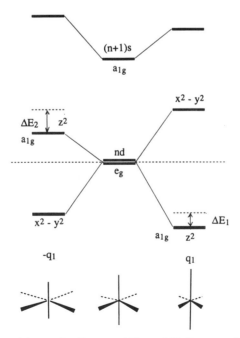

Figure 7.36 Effect of d–s mixing on the diagram of Figure 7.35. Because of the difference in the two z^2–s energy separations, $|\Delta E_2| > |E_1|$.

Table 7.3 The energies of d^9 and d^{10} systems from second-order Jahn–Teller considerations

	q_1	$-q_1$
d^9	$\Delta E + 2\,\Delta E_1$	$\Delta E + \Delta E_2$
d^{10}	$2\,\Delta E_1$	$2\,\Delta E_2$

the d^{10} configuration the best electronic stabilization occurs for the two short/four long case since here nz^2 lies closer to $(n+1)s$. For the d^9 configuration the lower-energy structure is determined on the model by the double occupation of the lower-energy orbital, nz^2 in the two long/four short case (Refs. 42, 137). Table 7.3 lists the stabilization energies associated with the two distortions as a function of d count. Similar arguments apply for the case (Ref. 3) of tetrahedrally coordinated atoms and are shown in Figure 7.37. The distortion that leads to three short and one lond bond is the one favored for the d^{10} configuration. This is just the motion of **7.6**, which leads to movement of the metal atom into a face of the tetrahedron.

The arguments describing the Jahn–Teller distortions of the cupric halides used a model that employed a localized description of the electronic state. Figure 7.36 showed a picture of individual d levels rather than of energy bands. An important feature of such distortions of solids is that they frequently demand a localized electronic description. This is seen experimentally in, for example (Ref. 309), the $La_{1-x}Sr_xMnO_3$ system. $LaMn(III)O_3$ itself is a localized, insulating material with a static Jahn–Teller distortion [Fig. 7.38(b)]. On doping with strontium, however, the

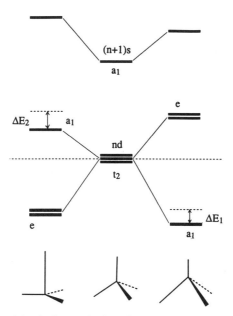

Figure 7.37 Effect of d–s mixing in the tetrahedron. Because of the difference in the two z^2–s energy separations, $|\Delta E_2| > |\Delta E_1|$.

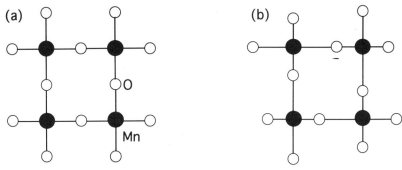

Figure 7.38 (a) The undistorted structure of $La_{1-x}Sr_xMnO_3$. (b) The in-plane structure of $LaMnO_3$.

appearance of metallic behavior (for $x > 0.1$) is associated with the loss of the the distortion and generation of an octahedral Mn coordination. Figure 5.38 shows the derivation of the $x^2 - y^2$ band of the CuO_2 sheet. In the doubled cell for the distorted structure of Figure 7.38 then this band must be folded back as in Figure 7.39(a), leading to a degeneracy at the zone edge. The z^2 band will also be important for this structure. In La_2CuO_4, with very long Cu–O distances perpendicular to the sheet the z^2 band lies below $x^2 - y^2$ and is filled. In the sheet of Figure 7.38, however, the Mn–O distances perpendicular to the sheet are short and the z^2 band needs to be included in the picture. Its two-dimensional behavior is shown in Figure 7.39(b). Although the model uses two separate bands, the z^2 and $x^2 - y^2$ orbitals will in fact mix at general points in the zone; so the picture is not quite this simple. (Dispersion perpendicular to the sheet for z^2 is also ignored. In terms of the parameters of the AOM, the bands are expected to fall energetically as in Figure 7.40, where e'_σ is used for the linkages perpendicular to the sheet and e_σ for linkages within the sheet. A p orbital only model is used for the oxygen levels. This means that the bottom of the $x^2 - y^2$ band is metal–oxygen non-bonding with an energy of e_d. The wavefunctions are also shown. On distortion Figure 7.38(b) → (a) the band structure of Figure 7.39 is found by calculation to

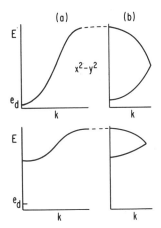

Figure 7.39 The bands of the CuO_2 sheet and how they are folded in the doubled cell. (a) The $x^2 - y^2$ band. (b) The z^2 band.

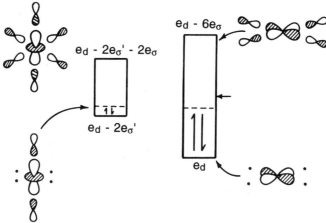

Figure 7.40   The energetic location of the top and bottom of the $x^2 - y^2$ and z^2 bands using the parameters of the AOM.

change very little. There is no dramatic splitting of the levels as seen in Figure 7.35, for example. There is no splitting of the obvious degeneracy at the zone edge, and indeed, even if there were, the orbital occupancy for both Cu(II) (two electrons half-filling this pair of bands) or Mn(III) (one electron) does not lead in general to a k_F that lies close to this point (shown in Fig. 7.40 by an arrow). The lack of splitting of the bands on distortion is easy to see in principle. Because the screw axis relating the copper atoms at (0,0) and $(\frac{1}{2},\frac{1}{2})$ is maintained during the distortion, as described in Section 3.3, there is a symmetry requirement for degenerate levels at the zone edge. Because of this combination of factors, there is no stabilization on the simplest model for any band filling during this distortion.

The distortion (two long/four short) observed in $La_{2-x}Sr_xCuO_4$ is present in both the undoped state ($x = 0$), where a localized description is appropriate, and for the metallic delocalized description, when $x \gtrsim 0.1$. Here the electronic situation is quite different. The electronic origin of the two long/four short distortion of the sheet evident in Figure 5.33 leads to a simple splitting of the z^2 and $x^2 - y^2$ orbitals (as shown at the left-hand side of Figure 5.38). The generation of an energy band from the $x^2 - y^2$ level follows quite naturally. Here there is no symmetry-enforced behavior comparable to that of Figure 7.39.

7.8   Distortions of the cadmium halide structure: trigonal prismatic coordination

The MoS_2 structure (Fig. 7.41) is found for some early transition metal chalcogenides. Its basic features may be derived from those of the cadmium halide structures by sliding every other chalcogen layer so that the metal atoms are located in trigonal prismatic rather than octahedral holes. Just as there are cadmium chloride/iodide stacking variants in the octahedral case (Figs. 7.19 and 7.20), so there are similar variants in the trigonal prismatic case. But these will not be of interest here. A clue to the electronic structure of MoS_2 (d^2) is that it is an insulator, suggesting that in

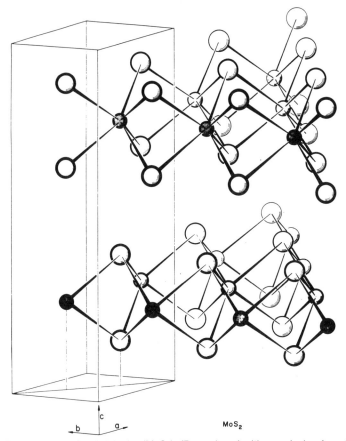

b   c   a   MoS$_2$

Figure 7.41 The structure of molybdenite (MoS$_2$). (Reproduced with permission from Ref. 280.)

this structure there is a single d-located energy band split off from the rest in this geometry. NbS$_2$ and TaS$_2$ (d^1) are metals, and their d count corresponds to half-filling of this band. The density of states of Figure 7.42 shows that this is indeed true. The bands are labeled according to the d-orbital splitting pattern expected for the local trigonal prismatic geometry shown in Figure 7.43. However, this is not quite the whole picture (Ref. 394 or 395). There is strong mixing between z^2 and the $\{x^2 - y^2, xy\}$ pair that ensures a gap between the two sets of energy bands. This means that the character of the lowest two bands is a mixture of all three orbitals. The mixing process has led to an orbital that is heavily involved in metal–metal bonding, as shown by the COOP curve (Ref. 394) of Figure 7.44. Indeed, using the method of orbital localization described in Section 4.3, the result is shown in Figure 7.45. Notice that this process gives rise to an orbital distributed over three adjacent sites (just as for the a_1' orbital of H$_3^+$) and contains two electrons. Metal–metal bonding of this type is a characteristic of the solid-state array and allows stabilization of the trigonal prismatic structure in a way that is not possible for molecular examples. There, the steric penalty associated with the eclipsed sulfur atoms of the trigonal prism is often sufficient to reader the trigonal prismatic structure energetically unfavorable. This steric effect in the solid is probably behind the plot

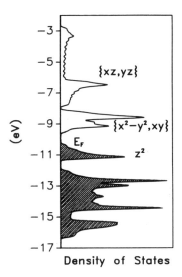

Figure 7.42 The densities of states in the *d* orbital region for 2*H*-MoS$_2$ containing trigonal prismatic metal coordination. Occupied levels are shaded. (Reproduced from Ref. 394 by permission of The American Chemical Society.)

(Ref. 150) of Figure 7.46. It shows how the radii of the metal and chalcogenide are also important. Thus the structural preference here is determined by a combination of electronic and nonbonded effects.

The MoS$_2$ structure is thus found for d^1 and d^2 electron counts, but not for higher ones. Figure 7.47 shows a calculated (Ref. 58) energy difference curve between the

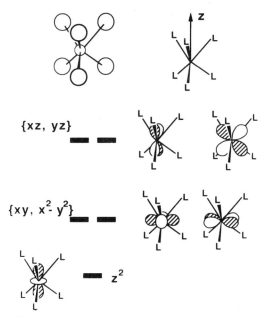

Figure 7.43 The ligand-field splitting for Mo in a trigonal prismatic environment. (Reproduced from Ref. 394 by permission of The American Chemical Society.)

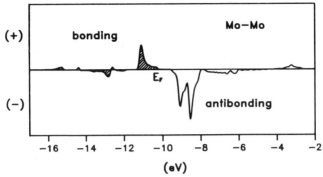

Figure 7.44 Crystal orbital overlap population curves for the Mo–Mo bonds in an MoS_2 layer. Notice that the Mo–Mo bonding levels are optimally occupied for MoS_2 itself. (Reproduced from Ref. 394 by permission of The American Chemical Society.)

Figure 7.45 Contour plots for the localized orbital in the layer plane for MoS_2. (Reproduced from Ref. 394 by permission of The American Chemical Society.)

two structures for an NbO_2 sheet. Interestingly, if MoS_2 (d^2) is intercalated with lithium to give Li_xMoS_2, at a critical lithium concentration the solid undergoes (Ref. 307) a first-order phase transition from the MoS_2 type to the $CdHal_2$ type, a result readily understandable from Figure 7.42. Indeed, the structural effect of electron doping, leading as it does to population of higher-energy orbitals, is in many ways similar to that resulting from an increase in temperature and thermal population of these higher-energy levels. Thus $MoTe_2$ on heating undergoes an analogous structural change from the MoS_2 type to the CdI_2 type. Although MoS_2 is an insulator, TaS_2, with a partially filled band, is a metal that undergoes a CDW distortion at lower temperatures.

7.9   Distortions of the cadmium halide structure: t_{2g} block instabilities

The cadmium halide structure is found for many transition metal chalcogenides and for compounds involving the heavier halides and pnictides. The $CdCl_2$ structure

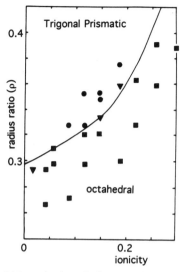

Figure 7.46 Structural sorting of trigonal prismatic from octahedral coordination in early transition metal chalcogenides. The radius ratio is self-explanatory. The ionicity is defined here as $1 - \exp(\Delta\chi^2/4)$, where $\Delta\chi$ is the difference in electronegativity between metal and chalcogen. Examples with the cadmium halide structure are marked with a square; those with the MoS_2 structure with a circle and those found in both structure types with a triangle. (Adapted from Ref. 150.)

(Fig. 7.20) may be regarded as a defect rocksalt (cubic) arrangement, with every other metal layer along the [111] direction of the parent removed and the CdI_2 structure (Fig. 7.19) as a defect NiAs (hexagonal) arrangement, with every other metal layer along the [0001] direction of the parent removed. Some ternary oxides also have this structure. For example, $LiVO_2$ may be regarded either as a derivative structure of rocksalt, with Li and V occupying the metal sites, or as a stuffed $CdCl_2$ structure

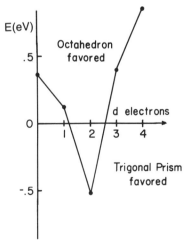

Figure 7.47 Relative energy of trigonal prismatic versus octahedral NbO_2 layers as a function of d count.

with the Li atoms occupying octahedral holes between the VO_2 sheets. Since the Li atom is more electropositive than the VO_2 unit, the latter description will be more useful and the material regarded electronically as VO_2^-. In many of these cases the structure is distorted in various ways that involve clustering of the metal atoms of the structure. Figure 7.48(a) shows (Ref. 374) the metal atom net of the undistorted structure, and Figures 7.48(b)–(g) show how the metal atoms cluster in a variety of systems that are all characterized by low metal d counts.

The important first step in the study of these distortions is to derive the electronic structure of the undistorted MX_2 layer (Ref. 58). Since each metal ion is in octahedral coordination, in broad terms for $X = O$ the density of states will be similar to that described for the rocksalt or TiO_2 structures, where the metal d bands may be split into "e_g" and "t_{2g}" sets. Recall that these are, respectively, associated with σ and π overlap with the orbitals of surrounding atoms. These t_{2g} orbitals are also able to take part in metal–metal interactions within the layer. The electronic description of these partially filled t_{2g} block systems is accessible using a simple approach and will lead to the introduction of the concept of hidden nesting (Ref. 374).

The metal atom is located at the origin of a local coordinate system so that the six ligands lie in the $\pm x$, $\pm y$, $\pm z$ directions. The three d orbitals of the "t_{2g}" set, xy, xz, and yz, then lie as shown in **7.8**, and are simply related by rotation around the threefold axis perpendicular to the MO_2 plane. The simplest model for describing

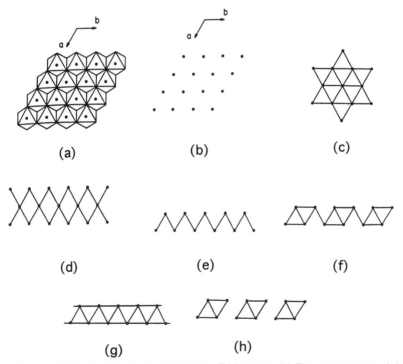

Figure 7.48 (a) Schematic view of an undistorted 1T-MX_2 layer. (b) The arrangement of the metal ions in such a layer. (c) $\sqrt{13} \times \sqrt{13}$ cluster found for d^1 metals. (d) Ribbon-chain cluster found for d^2 metals. (e) Zigzag chain cluster found for $d^{4/3}$ metals. (f) Diamond-chain cluster found for d^3 metals. (g) Zigzag double chains. (h) Rhombic clusters.

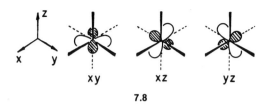

7.8

metal–metal bonding within the layer is one where each t_{2g} d orbital is allowed to interact only with orbitals on neighboring atoms that carry the same label; that is, only xy–xy, xz–xz, and yz–yz interactions are allowed. The band structure of the t_{2g} subblock is then easily described since it is entirely determined by translational symmetry and a few interaction integrals (β_σ, β_π, and β_δ of **4.10**) between the interacting d orbitals. **7.9** shows how the interaction integral between the xy orbitals

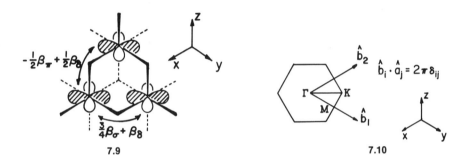

7.9   **7.10**

is derived. The coefficients are those determined simply from the local geometry and the angular behavior of the atomic orbitals. The threefold axis through the metal should make it clear that the xz–xz and yz–yz interactions will be identical.

The form of the wave functions for the "t_{2g}"-derived d bands follows directly from the translational symmetry requirements of Eq. (2.6). **7.10** shows the first Brillouin zone for the lattice, and it is easy to derive the energy

$$E_{xy}(\mathbf{k}) = \langle \psi_{xy} | \mathscr{H}^{\mathrm{eff}} | \psi_{xy} \rangle = \tfrac{3}{2}\beta_\sigma \cos k_1 - \beta_\pi[\cos k_2 + \cos(k_1 + k_2)]$$
$$+ \beta_\delta[\tfrac{1}{2}\cos k_1 + \cos k_2 + \cos(k_1 + k_2)] \qquad (7.5)$$

Overlap has been neglected in the normalization and $\mathbf{k} = k_1\mathbf{b}_1 + k_2\mathbf{b}_2$. Expressions for $E_{xz}(\mathbf{k})$ and $E_{yz}(\mathbf{k})$ can be obtained by permuting the variables k_1, k_2, and $k_1 + k_2$ in Eq. (7.5): For $E_{xz}(\mathbf{k})$ change k_1 to $(k_1 + k_2)$, k_2 to k_1, and $(k_1 + k_2)$ to k_2; for $E_{yz}(\mathbf{k})$ change k_1 to k_2, k_2 to $(k_1 + k_2)$, and $(k_1 + k_2)$ to k_1. To construct the band structure from these expressions it is assumed that the interaction integrals β_σ, β_π, and β_δ scale with the corresponding overlap integrals S_σ, S_π, and S_δ, as in the usual Wolfsberg–Helmholz approximation. From values calculated for Mo atoms 2.87 Å apart, $\beta_\delta = 0.153\beta$, $\beta_\pi = 0.783\beta$, $\beta_\sigma = \beta$. Figure 7.49(a) shows the calculated dispersion along high-symmetry lines in the Brillouin zone. (The heavy lines indicate doubly degenerate levels.) Figure 7.49(b) shows (Ref. 58) the result of an extended Hückel calculation on a full MoO_2 layer. The oxide p bands lie below the bottom of the

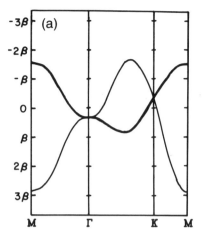 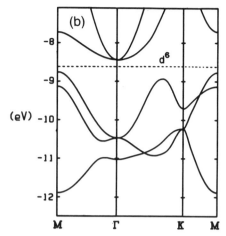

Figure 7.49 (a) Energy bands from a simple Hückel model for metal–metal interactions in the hexagonal layer. (b) Calculated Mo d bands for this layer using the extended Hückel tight-binding method. Notice the correspondence with (a).

energy range shown, and the metal "e_g" bands lie above the d^6 line shown at around -8.6 eV. The similarity between the approximate and "exact" band structure of these "t_{2g}" bands is quite striking. Quantitative agreement is best [in fact, that shown in Fig. 7.49(a)] for a value for $\beta = 0.73$ eV. The largest differences between the two figures occur where degeneracies are found in the idealized picture. This arises from the fact that the simple treatment outlined ignores any coupling between $\psi_{xy}(\mathbf{k})$, $\psi_{xz}(\mathbf{k})$, and $\psi_{yz}(\mathbf{k})$. Energy gap arguments suggest that the splittings induced by introducing any coupling will be greatest when the interacting levels are degenerate. One source of this coupling is direct metal–metal interaction [i.e., the basis functions $\psi_{xy}(\mathbf{k})$, $\psi_{xz}(\mathbf{k})$, and $\psi_{yz}(\mathbf{k})$ are not orthogonal). As an example **7.11** shows how two

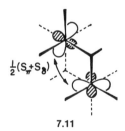

7.11

of the six xz orbitals adjacent to an xy orbital have a nonzero overlap. Another route is via through-bond coupling using the d–p π interaction with the surrounding oxide ligands. The basic ideas of the model (Ref. 58) will, however, suffice for the needs here. The correspondence between the computed results for the t_{2g} bands of the MoO$_2$ slab and the model where the xy, xz, and yz bands are generated independently is a useful approximation in the analysis of the distorted systems to be treated in the following. For the oxide layer, the approximate independence of the xy, xz, and yz orbital manifolds suggests a simple scheme for generating distorted structures similar to that used in Section 3.1 for the simple cubic structure, and developed further in Section 7.4.

Consider, for example, the distorted layer structure of $Na_{0.85}Mo_2O_4$ shown in Figure 7.48(g), where the metal atoms form zigzag double chains within the layers. This geometrical distortion can be envisaged as one where two of the bands (xz, yz) have been split by a pairing distortion while the third (xy) has been left unchanged. The change in the nature of the electronic structure is shown in **7.12**. For

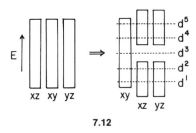

7.12

$Na_{0.85}Mo_2O_4$ with a d count of 2.43, the lower (bonding) xz and yz bands will be fully occupied and the xy band available to take up the extra 0.43 electrons per Mo atom.

Figure 7.50 shows the results (Ref. 58) of band structure calculations for $Na_{0.85}Mo_2O_4$ (the calculations were performed on a single two-dimensional MoO_2 layer) to probe the validity of this simple model. The left-hand panel of Figure 7.50 shows the total density of states (DOS). Only the Mo d bands are shown, the oxide s and p bands lying deeper in energy. As in the undistorted case, there is a separation of the e_g and t_{2g} bands, but the latter have changed considerably on distortion to form zigzag Mo chains. A gap has opened at the d^2 electron count. Projected densities of states are shown in the remaining panels of Figure 7.50. In the second panel are shown the contribution from the d orbitals that are directed for σ overlap along the short bonds of the zigzag chain, the xz and yz orbitals of **7.8**. The results mimic the scheme of **7.12** quite well in that there are large contributions at the top and bottom of the t_{2g} block with a much smaller contribution in the middle. The remaining metal contribution to the density of states in this energy range comes almost entirely from xy, the remaining t_{2g} orbital. The remaining two panels of Figure 7.50 show the xz and yz orbitals projected in a different way, this time as the σ and σ^* combinations illustrated in **7.13**. Notice that for the $d^{2.43}$ configuration, that appropriate for $Na_{0.85}Mo_2O_4$, the σ combination is virtually full while σ^* remains empty. Thus the

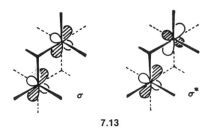

7.13

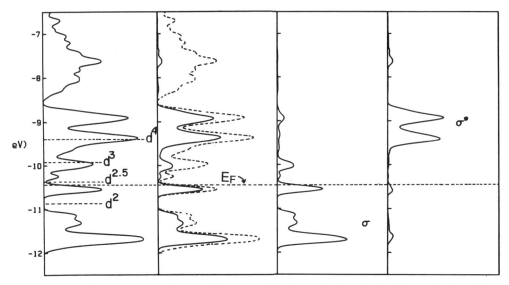

Figure 7.50 Density of states for the MoO_2 layer with zigzag chains. The far left panel shows the total density of states; the center left panel the partial density of states associated with the σ interactions along the short Mo–Mo bonds. The last two panels show the "σ" and "σ^*" contributions.

simple scheme of **7.12** is in reasonable accord with the results of the detailed calculation of Figure 7.50.

The superposition of these pairing distortions in three independent xy, xz, and yz band systems can be used quite generally to generate metal–metal-bonding networks. Half-filled bands will lead to alternating short–long–short–long distortions along the respective direction. Thus a d^1 center is expected to form just one short metal–metal bond, but a d^2 center may form two such linkages, 60° or 120° from each other (**7.14**).

7.14

On the model, clearly the arrangement where the two short bonds are 180° from each other is not allowed. A d^3 center will form three bonds, none of which will be located 180° from another. Possible two-dimensional metal–metal-bonded networks may be built by linking these units together. So the metal–metal bonding network in $Na_{0.85}Mo_2O_4$ may be described as the linking of the units in **7.14**. However, the network in $LiVO_2$ (regarded, as noted, as a VO_2 layer with intercalated Li) has been described as consisting of metal–metal-bonded triangles (**7.15**) at low temperatures. This pattern is just another way of linking the units possible for the d^2 configuration in **7.14** but cannot be described as a superposition of the pairing distortions in two dimensions.

7.15

For the d^3 materials ReX_2 (X = S, Se) and $M'Mo_2S_4$ (M' = V, Cr, Fe, and Co) alternate lengthening and shortening along all three directions is predicted, and indeed the diamond chains [Fig. 7.48(f)] found in these structures are understandable (Ref. 374) in this way. Figure 7.51(d) shows schematically the electronic picture. There is initially a problem with the zigzag chains [Fig. 7.48(e)] found for the d^2 systems β-$MoTe_2$, WTe_2, α-ZrI_2, and $M'Nb_2Se_4$ (M' = Ti, V, Cr). From Figure 7.51(c), if the electrons equally occupy all three t_{2g} bands, then a trimerization is expected in all three directions. The distortion observed is clearly a dimerization along two

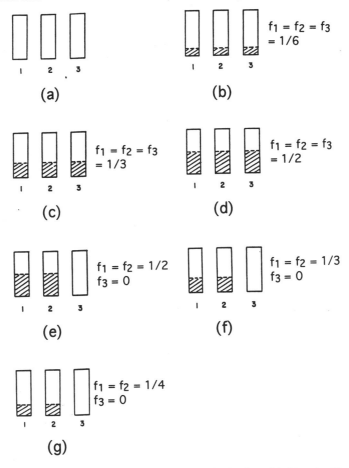

Figure 7.51 (a)–(g) Some of the possible band occupancies for the three "t_{2g}" bands of the hexagonal layer.

directions. This is understandable if two of the bands of the parent are half-filled and the other remains empty, as in Figure 7.51(e). A problem of a different type is found in $1T$-MTe$_2$ (M = V, Nb, Ta). Here [Fig. 7.48(a)] the distortion found is expected for the case [Fig. 7.51(f)] where two of the bands are $\frac{1}{3}$ full (a trimerization) and the other empty, one scenario for a $d^{4/3}$ configuration. However, here it has been argued (Ref. 374) that because of overlap between metal and Te bands (**7.16**) the true metal d count here is actually $\frac{4}{3}$ rather than 1.

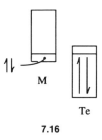

7.16

An interesting point concerning the preceding discussion is that it has moved from a delocalized picture of bonding in the undistorted MO$_2$ octahedral layer to a localized picture for the distorted structures. In the latter a close metal–metal contact can be associated with a two-center–two-electron bond. In principle, this is no different from the result of the Peierls distortion of hydrogen (Section 2.4), which led to H$_2$ molecules.

There are problems, however, with the distortions found for the d^1 configuration. Figure 7.48(c) shows the $\sqrt{13} \times \sqrt{13}$ distortion found for TaX_2 (X = S, Se). A $\sqrt{7} \times \sqrt{7}$ modulation is found for VSe$_2$ and a $\sqrt{19} \times \sqrt{19}$ cluster for NbTe$_2$. In order to understand these arrangements, recourse has to be made to the electronic structure of the cluster unit (Ref. 374), and the electronically satisfying relationship to the parent described is lost.

More insights into this structural problem come from study of the Fermi surface and the introduction of the concept of hidden nesting (Ref. 82). As described, one way of looking at the band structure of the material is as three orthogonal bands containing interactions that run in directions 120° with respect to each other. The complete band structure and Fermi surface, although complex, may be broken down into contributions from each interaction and the hidden nesting uncovered. Figure 7.52 shows the computed (Ref. 374) Fermi surfaces for $M$$X_2$ layers (X = S) with d^{1-3} electron counts. Figure 7.53 shows how these Fermi surfaces may be regarded in terms of the three, $\frac{1}{6}$- (d^1), $\frac{1}{3}$- (d^2), and $\frac{1}{2}$- (d^3) filled bands of Figure 7.51. The rounded features found in the computed surfaces, but not in the simple picture, come from avoided band crossings as in Figure 7.54. There is some resemblance between the two for the d^1 and d^3 counts but not for d^2. The results put into perspective the use of study of the Fermi surface of the parent as a guide to the nature of the distorted structure.

Figure 7.55 illustrates a general problem with approaches of this type. It shows how the energy of the selenium system varies on distortion away from the simple cubic structure. Certainly the observed distortion is a dimerization along two

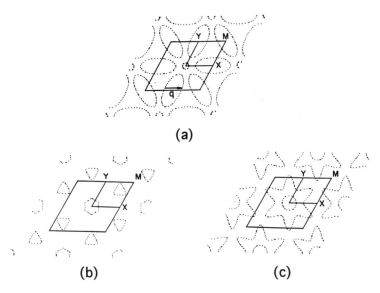

Figure 7.52 Computed Fermi surfaces for (a) d^1, (b) d^2, and (c) d^3 metals in the hexagonal layer. (Adapted from Ref. 374.)

directions with bond breaking completely in the third. The structural change where there is an identical distortion in all three directions is of higher energy. However, there is no obvious way, a priori, of predicting this. Thus the most stable distorted variant is formally derived from a higher-energy undistorted parent. A similar problem arises in energetically distinguishing between the rhombic cluster or zigzag distortions of Figure 7.48. Figure 7.56 shows a calculated energy difference plot (Ref. 58). At around 3–4 d electrons it is difficult to make predictions.

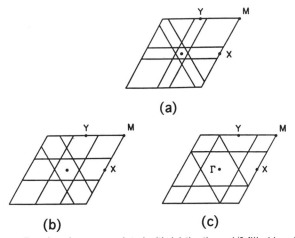

Figure 7.53 Hidden Fermi surfaces associated with (a) the three 1/6-filled bands of Figure 7.51(b) to be compared with the surface of Figure 7.52(a); (b) the three 1/3-filled bands of Figure 7.51(c) to be compared with the surface of Figure 7.52(b); (c) the three 1/2-filled bands of Figure 7.51(d) to be compared with the surface of Figure 7.52(c). (Adapted from Ref. 374.)

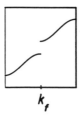

Figure 7.54 Avoided band crossings at some of the crossing points of Figure 7.53 to allow correlation with those of Figure 7.52. (Adapted from Ref. 374.)

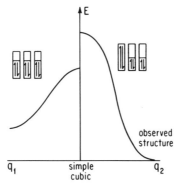

Figure 7.55 Distortions of two types ($q_{1,2}$) that are associated with different band fillings at the undistorted structure. (The specific case of selenium is used as an example here.)

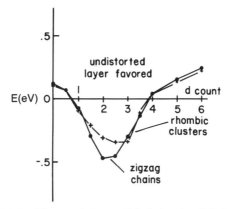

Figure 7.56 Computed energy difference between distorted and undistorted structures for the zigzag chain and embedded rhombic cluster distortions of the hexagonal layer as a function of d count.

7.10 The rutile versus cadmium halide versus pyrite structures

The structure map (Refs. 72, 73) of Figure 7.31 sorts many of the double-octet AB_2 systems into the two structural fields containing rutile (TiO_2) and cadmium halide types. Broadly speaking, the rutile structure is found for compounds containing first row anions such as oxides and fluorides, and the cadmium halide type for compounds

(a)   (b)

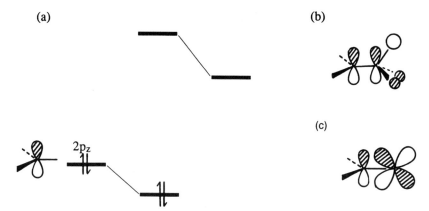

(c)

Figure 7.57   (a) The stabilization of a doubly occupied central atom z orbital by an acceptor orbital on one of the ligands. The nature of the acceptor orbital in (b) $N(SiH_3)_3$ and (c) rutile.

containing heavier halides and chalcogenides. What determines the preference of one structure over the other? Both arrangements contain metal atoms in octahedral six-coordination and anions in three-coordination, but the anions are pyramidal in the cadmium halide structure and planar in the rutile structure. A model (Ref. 44) based on this difference in anion geometry is a useful one, especially when comparison is made with the structures of small molecules. Whereas both NH_3 and PH_3 are pyramidal molecules, their inversion barriers differ considerably, ~ 5 and ~ 35 kcal/mole, respectively. It is, therefore, considerably easier to stabilize a planar three-coordinate nitrogen atom in some way than it is to stabilize a planar phosphorous atom. From these data it is not surprising to find planar NR_3 molecules when the R group is very large and steric interactions between the ligands minimized at the planar geometry. Electronic factors are of importance too (Ref. 5). $N(SiH_3)_3$ has a planar skeleton, but $P(SiH_3)_3$ is a pyramidal species. The SiH_3 unit has a low-lying σ^* orbital, which, as Figure 7.57(a) shows, may act as a π acceptor with respect to the HOMO of the trigonal planar central atom. This orbital in PR_3 gives rise to the acceptor nature of this ligand when coordinated to transition metals. (Earlier electronic descriptions used Si or P $3d$ orbitals, but this is not correct.) A similar stabilizing interaction (calculated to be around 10 kcal/mole) may occur between metal d and oxygen p in rutile [Fig. 7.57(c)]. The generation of π bonding and antibonding levels are clear to see in the density of states of Figure 4.7. An interaction of exactly this type was described in Section 4.4 for NbO.

There is a prediction concerning the d-count dependence of this interaction which controls the stability of the planar versus the pyramidal structure. As the electron count increases, then eventually the π antibonding level will be populated and this driving force removed. In fact, if overlap is included, the antibonding orbital goes up difference curve between the two structures containing planar (rutile) and pyramidal ($CaCl_2$) oxide ions. Notice that the $CaCl_2$ structure becomes stable for six d electrons, in accord with the simple argument. In fact, PtO_2 (d^6) does adopt the $CaCl_2$ structure.

A simple explanation for the origin of the large difference in inversion barrier between first row atoms and their heavier analogs is available from the second-order Jahn–Teller theorem (Ref. 5). Figure 1.18 shows a part of the molecular orbital

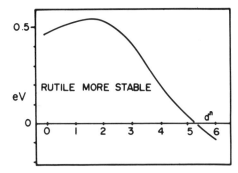

Figure 7.58 Computed energy difference between the rutile and $CaCl_2$ structures as a function of d count. The $CaCl_2$ structure is indeed found for the d^6 configuration appropriate for PtO_2.

diagram for a trigonal planar AB_3 molecule. Note that the HOMO and LUMO have, respectively, a_2'' and a_1' symmetry. Γ_q is thus a_2'' and is of the correct symmetry to take the planar to pyramidal geometry. The a_1' orbital is largely located on the central atom, and its energy will be sensitive to variations in its electronegativity. It should, therefore, lie deeper in planar NH_3 than in planar PH_3, to give a smaller energy gap, $\Delta E = [E_j(0) - E_0(0)]$ of Eq. (7.4), for the PH_3 case. The driving force for distortion away from the planar structure should then be larger for PH_3 than for NH_3.

Not shown on the structure map of Figure 7.31 are the well-known AX_2 structures, which contain $X-X$ pairs of atoms such as those of marcasite and pyrite (Figs. 7.7 and 7.8) described in Section 7.1 and the related, distorted löllingite structure. It is interesting to ask what the factors are that stabilize one of these structures over those of the cadmium halide or rutile types. FeS_2 in the pyrite structure would be written as $Fe^{2+}(S_2)^2$ (with 14 electrons), but in the cadmium halide or rutile structure as $Fe^{4+}(S^{2-})_2$ (sixteen electrons). Thus a part of the answer lies in the energetic cost of oxidation $Fe^{2+} \rightarrow Fe^{4+}$. Redox potentials in solids are considerably smaller than in the gas phase, as discussed in Section 5.3, but a plot (Ref. 66) of the sum of the third and fourth ionization potentials quite cleanly separates examples of solids with the two structures as shown in Figure 7.59.

7.11 Second-order structural changes

Equation (7.4) showed how the second-order Jahn–Teller mechanism allowed the mixing of higher-energy levels into lower-energy ones as a result of a geometrical distortion. In the delocalized picture an exactly analogous process allows energy bands to mix, or rather the orbitals that comprise them. This was described for the case of polyacene in Section 2.6. Another interesting example that is accessible (Ref. 381) in this way is the bond alternation found in many ABO_3 perovskites [Fig. 5.32(a)] and related structures. The undistorted perovskite structure consists of BO_6 octahedra (B is often a transition metal), vertex linked in all three directions, with a large central atom A (often a rare earth) in the large interstice so generated. (The cubic structure type that results of stoichiometry BO_3 is that of ReO_3.) The bond

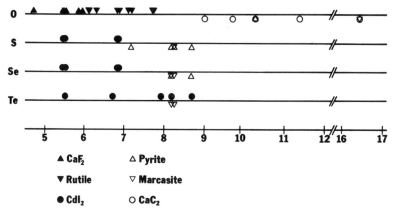

▲ CaF₂   △ Pyrite

▼ Rutile   ▽ Marcasite

● CdI₂   ○ CaC₂

Figure 7.59 Structures of the dioxides and dichalcogenides as a function of the sum of the third and fourth gas-phase ionization potentials of the atoms. Only about one-third of the known examples are plotted because of a lack of ionization potential data.

alternation (Fig. 7.60) takes place in either one, two, or three dimensions, leading to tetragonal, orthorhombic, and rhombohedral structures, respectively. The central A atom has been left out of these diagrams for clarity. $BaTiO_3$ is an interesting system in that all four structures are found depending on the temperature: cubic ($120 < T < 1460°C$), tetragonal ($5 < T < 120°C$), orthorhombic ($-80 < T < 5°C$), and rhombohedral ($T < -80°C$). The density of states for such a system looks very much like those of TiO_2 and rocksalt MO solids of Figures 4.4(a) and 4.7, understandable given the identical local coordination of the transition metal. The energetic behavior on distortion of the bands of interest is shown in Figure 7.61(a). Since in these ABO_3 perovskites the central A atom is usually quite electropositive, an electronic picture that focuses on the BO_3 skeleton is an appropriate one. Shown are the π ("t_{2g}") orbitals of the metal and the π oxygen orbitals that can interact with them. Notice how the mixing between the levels at the top of the oxygen p band and those at the bottom of the metal d band on distortion

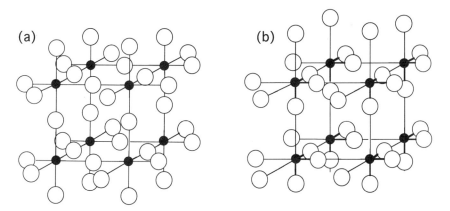

Figure 7.60 Bond alternation in the perovskite (a) structure. This may proceed along one, two, or [as shown in (b)] all three directions. The results are tetragonal, orthorhombic, and rhombohedral structures, respectively.

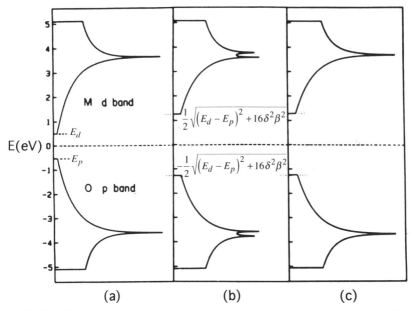

Figure 7.61 Densities of states in the perovskite structure and its distorted variants. (a) The Mo–O π bands of the undistorted structure; (b) the xz and yz bands of the tetragonally distorted structure; and (c) and xz bands for the orthorhombic distortion.

leads to an increase in the gap between oxide and metal levels. Figure 7.62 shows the orbitals at $\mathbf{k} = 0$, which are those at the top and bottom of these two bands. Both orbitals are largely nonbonding in the undistorted structure, but on distortion the upper orbital becomes metal–oxygen antibonding and the lower one metal–oxygen bonding.

The fact that the upper orbital increases in energy on distortion suggests a mechanism for switching off the distortion by increasing the electron count so that the upper orbital starts to fill with electrons. This is exactly the same idea behind

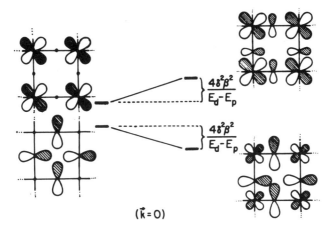

Figure 7.62 The change in the nature of the wave functions at the bottom of the d band and at the top of the p band on distortion.

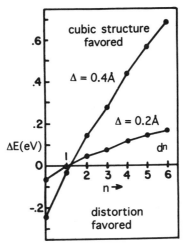

Figure 7.63   The energetics of the tetragonal distortion of the perovskite via bond alternation. The two plots correspond to two different bond length differences in the distorted structure.

the stabilization of planar ammonia by population of the LUMO as in its first excited state. The results of a numerical calculation are shown in Figure 7.63 and suggests (Ref. 381) that by the time the d^1 configuration is reached the driving force for distortion has disappeared. Such undistorted structures are found for compounds such as ReO_3 itself.

8 □ More about structures

There is no monopoly on ways to view chemical bonding and electronic and geometrical structure. The structures of the elements, for example, have been described in two different ways already in this book. This chapter develops the moments approach further and also introduces some of the ideas behind pseudo-potential theory.

8.1 The structures of the elements in terms of moments

The moments method (Refs. 45, 62, 105–108, 117, 118, 135) is a particularly useful tool with which to correlate structure with electron count (Ref. 114). Since the moments themselves may be directly expanded in terms of walks through the orbital topology of the lattice, there is a strong connection made immediately between structural and electronic features. [In theoretical terms the method has similarities to the recursion method (Refs. 163, 168, 297).] In terms of the electronic densities of states, the first moment locates the energy zero; the second is a measure, as noted earlier, of the width of the density of states; the third and higher odd moments define the asymmetry of the distribution; etc. The first step (Ref. 105) is to set equal the second moments of the energy densities of states of the structures being considered. The second is then to search for structural reasons behind the variation of relevant energy differences as a function of electron count, $\Delta E(x)$, in terms of the contributions from local structural features. Especially useful are the contributions to $\Delta E(x)$ from rings of various sizes (Ref. 45). Figure 1.13 showed some plots for the molecular case. Figure 8.1 shows in general the energy difference curves expected (Ref. 222) between two systems that differ at the nth moment. Note that there are n modes, that the amplitude of the curves drops off as n increases, and that it is the structure with the largest nth moment that is most stable at the earliest band fillings. The actual crossing points (Ref. 170) and the form of these curves vary a bit from system to system. Figure 8.1(b) shows the effect of a tail in the energy density of states often found for sp-bonded systems. Note that the third moment plot crosses the axis earlier than in Figure 8.1(a).

It is useful to first probe the origin of the very powerful criterion of second moment conservation (Refs. 179, 220, 223) when comparing the energies of structures. Its use lies behind the operation (Ref. 56) of the bond valence sum rule, as shown in Section 6.3. The equilibrium interatomic separation is set of course by the balance of attractive [$U(r)$] and repulsive [$V(r)$] terms in the energy. There are two basic assumptions that go into the structural energy difference theorem (Refs. 179, 289, 290, 292, 296) as it is called. The first is that the form of the attractive part of the energy is well represented by the orbital calculations of the type described throughout this book.

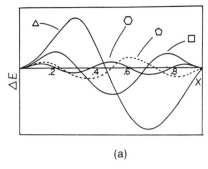
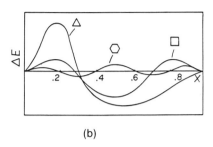

(a)   (b)

Figure 8.1   Qualitative energetic effect of different ring sizes as a function of electron count (fractional orbital occupancy, x). The energy differences plotted are between a structure containing a single ring of a given size and a structure containing no rings at all. The results in (a) correspond to the case where there is a uniform band, and in (b) the case where there is a strong bonding tail. (Reproduced from Ref. 221 by permission of The American Chemical Society.)

The second is that the repulsive part of the energy is of short range and may be written as a sum of pairwise repulsion energies over the Γ nearest neighbours coordinated to the atom under consideration. This repulsion includes the Coulombic repulsion of the nuclei and the requirements of the Pauli Principle as the atoms overlap. In Section 1.6 it was shown how the coordination number and second moment of the energy density of states are directly related, that is,

$$\Gamma = \eta \mu_2 = \eta \int_{-\infty}^{\infty} \rho(E)E^2 \, dE \tag{8.1}$$

Here η is a proportionality constant. The total energy of the system may then be written as

$$E_T = U(r_1, r_2, \ldots) + V(r_1, r_2, \ldots)$$
$$= \eta' \int_{-\infty}^{\infty} \rho(E)E^2 \, dE + \int_{-\infty}^{E_F} \rho(E)E \, dE \tag{8.2}$$

Now consider two systems, 1 and 2. The energy difference between the two is simply given by

$$\Delta E_T = U_1(r_1(\text{eq})) + V_1(r_1(\text{eq})) - U_2(r_2(\text{eq})) + V_2(r_2(\text{eq})) \tag{8.3}$$

Near the equilibrium geometry the energy is a slowly varying function of the geometry and so one can write

$$\Delta E_T = 0 = U_1(r_1(\text{eq})) + V_1(r_1(\text{eq})) - U_2(r_2(\text{eq}) + d) + V_2(r_2(\text{eq}) + d) \tag{8.4}$$

A value for d may be chosen such that $U_1(r_1(\text{eq})) = U_2(r_2(\text{eq}) + d)$, which implies that

$$\Delta E_T = \int_{-\infty}^{E_{F1}} \rho(E, r_1(\text{eq}))E \, dE - \int_{-\infty}^{E_{F2}} \rho(E, r_2(\text{eq})) + d)E \, dE \tag{8.5}$$

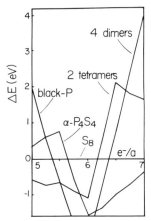

Figure 8.2 The calculated energy differences between black phosphorus, α-P_4S_4, S_8, $2S_4^{2-}$, and $4S_4^{2-}$ as a function of the number of electrons per atom. The calculations used the second moment scaling restriction with the observed bond lengths in S_8. (Reproduced from Ref. 220 by permission of The American Chemical Society.)

The value of r_1 comes from the experimental geometry for system 1, and the value of $r_2 + d$ comes from the relationship

$$\int_{-\infty}^{E_{F1}} \rho(E, r_1(\text{eq}))E^2 \, dE = \int_{-\infty}^{E_{F1}} \rho(E, r_2(\text{eq}) + d)E^2 \, dE \qquad (8.6)$$

Thus the energy difference between the two structures is given by the difference in the electronic energy, ΔU, evaluated for two structures with equal repulsive energies, that is, equal second moments. This is the structural energy difference theorem.

Figure 8.2 shows (Refs. 220, 222) some calculated energy difference as a function of electron count for some simple systems (**8.1**) using this second moment conservation idea. Black phosphorus (Fig. 7.2) contains three-coordinate atoms, but as the

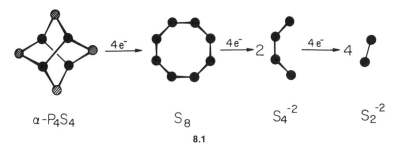

8.1

electron count increases, the coordination number decreases. Modeling the energy differences associated with geometry changes has long been a problem, but is one effectively solved by using this trick. Notice how the observed structures for a given electron count are faithfully reproduced by the calculation. Of course, the lowest-energy structure found in each case is the one expected using the ideas of the two-centre–two-electron bond.

Table 8.1 lists the structures of the elements (Ref. 222) from the posttransition

Table 8.1 The structures of some of the post-transition elements.

Element	Cu	Zn	Ga	Ge	As	Se
Structure type	fcc	hcp	Ga	diamond	As	Se
Geometric features	tetrahedra	tetrahedra	triangles	hexagons	hexagons, squares	chains, squares

Element	Ag	Cd	In	Sn	Sb	Te
Structure type	fcc	hcp	In	w-Sn, diamond	As	Se
Geometric features	tetrahedra	tetrahedra	tetrahedra	hexagons, squares, hexagons	hexagons, squares	chains, squares

Element	Tl	Pb	Bi	Po
Structure type	hcp	fcc	As	simple cubic
geometric features	tetrahedra	tetrahedra	hexagons, squares	squares

Groups 11 to 17 with some of the structural features of interest. Figure 8.3 shows the structures themselves. The heavier lines correspond to the short distances. As noted in Table 8.1, although these structures often contain short bonds connecting the atoms within a slab, sheet, or chain, there are usually nonbonded distances that are often not very much longer. The table also lists the types of rings present, data of use in the moments analysis. Since the importance of the orbital walks that determine the moments depend upon the relevant hopping integrals from Eq. (1.52), and **1.7** and **1.8**, the shorter internuclear distances will be the ones of greatest importance. Figure 8.4 shows (Refs. 63, 222) the results of some Hückel tight-binding calculations on the heavy elements, Tl–Po, from the fifth row of the periodic table. The calculations, which only used the valence $6p$ orbitals (vide infra), reproduce the observed structural trends with electron count, and it is interesting to note that the form of these energy difference curves are very similar to those of Figures 1.13 and 8.1. These describe the energy difference between two structures as a function of the fractional band or orbital occupancy, $\Delta E(x)$, and are interpreted in terms of differences in the moments of the energy density of states. From the form of the curves of Figure 8.4, the energies of the fcc and hcp structures relative to that of the simple cubic reference structure are determined by the third moment, and that of the α-arsenic structure relative to simple cubic by the fourth moment. Using the correlation between the order of the first disparate moment and the difference in the orbital walks around the structure, these energy differences are easy to understand. The fcc and hcp structures have a very high density of three-membered rings that stabilize these structures at low electron counts (Fig. 8.1). The energy difference between simple cubic and α-arsenic is a fourth moment problem, set by the squares of atoms in the former. Since from Figure 8.1 the structure with the largest fourth moment is the one that is more stable at early band fillings, the structure of α-arsenic is the one most stable at the half-filled point. At higher electron counts the simple cubic structure, with its four-rings, becomes more stable.

The energy difference plots also correctly reproduce the observed change in the nature of the close packing on moving from one to two electrons (Ref. 3). This may also be studied in terms of the moments method but not quite so usefully. On drawing out the walks associated with the two structures, the first walk that is different is one of length five. Indeed, from **8.2** there are five nodes in the calculated $\Delta E(x)$ curve

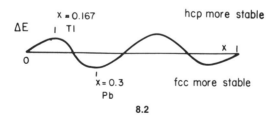

8.2

for the pair of structures. However, it is difficult to see this walk easily and to associate it with the difference in the mode of packing. A similar geometrical problem arises with the close-packing variants of the lanthanide elements mentioned in Section 4.6. This series of main group structures, however, highlights the importance of straight-

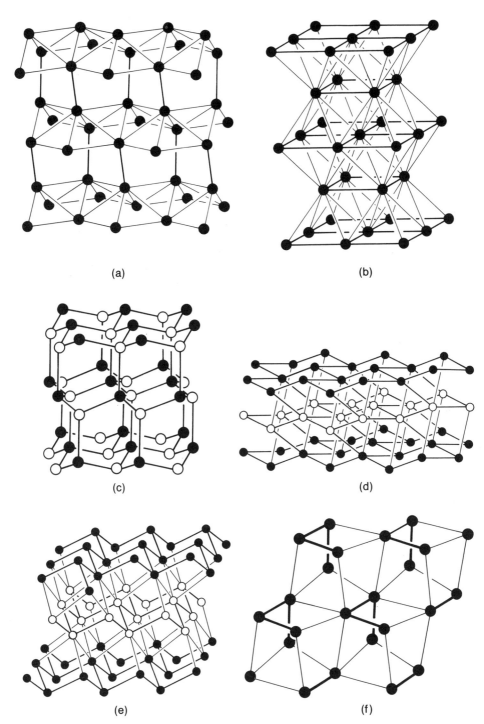

Figure 8.3 The crystal structures of (a) Ga, (b) In, (c) Ge and gray Sn, (d) white Sn, (e) Bi (As structure), (f) Te. Thick lines indicate shorter bonds. (Reproduced from Ref. 222 by permission of The American Chemical Society.)

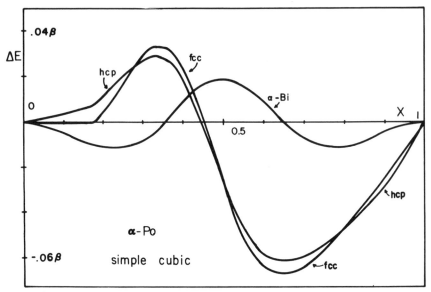

Figure 8.4 Computed energy difference curves for the structures of the heavy main group elements as a function of fractional orbital occupancy, x, relative to the simple cubic structure of α-Po. (A p orbital only model was used here.)

forward topological ideas. The calculations of Figure 8.4 are quite simple ones. At the bottom of the periodic table the s orbitals are quite contracted due to relativistic effects, which result in the generation of a lone pair of electrons. The calculations therefore only used the valence 6p orbitals. Since the atoms in the structures considered have different coordination numbers, the second moments of their energy densities of states need to be kept fixed, just as for the case of the H_3 ions discussed in Section 1.6.

Figure 8.5 shows the calculated (Ref. 63) energy difference curve, using both s and p orbitals now, for the two first row elements from this part of the periodic table that form solids at ambient pressure and temperature, carbon and boron. Notice the third moment form of this curve, set of course by the presence of three-membered rings in the rhombohedral boron structure (Fig. 8.6). Within the moments language rhombohedral boron is more stable than diamond for three electrons per atom because it contains three-membered rings. Diamond is more stable than rhombohedral boron at four electrons per atom because it does not. The fact that diamond has six-membered rings does not enter into the argument. [Parenthetically, it is interesting to remember that the rhombohedral boron structure may be understood by use of Wade's rules (Ref. 359).] The two curves shown differ only in the bond lengths used. Curve I shows the energy difference curve for the (unrealistic) case where the interatomic distances are the same for both structures, but curve II is for the case when the observed distances are used for both structures.

Figure 8.7 shows a more complete plot (Refs. 220, 222), which compares the energies of the hcp, rhombohedral boron, graphite, and diamond structures using second moment scaling. The plots are in complete agreement with experiment: Li and Be are hcp, B is rhombohedral, and carbon adopts the graphite structure. Again

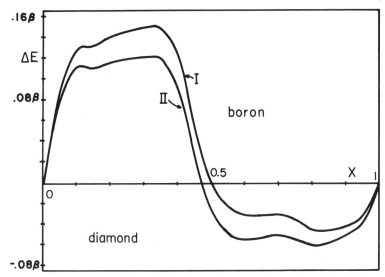

Figure 8.5   The computed energy difference between the cubic diamond and rhombohedral boron structures as a function of fractional orbital occupancy, x. Curve I is from calculations where the same internuclear distances are used for the two structures. The calculations leading to curve II used experimentally determined distances for the two structures.

the hcp plot shows a dominant third moment shape. There are more three-membered rings per atom for the hcp structure than there are in the structure of rhombohedral boron. As a result, the hcp arrangement has the larger third moment and the energy difference curve has the shape it does. Both diamond and graphite contain six-membered rings, and thus their energy difference curve with rhombohedral boron is set by third moment considerations. However, since there are more three-rings in rhombohedral boron than in diamond (where there are none, of course), the diamond structure is less stable at earlier band fillings. Notice that graphite is predicted (and observed) to be the more stable form of carbon. This is a result that only arises through the use of the second moment scaling requirements. Otherwise the lowest-energy structure is always calculated to be the one with the larger second moment, or larger coordination number.

Figure 8.8 shows analogous energy difference plots (Ref. 222) for the third row elements relative to that of the diamond structure found for germanium. It reproduces the observed trend: fcc (Cu), hcp (Zn), Ga (Ga), diamond (Ge), α-arsenic (just) (As), followed by Se. Here there is a crossing of fcc and hcp curves at early $(s + p)$ electron counts, corresponding to the elements Cu and Zn. The behavior, however, is different from that noted for Tl and Pb with early s electron counts. For Cu and Zn the fcc structure is the one stable at earliest fillings. The singular structure of gallium becomes stable just where it should be at three electrons per atom. The structure has fewer three-membered rings than the close-packed structures. This leads to a $\Delta E(x)$ curve (relative to the diamond structure of germanium) that has a smaller amplitude at low x and a crossing point at higher x when compared to those of the close-packed structures.

Figure 8.9 shows the energy difference plots (Ref. 222) for the fourth row elements relative to that of the white tin structure found for tin. Figure 8.9(a) shows the plots

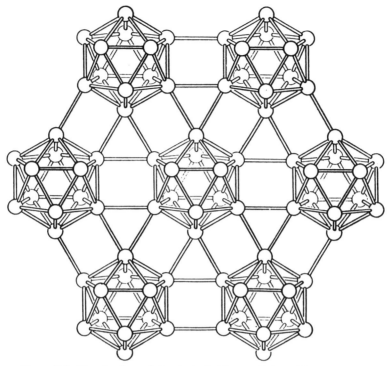

Figure 8.6 The structure of rhombohedral boron. (Adapted from Ref. 114.)

calculated assuming no contraction of the valence s orbital, expected to be of importance for these heavier elements. It is in quite poor agreement with experiment. Importantly, the white tin structure is not computed to be the lowest-energy arrangement for four electrons. Figure 8.9(b) shows analogous plots but for calculations where a contracted s orbital was used. It reproduces, with the exception

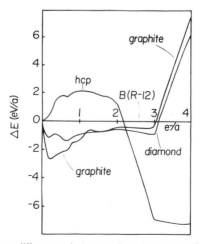

Figure 8.7 Computed energy difference between the structures of Li and Be (both hcp), the rhombohedral boron structure, and two carbon structures. (Reproduced from Ref. 220 by permission of The American Chemical Society.)

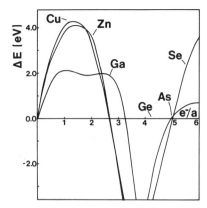

Figure 8.8 The computed energy difference between the Cu, Zn, Ga, Ge, As, and Se structures as a function of electron count. (Reproduced from Ref. 222 by permission of The American Chemical Society.)

of the In structure (found to be indistinguishable energetically from the fcc structure), the observed trend: fcc (Ag), hcp (Cd), white tin (Sn), α-arsenic (Sb), followed by Te. There is again the crossing of fcc and hcp curves at early $(s + p)$ electron counts.

The calculations for the second and third rows of the periodic table used the valence s and p orbitals, but those for the fifth row used s orbitals only. The gradual change in structure type on moving down the table, for example, from those of gallium and germanium, to the close-packed structures for thallium and lead is a result of the generation of an inert s valence pair. Thus $x \sim 0.14$ corresponds to Cu for the $(s + p)$ band of the third row but to Tl for the s band of the fifth. Figure 8.10 shows how

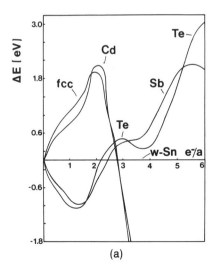
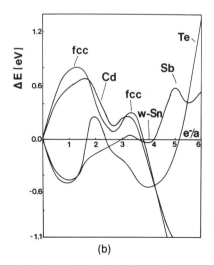

(a)   (b)

Figure 8.9 The computed energy difference between the fcc, Cd, In, white Sn, Sb, and Te structures. (The energetic difference between the fcc and In structures is so small that only one line is shown.) (a) The results of a calculation where there is no s orbital contraction. (b) The results of calculations where this has been taken into account. (Reproduced from Ref. 222 by permission of The American Chemical Society.)

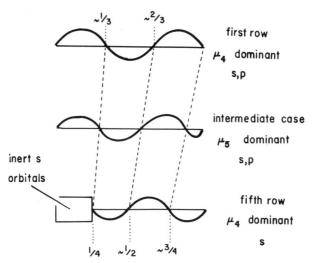

Figure 8.10   Transition from an $s + p$ orbital model at the top of the periodic table to a p-only one at the bottom.

the filling pattern varies and how the number of nodes changes on moving down the periodic table.

It is pertinent to make contact here with the structural principles of Laves (Ref. 218) in terms of the stability of crystal structures with high symmetry, high coordination numbers, and dense atomic packing. These criteria are satisfied for the hcp and fcc structures found for some of the metallic elements. It is important to note, however, that such structures are those that contain a high density of three-membered rings and thus are those structures that will be favored from the electronic point of view at low electron counts, which is indeed where they are found. This is then the electronic origin of Laves' comments in these systems.

8.2   The structures of some main group intermetallic compounds

The structures of the set of $XA_{2-x}B_x$ compounds, where X is an electropositive element and A and B are elements from Groups 8–16, are understandable (Ref. 221) using the ideas described in the previous section. There are eleven structural families here, namely, $MgZn_2$, $MgCu_2$, Cu_2Sb, $MoSi_2$, $CeCu_2$, $MgAgAs$, $CeCd_2$, $CaIn_2$, AlB_2, $ThSi_2$, and $ZrSi_2$. Some of the structures are very similar. Thus $CaIn_2$ is a stuffed hexagonal diamond (In) structure and $MgAgAs$ a stuffed derivative structure of cubic diamond (AgAs). These two structures are, of course, Zintl phases. $MoSi_2$ and Cu_2Sb are also structurally quite similar. This set of structures is shown in Figure 8.11. Some of them, such as that of $CaIn_2$, are only found for rather specific electron counts, whereas some, such as that of AlB_2, show a wider variation. The structures found for these solids are reported in Table 8.2. In terms of their prominent geometrical features, the $MgCu_2$ structure contains triangles of A and B atoms. $MoSi_2$ contains both triangles and squares. $CeCu_2$ and $CeCd_2$ contain squares and hexagons. $MgAgAs$, $CaIn_2$, and AlB_2 contain hexagons only. $ThSi_2$ has no small rings at all.

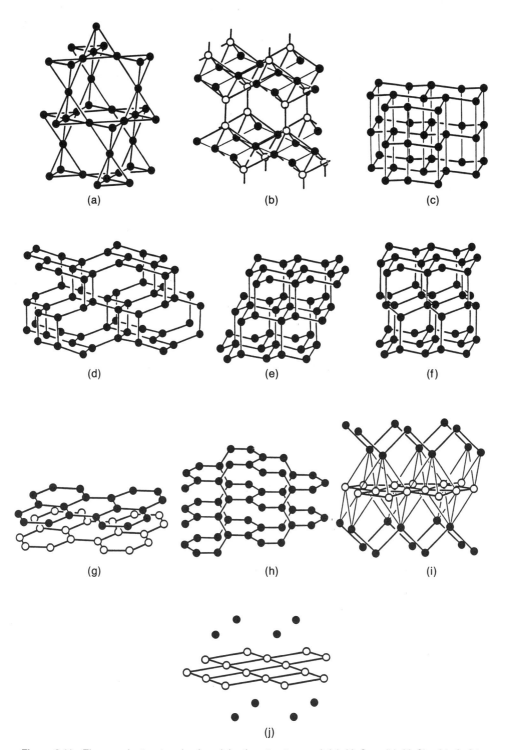

Figure 8.11 The covalent networks found in the structures of (a) MgCu$_2$, (b) MoSi$_2$, (c) CeCd$_2$, (d) CeCu$_2$, (e) MgAgAs, (f) CaIn$_2$, (g) AlB$_2$, (h) ThSi$_2$, (i) ZrSi$_2$, and (j) Cu$_2$Sb. (Reproduced from Ref. 221 by permission of The American Chemical Society.)

Table 8.2 A comparison between experiment and theory of the electron count at which $ZA_{2-x}B_x$ compounds are stable

No. of electrons $ZA_{2-x}B_x$ unit	Theory[a]	Expt[b]
1	$MgCu_2$	$MgCu_2$
2	$MgCu_2$	$MgCu_2$
3	$MgCu_2$	$MgCu_2$
4	$MoSi_2 = (Cu_2Sb) > $ $MgCu_2 > CeCd_2$	Cu_2Sb, $MgCu_2$
5	$MoSi_2 = (Cu_2Sb)$	$MoSi_2$, Cu_2Sb, $CeCu_2$
6	$MoSi_2 = (Cu_2Sb) = $ $CeCd_2 > CeCu_2$	complex
7	$CeCd_2 = CeCu_2 = $ $MgAgAs = (CaIn_2)$	$CeCd_2$, $CeCu_2$, Cu_2Sb, AlB_2
8	$MgAgAs = (CaIn_2)$	$MgAgAs$, $CaIn_2$
9	$AlB_2 > ThSi_2$	AlB_2, $MgCu_2$
10	$ThSi_2 > AlB_2$	$ThSi_2$
11	[c]	$ZrSi_2$, $ThSi_2$
12	[c]	$ZrSi_2$
13	$ZrSi_2 = Cu_2Sb$	$ZrSi_2$
14	Cu_2Sb	Cu_2Sb
15	Cu_2Sb	Cu_2Sb

[a] Parentheses indicate a structure for which no calculation has been carried out but that is geometrically similar to another structure type. Equal signs indicate the structure types are within 0.05 eV/atom of one another. Greater than signs indicate alternative structures that have energies within 0.40 eV/atom of the most stable structure.
[b] Italicized compounds indicate the most common electron count for that structure type.
[c] See original paper for further discussion concerning $ZrSi_2$ and $ThSi_2$ structures.

$ZrSi_2$ has two structural features, chains (which have no rings) and a square lattice. So just as described previously for the main group elements, the lowest-electron-count structures have triangles, then come structures with triangles and squares, followed by hexagons around the half-filled point, with squares at the highest fillings.

Figure 8.12 shows calculated energy differences (Ref. 221) between some of the structures and the $MoSi_2$ structure type. Their shape is as expected from the moments approach. The $MgCu_2$ structure with its many triangles of atoms is stable at the earliest fillings. $MoSi_2$ has few triangles but is more stable than the $CeCu_2$ and $CeCd_2$ structures, which contain squares and hexagons. A similar series of structural features is seen in molecular structures described by Wade's rules (Ref. 359). Consider the set of five-atom molecules of **8.3**. The closo species $B_5H_5^{2-}$ contains seven triangles only. Since there are a total of 15 skeletal orbitals in this molecule, with six bonding pairs the fractional orbital occupancy of the set is $x = \frac{6}{15}$ or 0.4. The nido compound has seven pairs, and thus $x = 0.47$. Here there are four triangles and one square. For the

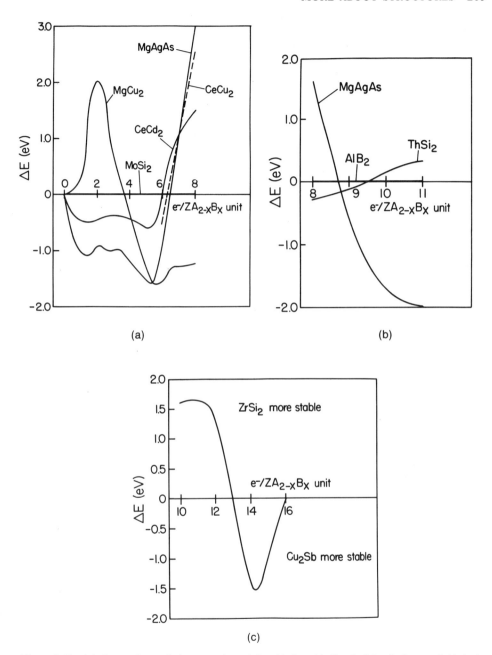

Figure 8.12 (a) Comparison of the energies of the $MgCu_2$, $MoSi_2$, $CeCd_2$, $CeCu_2$, and MgAgAs structure types. The curves are plotted relative to the energy of $MoSi_2$. (b) Comparison of the energies of the MgAgAs, AlB_2, and $ThSi_2$ structure types. (The energy of $CaIn_2$ is very close to that of MgAgAs and is not shown separately.) (c) Comparison of the energies of the $ZrSi_2$ and Cu_2Sb structure types. (Reproduced from Ref. 221 by permission of The American Chemical Society.)

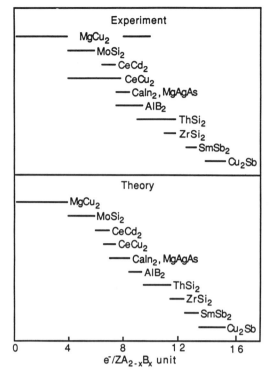

Figure 8.13 Theoretically calculated and experimentally observed electron count ranges for stability of the $ZA_{2-x}B_x$ phases. (Reproduced from Ref. 220 by permission of The American Chemical Society.)

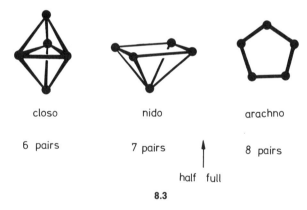

arachno compound there are neither squares not triangles, but one five-membered ring. Here $x = 0.53$.

Past the half-filled band at eight electrons Figure 8.12(b) shows how the MgAgAs (stuffed diamond derivative) structure rapidly becomes unstable relative to structures that are much more open (cf. Section 7.2). Figure 8.12(c) shows the eventual stability of the Cu_2Sb structure. The agreement between theory and experiment (Table 7.2 and Fig. 8.13), although not perfect, is quite good.

8.3   The Hume-Rothery rules

Hume-Rothery noted many years ago (Ref. 189) that the structure adopted by noble metal alloys, such as the brasses, appeared to be controlled by the electron concentration, or the average number of electrons per atom. He called such systems "electron compounds," although today it is well known that there are many areas of chemistry where electron count is a crucial parameter. The effective atomic number rule [in solids the Grimm–Sommerfeld rule (Ref. 274) of Section 7.1], Wade's rules (Ref. 359), and Hückel's rule (Ref. 183, 335) are examples that spring to mind. The discussion of Section 4.7 showed that zone-face touching by the Fermi surface (Refs. 254, 336) was not a good model to understand these observations. The electronic structure of the transition metals and their crystal structures were well described in terms of d-orbital interactions between adjacent atoms. The discussion (Ref. 178) in this section continues that orbital philosophy.

There are five phases whose stability Hume-Rothery identified as being controlled by electron count: hcp, bcc, fcc, β-Mn (Fig. 4.30), and γ-brass (**8.4**). The ranges of

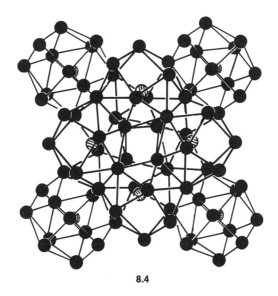

8.4

electron count for which these structures are found are shown in Fig. 8.14(a). The fcc structure is found for an average of 11 $(s + p + d)$ electrons per atom and the hcp structure for 12 electrons, a result in accord with the discussion of Section 8.1. The other structure types lie in between, in this narrow range. The structures of β-Mn and γ-brass are complex ones. The hcp structure comes in three variants that differ only in their c/a values, ζ (close to the ideal value of 1.633), ε (1.55–1.58), and η (1.77–1.88). The structure of the η phase for elemental zinc was discussed in Section 5.2. The three structures found for the hcp structure correspond to three distinctly different phases. Biphasic regions are known in the phase diagrams of Ag–Zn, Au–Zn, Cu–Zn, and Ag–Cd, for example. Figure 8.15 shows the results of a series of calculations (Refs. 178, 220) using the second moment scaling method on these

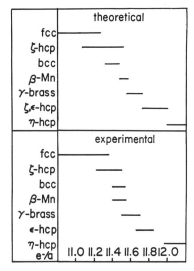

Figure 8.14 Theoretically calculated and experimentally observed electron count ranges for stability of the Hume–Rothery phases. (Reproduced from Ref. 221 by permission of The American Chemical Society.)

structure types, and Figure 8.14(b) a comparison between theory and experiment. The agreement is very good indeed.

A similar approach allows (Ref. 178) resolution of the electron count dependence for binary alloys of the transition elements themselves. Structures that are based on the fcc, hcp, and bcc structures are found, in addition to the much more complex structures observed for the χ and σ phases. An important result is that it is orbital factors that are involved in determining structure type. In this sense the origin of the Hume-Rothery rules and similar counting rules for the binary transition metal alloys

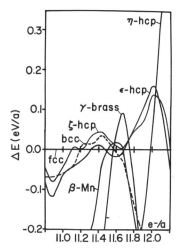

Figure 8.15 Calculated energy differences for the seven Hume–Rothery phases as a function of the number of $(s + p + d)$ electrons per atom. (Reproduced from Ref. 221 by permission of The American Chemical Society.)

is just a (considerably) more complex version of the Hückel rules for cyclic hydrocarbons. In Section 8.4 the use of pseudopotentials gives some added insights from a completely different perspective.

8.4 Pseudopotential theory

Pseudopotential theory (Refs. 97, 167, 171) and its application to the structure of solids has its origins in the results of Section 4.7, where band gaps opened up in the nearly-free-electron (NFE) model, as a result of including the effect of the nuclei as a perturbation of the free-electron bands. In principle, the secular determinant of Eq. (4.33) places no restrictions on the strength of the perturbation. However, in general the band gaps that would be generated if real atomic potentials were used are much too large. (Exceptions are the light elements H, He, and perhaps Li and Be.) But, as noted in Section 4.7, the band structures of many systems are well described by free-electron bands perturbed by a weak potential, the pseudopotential (Refs. 167, 302). Some of its properties were described in Section 6.2. There are several forms for this potential, one of which was shown in Figure 6.5. The model potential shown in **8.5** is defined by the two parameters R_m and A_l. Outside R_m the potential

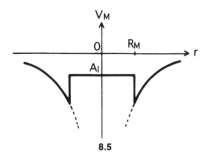

8.5

is just a Coulombic one; inside it is constant. A_l is adjusted to give the correct values of the observed l ionization energies of the ion. Often a one-parameter form is used (Ref. 167) that sets A_l to zero (the empty-core pseudopotential). This results in $R_m = R_c$, the radius of the "core" of Section 6.2. Irrespective of the form of the pseudopotential, the main point is that it is weak and leads to gaps of the correct magnitude. Importantly, the places where the gaps lie are determined by the symmetry of the crystal lattice. Much of the literature that describes the application of pseudopotential theory to the structures of solids involves the Fourier transform of the pseudopotential (Ref. 97, 167, 171), and the electronic arguments lie in reciprocal space. The observed crystal structure is suggested to be the one where the lattice vectors have moved so as to open gaps wherever possible at the Fermi level. There is an alternative, and exactly analogous, description in terms of real-space ideas (Refs. 154, 155).

The results (which will not be proved) of transforming the band structure energy into real space is an interesting one. The total energy may be written in a particularly

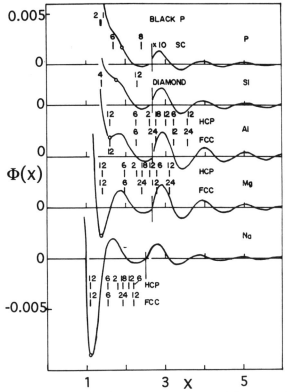

Figure 8.16   A typical interatomic potential, $\Phi(x)$, for rearrangement of the atoms at constant volume ($x = Rk_F/\pi$). The set shown is for the second (long) row of the periodic table. There is a strong repulsion at short distances, followed by a potential energy minimum, followed at larger x by weaker oscillations. For the system where the atoms are located at the point shown, the energy may be lowered by moving some atoms in closer and some atoms farther away. (Adapted from Ref. 155.)

simple form as

$$E_T = E_\Omega + \tfrac{1}{2} \sum_{ij} \Phi(R_{ij}), \qquad i \neq j \tag{8.7}$$

where Φ is the effective two-body potential between the atoms i and j, a distance R_{ij} apart. E_Ω is a volume-dependent but structure-independent term. Thus $\Phi(R_{ij})$ is not a complete interaction potential, but one for the atoms at constant volume. The pair potentials consist of three main regions (Fig. 8.16). There is a repulsive contribution at short R of course. At long distances there are oscillations (so-called Friedel oscillations), but it is the behavior at intermediate distances that is important. (The reduced ordinate $x = Rk_F/\pi$ is used in Fig. 8.16, which highlights the form of the ripples that are a feature at large R.) In considering the structural predictions of Eq. (8.7) it is necessary to work at constant volume so that only the variations in the two-body terms are present. The philosophy is quite simple. If the nearest neighbors of a particular structure fall in a minimum in the $\Phi(R)$, then that structure is expected to be stable. If they fall on a peak, then the energy will be lowered by moving some atoms closer and some atoms farther away. Such motion has to be one that conserves the volume of the solid. In terms of the reciprocal-space arguments, such movement increases the band gaps, D_{cp} is the value of R expected for the material if it were closed packed (indicated by a circle in Fig. 8.16). This figure shows that indeed for Na, Mg, and Al the value of D_{cp} is such that the nearest neighbors lie in the first minimum (R_{min}) in the $\Phi(R)$ plot, and a close-packed structure is expected. For Si and P this is not the case, and more open structures are expected with a smaller number of closer neighbors. For silicon, for example, some atoms move closer to make four contacts, and some move outward in accord with the location of the first minimum at the position close to that expected for 12 second-nearest neighbors. Recall that the general trends of the main group elements with increasing electron count are from close-packed structures (initially hcp but then fcc) to more open structures. (Associated with this is the change from metallic to nonmetallic properties, as discussed in Section 5.3.) This observation comes directly from the form of the pair potential. There is also information concerning the preference for the hcp or fcc structure. For Na and Mg there are more neighbors in the second hcp attractive well than for fcc, and so the hcp structure is found. For Al the observed fcc arrangement is consistent with the fact that the fifth-nearest neighbors lie at the maximum of the second repulsive wiggle for hcp.

The behavior of the Group 12 metals Zn and Cd was discussed in Section 5.2, and their structures reproduced numerically in Figure 8.8. A pseudopotential argument exists (Ref. 167) to understand their structure, shown in Figure 8.17. For Mg D_{cp} lies at R_{min}, and a close-packed arrangement is expected. For Cd this is not true, and the structure may distort (increased c/a) so as to bring six in-plane neighbors closer together and six out-of-plane neighbors (along c) farther apart. A similar picture holds for gallium. In neither case does the theory make a prediction concerning the identity of the lowest-energy structure. It does give information, however, about the instability of the close-packed arrangement.

The variation in the form of the curves of Figure 8.16 on moving across the periodic table is not quite so easy to understand. It depends in a complex way on Z and the parameters R_c and R_a, the core and atomic radii, respectively. It is clear to see, however, that as Z increases along a row of the periodic table, the amplitude of the oscillations decreases and the first minimum is covered up by the repulsive core

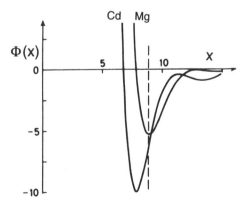

Figure 8.17 Interatomic potential, $\Phi(x)$, for Mg and Cd. The dashed line indicates the distance of the nearest neighbors of ideal c/a ratios. (Adapted from Ref. 167.)

potential. Recent studies have attempted to build in some structural dependence of the pair potentials (Refs. 169, 295).

8.5 The structures of the first row elements

Because of their chemical importance the elements of the first (long) row of the periodic table (Li–Ne) receive much of the attention of chemists. The structures and chemistry of their compounds are usually regarded as the yardstick with which those of the other elements are compared. A part of this attitude is of course set by the domination of a large part of chemistry by the properties of carbon-containing compounds. However, on viewing the properties of the main group elements in general, one can readily see that it is the properties of the first row elements boron to fluorine that are unusual. Thus, although CO_2 is a gas under ambient conditions, all the other Group 14 AO_2 systems are solids. N_2 is a gas, but the heavier elements are found as extended solids at room temperature. Traditional explanations of these differences focus on the importance of π bonding for N_2, acetylene, etc., and point out that the triple bond (one σ, two π) between the two nitrogen atoms is energetically worth more than the three single bonds found in the structure of black phosphorus (Fig. 7.2), for example. In comparing the relative importance of σ and π bonding on moving down a group, the effectiveness of π bonding drops off rapidly beyond the first row. Of course, this σ/π ratio will depend rather critically on the ratio of orbital "size" to the internuclear separation. [A quantitative idea of the relative importance of σ and π interactions comes (Ref. 228) from experimental measurements of bandwidths and gaps and identification with the theoretical values using β_σ and β_π.] One could present a model that rationalized much of the difference in behavior between the first row and heavier elements based on the much shorter internuclear separations than would be expected by extrapolation from the heavier elements. Such shortening would dramatically stabilize multiple bonds between the atoms and thus lead to generation of small molecules rather than extended arrays. Table 5.2 shows that the internuclear separations for the first row element are distinctly shorter than expected by such an extrapolation from those of the heavier elements.

Figure 8.18 shows a suggested $\Phi(R)$ function (Section 8.4) for boron (Ref. 155).

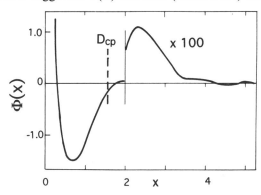

Figure 8.18 A suggested interatomic potential for boron. (Adapted from Ref. 155.)

Notice that although it has the same shape as that found for Na or Mg of Figure 8.16, the value of R_{min} lies to considerably shorter R than does R_{min} for these elements. In fact, $R_{min} < D_{cp}$, a result that implies (just as for the discussion concerning silicon and phosphorus) that some atoms will move in to closer distances but others will move farther away from the close-packed arrangement to give an open structure. The effect here though will be much more pronounced, since the disparity between R_{min} and D_{cp} is larger. The relationship of the following equation has been suggested

$$R_{min}/D_{cp} = [(\Gamma + 1)/13]^{1/3} \qquad (8.8)$$

(Ref. 155) to describe the dependence of coordination number (Γ) on these two parameters, and using the values for boron a figure of $\Gamma \approx 0$ results. The exact value is not of great importance, but note that it is small and very different from that (12) found for its congener, Al. Similar results are found for the first row elements in general. The pair potential, therefore, is such that a few very close contacts are preferred. Strong π bonding for the first row elements is a natural consequence of this.

What is the origin of this tremendous and chemically dramatic shift in the nature of Φ? The answer is the lack of a filled $1p$ core to provide a repulsion (Fig. 6.4) of the $2p$ valence electrons. This leads to a larger attractive potential than might have been expected by extrapolation from the heavier elements (Ref. 214). A related result comes from study of the functional form of the repulsive potential in Eq. (8.2). It may be shown (Ref. 1) that if the repulsive potential is somewhat softer than H_{ij}^2, then smaller coordination numbers result. A similar effect is found for the three transition metal series. Here the lack of a $2d$ orbital leads to quite contracted $3d$ orbitals and the interesting properties concerning metal–insulator transitions (of Section 5.5) and high- and low-spin compounds (of Section 1.7) containing such elements. Such electronic questions are usually less prevalent for the $4d$ and $5d$ series.

8.6 The coloring problem

One question of interest concerning the structures of binary compounds that adopt a derivative structure of some type is how the atoms are ordered over the sites of the parent. This may be called the coloring problem (Ref. 70), since it asks initially how many ways there are to color the sites of a parent structure in two, three, or in general the number of colors given by the number of different atoms involved. Some of these coloring patterns may be ordered ones, leading to a crystalline material; some may be disordered. Three examples of the coloring problem are shown in Fig. 8.19. An important question is the identification of the electronic reasons behind the adoption of one structure over its alternative. A calculated $\Delta E(x)$ plot for the two isomers of substituted cyclobutadiene (**8.6**) using simple Hückel theory is shown in **8.7**. It is in excellent agreement with experiment. The square cyclobutadiene molecule is best stabilized by attaching electron-donating and -withdrawing groups alternately around the ring, as found in the so-called push–pull butadienes (Ref. 175) and molecules such as $B_2X_2R_2$ (X = N, P). Stabilization of this pattern is also found in the solid-state structure (Ref. 9) of CaB_2C_2. Figure 8.20 shows the energy density of

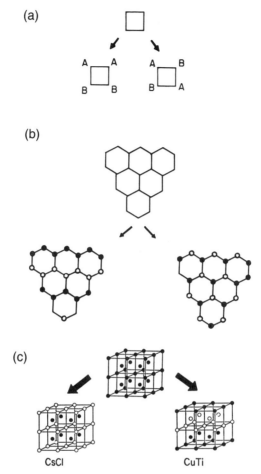

Figure 8.19 Some coloring problems. (a) Cyclobutadiene to $B_2N_2H_4$, (b) graphite to BN, (c) the bcc structure to CsCl and CuTi.

states for two possible coloring patterns. It is clear to see that the alternating pattern in this 48^2 structure is energetically favored for this electron count.

A $\Delta E(x)$ curve from a band structure calculation (Ref. 45, 69) on the coloring patterns of the bcc structure is shown in Fig. 8.21. There are about 40 binary transition alloys with either the CsCl or CuTi structure, as shown in Table 8.3. Notice the sharp dividing line between the two types at about 7.25 electrons per atom, which

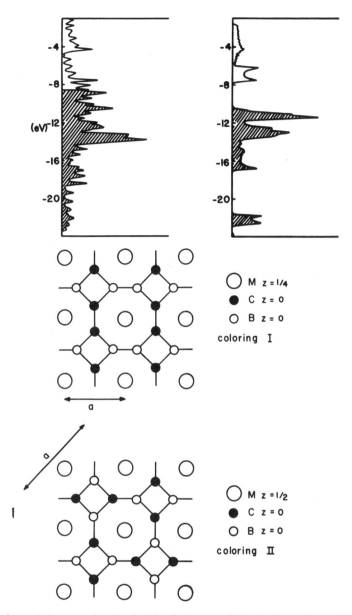

Figure 8.20 Computed energy density of states for two possible coloring patterns of the nonmetal net (48^2) in CaB_2C_2.

is reproduced by the calculated energy difference curve. (Of the four examples of Table 8.3 with electron counts of 8.5 and above, two are magnetic phases with electrons in higher-energy orbitals, and two are not the lowest-energy phase.) Figure 8.22 shows a similar plot for two colorings of the hcp structure (**8.8**). Here all of the known examples are of one type, and, although in agreement with the calculated $\Delta E(x)$ curve, there is no test of the location of the crossing point.

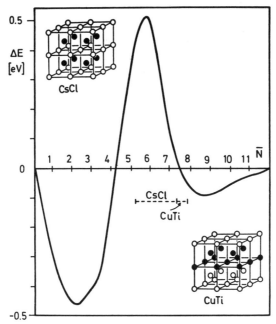

Figure 8.21 Computed energy difference between the CsCl and CuTi structures found for transition metal alloys as a function of the number of $(d + s)$ electrons. The dashed lines show where examples are found experimentally.

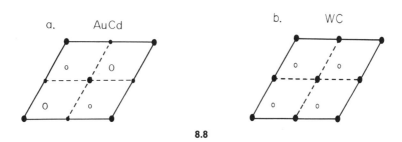

8.8

The most obvious feature of these energy difference curves is the set of four nodes found for the region between zero and twelve $(s + d)$ electrons, which identifies the electronic problem as a fourth moment one. It is also interesting to note that the structures that are calculated to be most stable at the half-filled point are the ones containing A–B contacts, whereas the structure stable in the other two regions is the one where there are both A–A and B–B contacts. A little counting allows confirmation of the fourth moment nature of the structural problem controlled by the nature of the nearest-neighbor contacts. Table 8.4 shows the moments of two model structures, the tetranuclear four-atom structures ($ABAB$ and $AABB$) and the infinite chains ($\cdots ABAB \cdots$ and $\cdots AABB \cdots$). If values of the hopping integrals are kept equal for all contacts, irrespective of their nature, then the only differences between the two arrangements arise as a result of the walks in place. From Table 8.4 it is

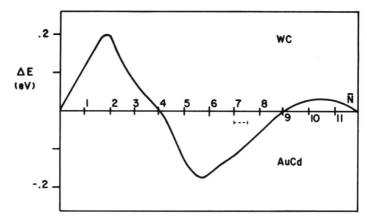

Figure 8.22 Computed energy difference between the WC and AuCd structures found for transition metal alloys as a function of the number of $(d + s)$ electrons. The dashed lines show where examples are found experimentally.

clear that the first disparate moment is the fourth, and thus four nodes are expected in $\Delta E(x)$ as observed. It is the structure with the smaller fourth moment that should be observed a half-filling, namely the "symmetrical" structure $\cdots ABAB\cdots$.

Another way to look at the structural preferences, which simply rephrases the ideas of the previous paragraph, is in terms of pair potentials between the atoms concerned

Table 8.3   Examples of transition metal alloys in the CsCl, CuTi structure and AuCd types

Av. no. of $s + d$ electrons	CsCl type	γ-CuTi type	AuCd type
5.5	TcTi, RuSc, ReTi, HfTc		
6.0	RhY, TcV, TaTe, RuZr, RuTi, RhSc, OsZr, OsTi, MnV, IrSc, Hf, Ru, CoSc		
6.5	RuV, PtSc, PdSc, OsV, NiSc, HfRh, CoZr, CoTi, CoHf		
7.0	AuLa, CuY, CuSc, AuY, AuSc, AgY, AgSc, AgLa		PtTi, PdTi
7.5		AgZr, AuHf, Au,Ti, CuTi, PdTa, AgTi	IrMo, IrW, NbPt, PtV, MoRh
8.0			
8.5	CoFe, FeRh, β-MnPd		
10.5	β-CuPd		

Table 8.4 The moments of some coloring patterns

$$\cdots ABAB \cdots$$

(a) All β equal

$\mu_1 = 2(\alpha_A + \alpha_B)$
$\mu_2 = 2(\alpha_A^2 + \alpha_B^2 + 4\beta^2)$
$\mu_3 = 2[\alpha_A^3 + \alpha_B^3 + 6\beta^2(\alpha_A + \alpha_B)]$
$\mu_4 = 2(a_A^4 + \alpha_B^4 + 12\beta^2 + 8\beta^2(\alpha_A^2 + \alpha_B^2 + 8\alpha_A\alpha_B\beta^2)$

(b) Unequal β

$\mu_= = (2\alpha_A + \alpha_B)$
$\mu_2 = 2(\alpha_A^2 + \alpha_B^2 + 4\beta_{AB}^2)$
$\mu_3 = 2[\alpha_A^3 + \alpha_B^3 + 6\beta_{AB}^2(\alpha_A + \mathcal{A}_B)]$

$$\cdots AABB \cdots$$

(c) All β equal

$\mu_1 = 2(\alpha_A + \alpha_B)$
$\mu_2 = 2(\alpha_A^2 + \alpha_B^2 + 4\beta^2)$
$\mu_3 = 2[\alpha_A^3 + \alpha_B^3 + 6\beta^2(\alpha_A + \alpha_B)]$
$\mu_4 = 2[\alpha_A^4 + \alpha_B^4 + 12\beta^2 + 10\beta^2(\alpha_A^2 + \alpha_B^2) + 4\alpha_A\alpha_B\beta^2]$

(d) Unequal β

$\mu_1 = 2(\alpha_A + \alpha_B)$
$\mu_2 = 2(\alpha_A^2 + \alpha_B^2 + 2\beta_{AB}^2 + \beta_{AA}^2 + \beta_{BB}^2)$
$\mu_3 = 2[\alpha_A^3 + \alpha_B^3 + 3\beta_{AB}^2(\alpha_A + \alpha_B) + 3(\alpha_A\beta_{AA}^2 + \alpha_B\beta_{BB}^2)]$

(e) Unequal β^a

$\Delta\mu_2(ABAB - AABB) \propto -(\alpha_A - \alpha_B)^2$
$\Delta\mu_3(ABAB - AABB) \propto -9(\alpha_A - \alpha_B)(\alpha_A^2 - \alpha_B^2)$

---

^{a} Assuming $\beta_{ij} \propto (\alpha_i + \alpha_j)$.

(Ref. 231). If the pair potential between two nearest-neighbor atoms i and j is written as Φ_{ij}, then the CsCl structure (or $\cdots ABAB \cdots$ structure) will be preferred over the CuTi structure (or $\cdots AABB \cdots$ structure) if $|2\Phi_{AB}| > |\Phi_{AA}| + |\Phi_{BB}|$ and the structural stability will be reversed if $|2\Phi_{AB}| < |\Phi_{AA}| + |\Phi_{BB}|$. Of course, one way of evaluating the pair potentials is through the energetics of the various structures, and so there is no new numerical information, but just some new concepts. In an analogous way, the electron count dependence of $2\Phi_{AB} - \Phi_{AA} - \Phi_{BB}$ will also be a fourth moment problem.

Such ideas lead to insights into other structural results without the use of a calculation. The coloring patterns for two possible inorganic analogs of polyacene (Ref. 45) are shown in **8.9** and **8.10** (open circles are Al atoms). Chains of this type are found in the structure of $Ba_3Al_2Sn_2$, the actual structure being the one of **8.10**.

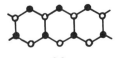

8.9

8.10

This contains homoatomic contacts. The structure of the graphite derivative structure of boron nitride of **3.7**, however, contains solely heteroatomic contacts. For BN the π band is half-full, and so from the preceding arguments the $\cdots ABAB \cdots$ structure should indeed be more stable. For the $Ba_3Al_2Sn_2$ example electron counting leads to a description of the sheet as $(Al^-Sn^-)^-$ and thus a value for the filling of the π manifold of 0.75. At this configuration an $\cdots AABB \cdots$ structure with its homonuclear contacts should be more stable, as observed.

From this discussion a general result emerges concerning the structures of binary and more complex compounds. At the half-filled point, the structure where nearest-neighbor heteroatomic contacts are maximized is predicted to be the most stable alternative of all of the coloring patterns. This leads to a certain symmetry in the most stable structure, and apparent from Figure 8.19. Thus the CsCl arrangement is cubic, but the CuTi structure tetragonal. For this reason, too, octet compounds tend to adopt those structures where the electronegativities of the atoms alternate. Zincblende (ZnS) contains alternating zinc and sulfur atoms, just as in the more complex diamond derivative structure such as chalcopyrite, stannite, and nowackiite. In all of these solids with four electrons per atom, all of the bonding orbitals are occupied and all of the antibonding ones unoccupied. The fractional orbital occupation is thus 0.5. Away from the half-filled point the $\cdots AABBAA \cdots$ pattern is expected. The structure of GaSe (Fig. 5.3) contains sheets ordered $\cdots SSGaGaSS \cdots$ with strong Ga–Ga linkages, as described in Section 7.1, and low-coordinate, electronegative (Se in this case) atoms. The structures of ZrCl and ZrBr are interesting ones (Fig. 8.23) in that they contain adjacent Zr sheets, $\cdots ClClZrZrClCl \cdots$. This is also understood in terms of the metal–metal bonding between adjacent zirconium sheets made possible by the d^3 electron configuration for Zr(I). Similar like-atom

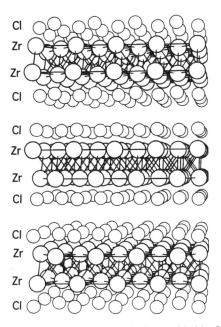

Figure 8.23   The structure of ZrCl. (Drawing kindly provided by Prof. J. D. Corbett.)

pairs are found in the (14 electron) pyrite structure in contrast to the alternation in the 16 electron $CdHal_2$ structure.

These ideas concerning the stability of nearest-neighbor contacts as a function of electron count may be extended to second-nearest neighbors. As **8.11** shows, a walk

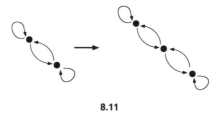

8.11

of length n involving first-nearest neighbors (a walk of length 4 is shown) is converted into a walk of length $n + 2$ for second-nearest neighbors. This means that the energetics determining the identity of next-nearest neighbors is determined by the sixth moment. Figure 8.1 compares energy difference curves for fourth and sixth moment problems. Because of the number of nodes in the plots, notice that like second-nearest neighbors are energetically favored at the half-filled point. This, of course, is found for all coloring patterns in which the atomic identity alternates, such as the structures of BN, ZnS, and indeed in virtually all mineral structures where there are anion–cation linkages. In general for this rather special but frequently found electronic configuration, that of the half-filled band, the alternating arrangement of atoms of different electronegativity is the most stable one for all orders. However, recall that the amplitude of the moments curves drop off as the order increases. This implies, as does chemical intuition, that it is the nearest-neighbor interactions that are the most important. Notice in a general way that these *orbital* arguments for the half-filled band give exactly the same result as does the electrostatic model. Isomorphisms between "ionic" and "covalent" (orbital) models have been frequent ones in this text. These results do not hold away from the half-filled band. As a consequence much more complex structures are found at these electron counts.

8.7   Structural stability and band gap

The idea that a large HOMO–LUMO gap leads to structural (Ref. 52) and kinetic stability is one that is frequently used in the molecular realm. In Section 7.7 the small d/s separation at the right-hand end of the transition metal series was identified as the reason behind the distortions of the octahedral geometry of d^9 and d^{10} systems via the second-order Jahn–Teller approach. In Section 7.9 it was shown how the relative stability of the cis- and trans-substituted cyclobutadienes varied with electron count. At the half-filled point the trans isomer is more stable, and this is the arrangement that has the larger HOMO–LUMO gap. A similar viewpoint should apply to solids.

Table 8.5 shows the results of some calculations (Ref. 52) on solids that differ in the ordering patterns of the anions, the solid-state analog of the cis/trans isomerism in molecular cyclobutadiene. In the language of the previous section these represent

Table 8.5 Computed band gaps and relative stability
for rocksalt and sphalerite solids with different anion
ordering patterns

System		E (eV)	E_g (eV)
Sphalerite	(i)	0	1.8
	(ii)	0.3	1.5
Rocksalt	(iii)	0	3.05
	(iv)	0.5	2.40

different colorings of the anion or cation arrangement. The two classic eight-electron
AB structures, rocksalt and sphalerite, were chosen for this purpose. In structure (i)
a derivative sphalerite structure with stoichiometry A_2PS, the arrangement of the
anions is just as in the antistructure of chalcopyrite ($CuFeS_2$). Each metal is
coordinated by two anions of each type $(2 + 2)$. In structure (ii) there are two different
metal atoms, coordinated, respectively, by 1P and 3S and by 3P and 1S $(3 + 1, 1 + 3)$.
The rocksalt derivative $MgZrX_2$ structure (iii) is one where each anion is coordinated
by three cations of each type $(3 + 3)$ and structure (iv), one where there are two
different anions with $4 + 2$ and $2 + 4$ coordination. Notice the correlation between
stability and band gap for both systems.

The simplest model for this problem is shown in **8.12**. It ignores the presence of s
and p orbitals on the atoms concerned, and represents the density of states by a
rectangular distribution. The two bands are each of width W ($|a - b|$) and sym-

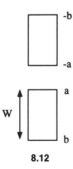

8.12

metrically situated around an arbitrary energy zero. The lower filled band is largely
anionic in character, while the empty band is largely cation located. The total number
of states, M, is defined by $M = \int_{-\infty}^{\infty} \rho(E) \, dE$, and the total electronic energy is
$E(\text{total}) = 2 \int_{-\infty}^{E_F} \rho(E)E \, dE$. In terms of a and b, the second moment of the density
of states, $\mu_2 = \frac{2}{3}M(a^2 + ab + b^2)$, $E(\text{total}) = \frac{1}{2}M(a + b)$ when half of these orbitals
are occupied, and $E_g = |2a|$. A little algebra shows that the total energy and
bandwidth are related by an expression including the second moment as

$$4E_T^2 = 2M\mu_2 - M^2W^2/3 \tag{8.9}$$

The band gap is given by

$$E_g = (2\mu_2 M - W^2/3)^{1/2} - W \qquad (8.10)$$

Setting the second moment of the electronic density of states equal for each pair of systems to be compared, both the stabilization energy and the band-gap increase as W increases, a result in accord with the idea that μ_2 represents the coordination strength of the atoms concerned. Thus the larger E_T, the larger the band gap E_g. Numerical calculations that support this view are shown in Table 5. Figure 8.20 showed the density of states for two possible coloring patterns (Ref. 49) for CaB_2C_2, which show the same effect. Here the less stable system has no gap.

This result is related to ideas from the molecular area in terms of the concept of hardness. Specifically, hardness can be directly associated with the HOMO–LUMO gap in molecules, a result that comes from density-functional theory (Refs. 278, 273). A well-established principle, that of maximum hardness (Ref. 279), which argues that molecules rearrange themselves so as to be as hard as possible, is just the molecular analog of that for the solid derived previously. The larger E_T, the larger the HOMO–LUMO gap (E_g).

References

1. Abell, G. C., *Phys. Rev. B* **31**, 6184 (1985).
2. Adams, D. M., *Inorganic Solids*, Wiley (1974).
3. Adler, D., in *Solid State Chemistry*, Hannay, N. B., editor, Vol. 2, Plenum (1967).
4. Adler, D., *Solid State Physics* **21**, 1 (1968).
5. Albright, T. A., Burdett, J. K., Whangbo, M.-H., *Orbital Interactions in Chemistry*, Wiley (1985).
6. Alcacer, J., Novais, H., in *Extended Linear Chain Compounds*, Miller, J. S., editor, Vol. 3, Plenum (1982).
7. Allan, G., *Annales Phys. (Paris)* **5**, 169 (1970).
8. Allen, L. C., *J. Amer. Chem. Soc.* **114**, 1510 (1992).
9. Alonso, J. A., March, N. H., *Electrons in Metals and Alloys*, Academic (1989).
10. Altmann, S. L., *Band Theory of Metals*, Pergamon (1970).
11. Altmann, S. L., *Induced Representations in Crystals and Molecules*, Academic (1977).
12. Altmann, S. L., *Band Theory of Solids; An Introduction from the Point of View of Symmetry*, Oxford (1991).
13. Andersen, O. K., in *The Electronic Structure of Complex Systems*, Phariseau, P., Temmerman, W. M., editors, Plenum (1984).
14. Andersen, O. K., in *Highlights of Condensed Matter Physics*, Bassani, F., Fumi, F., Tosi, M., editors, North-Holland (1985).
15. Anderson, P. W., *Phys. Rev.* **109**, 1492 (1958).
16. Ashcroft, N. W., Mermin, N. D., *Solid State Physics*, Saunders (1976).
17. Bacon, G. E., *The Architecture of Solids*, Wykeham (1981).
18. Baldereschi, A., *Phys. Rev. B* **7**, 5212 (1973).
19. Barbee, T. W., Cohen, M. L., *Phys. Rev. B* **43**, 5269 (1991).
20. Barisic, S., Bjelis, A., in Kamimura, H., editor, *Theoretical Aspects of Band Structures and Electronic Properties of Pseudo-One-Dimensional Solids*, Reidel (1985).
21. Barratt, C., Massalski, T. B., *Structure of Metals*, 3rd edition, Pergamon (1980).
22. Bartell, L. S., *J. Chem. Educ.* **45**, 754 (1968).
23. Bednorz, J. G., Müller, K. A., *Z. Phys. B* **64**, 189 (1986).
24. Blase, W., Cordier, G., Peters, K., Somer, M., von Schnering, H. G., *Z. Anorg. Allgem. Chem.* **474**, 221 (1981).
25. Bloch, A. N., Schatteman, G. C., in *Structure and Bonding in Crystals*, Vol. 1, O'Keeffe, M., Navrotsky, A., editors, Academic (1981).
26. Boom, R., de Boer, F. R., Miedema, A. R., *J. Less Common Metals* **45**, 237 (1976).
27. Boom, R., de Boer, F. R., Miedema, A. R., *J. Less Common Metals* **46**, 271 (1976).
28. Born, M., Landé, A., *Sitzungsber. Preuss. Akad. Wiss. (Berlin)* **45**, 1048 (1918).
29. Born, M., Mayer, J. E., *Z. Phys.* **75**, 1 (1932).
30. Bradley, C. J., Cracknall, A. P., *The Mathematical Theory of Symmetry in Solids*, Oxford (1972).
31. Bragg, W. H., *Z. für Krist.* **74**, 237 (1930).
32. Brennan, T. L., Burdett, J. K., *Inorg. Chem.* **32**, 750 (1993).

33. Brown, I. D., in *Structure and Bonding in Crystals*, Vol. 2, O'Keeffe, M., Navrotsky, A., editors, Academic (1981).
34. Brown, I. D., *J. Solid State Chem.* **90**, 155 (1991).
35. Brown, I. D., *Acta Cryst. B* **48**, 553 (1992).
36. Brown, I. D., Altermatt, D., *Acta Cryst. B* **41**, 244 (1992).
37. Brown, I. D., Shannon, R. D., *Acta Cryst. A* **29**, 266 (1973).
38. Bube, R. H., *Electrons in Solids*, Academic (1992).
39. Bullett, D. W., *Solid State Physics* **35**, 129 (1980).
40. Burdett, J. K., *Nature* **279**, 121 (1979).
41. Burdett, J. K., *Molecular Shapes: Theoretical Models of Inorganic Stereochemistry*, Wiley (1980).
42. Burdett, J. K., *Inorg. Chem.* **20**, 1959 (1981).
43. Burdett, J. K., *Prog. Sol. State Chem.* **15**, 173 (1984).
44. Burdett, J. K., *Inorg. Chem.* **24**, 2244 (1985).
45. Burdett, J. K., *Structure and Bonding* **65**, 30 (1987).
46. Burdett, J. K., *Chem. Rev.* **88**, 3 (1988).
47. Burdett, J. K., in *The Stability of Minerals*, Price, G. D., Ross, N. L., editors, Chapman and Hall (1992).
48. Burdett, J. K., *Adv. Chem. Phys.* **83**, 207 (1993).
49. Burdett, J. K., Canadell, E., Hughbanks, T., *J. Amer. Chem. Soc.* **108**, 3971 (1986).
50. Burdett, J. K., Canadell, E., Miller, G. J., *J. Amer. Chem. Soc.* **108**, 6561 (1986).
51. Burdett, J. K., Coddens, B., *Inorg. Chem.* **27**, 418 (1988).
52. Burdett, J. K., Coddens, B., Kulkarni, G., *Inorg. Chem.* **271**, 3259 (1988).
53. Burdett, J. K., Eisenstein, O., *Inorg. Chem.* **31**, 1758 (1992).
54. Burdett, J. K., Fässler, T., *Inorg. Chem.* **30**, 2859 (1991).
55. Burdett, J. K., Haaland, P., McLarnan, T. J., *J. Chem. Phys.* **75**, 5774 (1981).
56. Burdett, J. K., Hawthorne, F. C., *Amer. Min.* **78**, 884 (1993).
57. Burdett, J. K., Hughbanks, T., *J. Amer. Chem. Soc.* **106**, 3101 (1984).
58. Burdett, J. K., Hughbanks, T., *Inorg. Chem.* **24**, 1741 (1985).
59. Burdett, J. K., Kulkarni, G. V., *Chem. Phys. Lett.* **160**, 350 (1989).
60. Burdett, J. K., Kulkarni, G. V., *Phys. Rev. B* **40**, 8908 (1989).
61. Burdett, J. K., Lee, S., *J. Amer. Chem. Soc.* **105**, 1079 (1983).
62. Burdett, J. K., Lee, S., *J. Amer. Chem. Soc.* **107**, 3050 (1985).
63. Burdett, J. K., Lee, S., *J. Amer. Chem. Soc.* **107**, 3063 (1985).
64. Burdett, J. K., Lin, J.-H., *Acta Cryst. B* **28**, 408 (1982).
65. Burdett, J. K., McLarnan, T. J., *J. Chem. Phys.* **75**, 5764 (1981).
66. Burdett, J. K., McLarnan, T. J., *Inorg. Chem.* **21**, 1119 (1982).
67. Burdett, J. K., McLarnan, T. J., *J. Amer. Chem. Soc.* **104**, 5229 (1982).
68. Burdett, J. K., McLarnan, T. J., *American Mineralogist* **69**, 601 (1984).
69. Burdett, J. K., McLarnan, T. J., *J. Solid State Chem.* **53**, 382 (1984).
70. Burdett, J. K., McLarnan, T. J., Lee, S., *J. Amer. Chem. Soc.* **107**, 3083 (1985).
71. Burdett, J. K., Price, S. L., *J. Phys. Chem. Solids* **43**, 521 (1982).
72. Burdett, J. K., Price, S. L., Price, G. D., *Phys. Rev. B* **24**, 2903 (1981).
73. Burdett, J. K., Price, S. L., Price, G. D., *Solid State Commun.* **40**, 923 (1981).
74. Burdett, J. K., Price, S. L., Price, G. D., *J. Amer. Chem. Soc.* **104**, 92 (1982).
75. Bürgi, H.-B., *Angew. Chem. Int. Ed.* **14**, 462 (1974).
76. Burns, G., *Solid State Physics*, Academic (1985).
77. Burns, G., Glazer, A. M., *Space Groups for Solid State Scientists*, 2nd edition, Academic (1990).
78. Callaway, J., *Solid State Physics* **7**, 100 (1958).
79. Callaway, J., *Energy Band Theory*, Academic (1964).

80. Callaway, J., *Quantum Theory of the Solid State*, Academic (1974).
81. Canadell, E., Ravy, S., Pouget, J. P., Brossard, L., *Solid State Comm.* **75**, 633 (1990).
82. Canadell, E., Whangbo, M.-H., *Chem. Rev.* **91**, 965 (1991).
83. Catlow, C. R. A., *Proc. Roy. Soc. A* **353**, 533 (1977).
84. Catlow, C. R. A., *Ann. Rev. Mat. Sci.* **16**, 517 (1986).
85. Catlow, C. R. A., Mackrodt, W. C., editors, *Computer Simulation of Solids*, Springer-Verlag (1982).
86. Catlow, C. R. A., Stoneham, A. M., *J. Phys. C* **16**, 4321 (1983).
87. Cava, R. J., *Science* **247**, 656 (1990).
88. Cava, R. J., Hewat, A. W., Hewat, E. A., Batlogg, B., Marezio, M., Rabe, K. M., Krajewski, J. J., Peck, W. F., Rupp, L. W., *Physica C* **165**, 419 (1990).
89. Chadi, D. J., Cohen, M. L., *Phys. Rev. B* **8**, 5747 (1973).
90. Chadi, D. J., Cohen, M. L., *Phys. Stat. Sol.* **68b**, 405 (1975).
91. Chelikowsky, J. R., Cohen, M. L., *Phys. Rev. B* **14**, 556 (1976).
92. Chen, M. M. L., Hoffmann, R., *J. Amer. Chem. Soc.* **98**, 1647 (1976).
93. Chien, J. C. W., *Polyacetylene: Chemistry, Physics and Materials Science*, Academic (1984).
94. Clark, G. M., *The Structures of Non-Molecular Solids*, Applied Science (1972).
95. Cohen, M. L., *Physics Today* **32**, 40 (1979).
96. Cohen, M. L., in *Structure and Bonding in Crystals*, Vol. 1, O'Keeffe, M., Navrotsky, A., editors, Academic (1981).
97. Cohen, M. L., Heine, V., *Solid State Phys.* **24**, 37 (1970).
98. Coles, B. R., Caplin, A. D., *Electronic Structures of Solids*, Arnold (1976).
99. Cordier, G., Hohn, P., Kniep, R., Rabenau, A., *Z. Anorg. Allgem. Chem.* **591**, 58 (1990).
100. Cotton, F. A., *Chemical Applications of Group Theory*, 3rd edition, Wiley (1990).
101. Coulson, C. A., *Trans. Faraday Soc.* **38**, 433 (1942).
102. Coulson, C. A., Fischer, I., *Phil. Mag.* **40**, 386 (1949).
103. Cox, P. A., *The Electronic Structure and Chemistry of Solids*, Oxford (1987).
104. Cox, P. A., in *Solid State Chemistry: Compounds*, Cheetham, A. K., Day, P., editors, Oxford (1992).
105. Cyrot, M., Cyrot-Lackman, F., *J. Phys F* **6**, 2257 (1976).
106. Cyrot-Lackman, F., *Adv. Phys.* **16**, 393 (1967).
107. Cyrot-Lackman, F., *J. Phys. Chem. Solids* **29**, 1235 (1968).
108. Cyrot-Lackman, F., *Surf. Sci.* **15**, 535 (1969).
109. Day, P., *Int. Rev. Phys. Chem.* **1**, 149 (1981).
110. Day, P., in *Solid State Chemistry: Compounds*, Cheetham, A. K., Day, P., editors, Oxford (1992).
111. Dewar, M. J. S., *Theory of Molecular Orbitals*, McGraw-Hill (1969).
112. DiSalvo, F. J., in *Electron–Phonon Interactions*, Riste, T., editor, Plenum (1977).
113. Dixon, D. E., Parry, G. S., *J. Phys. C* **2**, 1732 (1962).
114. Donohue, J., *The Structures of the Elements*, Krieger (1982).
115. Doublet, M.-L., Canadell, E., Pouget, J. P., Yagubskii, E. B., Ren, J., Whangbo, M.-H., *Solid State Comm.* **88**, 699 (1993).
116. Ducastelle, F., *J. Physique* **31**, 1055 (1970).
117. Ducastelle, F., Cyrot-Lackman, F., *J. Phys. Chem. Solids* **31**, 1295 (1970).
118. Ducastelle, F., Cyrot-Lackman, F., *J. Phys. Chem. Solids* **32**, 285 (1971).
119. Duke, C. B., in *Extended Linear Chain Compounds*, Miller, J. S., editor, Vol. 2, Plenum (1982).
120. Dunitz, J. D., Orgel, L. E., *J. Phys. Chem. Solids* **20**, 318 (1957).
121. Dunitz, J. D., Orgel, L. E., *Adv. Inorg. Chem. Radiochem* **2**, 1 (1960).
122. Dunn, T. M., McClure, D. S., Pearson, R. G., *Some Aspects of Crystal Field Theory*, Harper and Row (1965).

123. Duthie, J. C., Pettifor, D. G., *Phys. Rev. Lett.* **38**, 564 (1977).
124. Eisenmann, B., Janzon, K. H., Schäfer, H., Weiss, A., *Z. Naturf.* **24b**, 457 (1969).
125. Ellison, A. J. G., Navrotsky, A., *J. Solid State Chem.* **94**, 24 (1991).
126. Evain, M., Whangbo, M.-H., Beno, M. A., Geiser, U., Williams, J. M., *Inorg. Chem.* **109**, 7917 (1987).
127. Evans, R. C., *An Introduction to Crystal Chemistry*, 2nd edition, Cambridge (1964).
128. Fischer, J. E., Heiney, P. A., Smith, A. B., *Accts. Chem. Res.* **25**, 112 (1992).
129. Friedel, J., *Trans. AIME* **230**, 616 (1964).
130. Friedel, J., in *The Physics of Metals*, Vol. 1, *Electrons*, Ziman, J. M., editor, Cambridge (1969).
131. Frost, A. A., Musulin, B., *J. Chem. Phys.* **21**, 572 (1953).
132. Gallup, R. F., Fong, C. Y., Kauzlarich, S. M., *Inorg. Chem.* **31**, 115 (1992).
133. Gaspard, J.-P., in *Structures et Instabilités*, Godrèche, C., editor, Editions de Physique, Les Ullis (1986).
134. Gaspard, J.-P., Céolin, R., *Solid State Comm.* **84**, 839 (1992).
135. Gaspard, J.-P., Cyrot-Lackmann, F., *J. Phys. C* **6**, 3077 (1973).
136. Gelatt, G. D., Williams, A. R., Moruzzi, V. L., *Phys. Rev. B* **27**, 2005 (1983).
137. Gerloch, M., *Inorg. Chem.* **20**, 638 (1981).
138. Gerloch, M., Slade, R. C., *Ligand Field Parameters*, Cambridge (1973).
139. Gerstein, B. C., *J. Chem. Educ* **50**, 316 (1973).
140. Gibbs, G. V., Lasaga, A., *Phys. Chem. Min.* **14**, 107 (1987).
141. Gibbs, G. V., Meagher, E. P., Newton, M., Swanson, D. K., in *Structure and Bonding in Crystals*, Vol. 1, O'Keeffe, M., Navrotsky, A., editors, Academic (1981).
142. Gimarc, B. M., *J. Amer. Chem. Soc.* **105**, 1979 (1983).
143. Goodenough, J. B., *Magnetism and the Chemical Bond*, Wiley (1963).
144. Goodenough, J. B., *Prog. Solid State Chem.* **5**, 143 (1971).
145. Goodenough, J. B., in *Inorganic Compounds With Unusual Properties*, Johnson, M. K., King, R. B., Kurtz, P. M., Kutal, C., Norton, M. L., Scott, R. A., editors, American Chemical Society (1990).
146. Goodenough, J. B., *Phase Transitions* **22**, 79 (1990).
147. Griffiths, J. S., *Theory of Transition Metal Ions*, Cambridge (1961).
148. Grimes, R. W., Catlow, C. R. A., Shluger, A. L., editors, *Quantum Mechanical Calculations in Solid-State Studies*, World Scientific (1992).
149. Gschneidner, K. A., *Solid State Phys.* **16**, 275 (1964).
150. Haas, C., in *Crystal Structure and Chemical Bonding in Inorganic Chemistry*, Rooymans, C. J. M., Rabenau, A., editors, North-Holland (1975).
151. Haas, C., *Solid State Commun.* **26**, 709 (1978).
152. Haas, C., *J. Solid State Chem.* **57**, 82 (1985).
153. Haddon, R. C., *Accts. Chem. Res.* **25**, 127 (1992).
154. Hafner, J., *From Hamiltonians to Phase Diagrams*, Springer-Verlag (1987).
155. Hafner, J., Heine, V., *J. Phys. F* **13**, 2479 (1983).
156. Hall, G. G., *Proc. Roy. Soc., A* **202**, 366 (1950).
157. Hall, G. G., Lennard-Jones, J. E., *Proc. Roy. Soc. A* **202**, 155 (1950).
158. Hall, G. G., Lennard-Jones, J. E., *Proc. Roy. Soc. A* **205**, 357 (1951).
159. Harrison, W. A., *Solid State Theory*, Dover (1979).
160. Harrison, W. A., *Electronic Structure and the Properties of Solids*, Freeman (1980).
161. Hatfield, W. E., editor, *Molecular Metals*, Plenum (1979).
162. Haug, A., *Theoretical Solid State Physics*, Pergamon (1972).
163. Haydock, R., Heine, V., Kelly, M., *J. Phys. C* **5**, 2845 (1972).
164. Heeger, A. J., Macdiarmid, A. G., *Mol. Cryst. Liq. Cryst.* **74**, 1 (1980).
165. Heilbronner, E., Bock, H., *The HMO Model and its Applications*, Wiley (1976).

166. Heine, V., *Group Theory in Quantum Mechanics*, Pergamon (1960).
167. Heine, V., *Solid State Physics* **24**, 1 (1970).
168. Heine, V., Haydock, R., Bullett, D., Kelly, M., *Solid State Physics* **35**, 1 (1980).
169. Heine, V., Robertson, I. J., Payne, M. C., in *Bonding and Structure of Solids*, Haydock, R., Inglesfield, J. E., Pendry, J. B., editors, The Royal Society of London (1991).
170. Heine, V., Samson, J. H., *J. Phys. F* **10**, 2609 (1980).
171. Heine, V., Weaire, D., *Solid State Physics* **24**, 250 (1970).
172. Hodges, L., Ehrenreich, H., Freemann, A. J., *J. Appl. Phys.* **37**, 1449 (1960).
173. Hoffman, A. J., *Studies in Graph Theory*, Vol. 12, Fulkerson, D. R., editor, Math. Assoc. America (1975).
174. Hoffmann, R., *J. Chem. Phys.* **39**, 1397 (1962).
175. Hoffmann, R., *Chem. Comm.*, 240 (1969).
176. Hoffmann, R., *Accts. Chem. Res.* **4**, 1 (1971).
177. Hoffmann, R., *Solids and Surfaces*, VCH (1988).
178. Hoistad, L. M., Lee, S., *J. Amer. Chem. Soc.* **113**, 8216 (1991).
179. Hoistad, L. M., Lee, S., Pasternak, J., *J. Amer. Chem. Soc.* **114**, 4790 (1992).
180. Hoppe, R., *Adv. Fluorine Chem.* **6**, 387 (1970).
181. Hor, P. H., Meng, R. L., Wang, Y. Q., Gao, L., Huanh, Z. J., Bechtold, J., Forster, K., Chu, C. W., *Phys. Rev. Lett.* **58**, 1891 (1987).
182. Hubbard, J., *Proc. Roy. Soc. A* **276**, 238 (1963); **277**, 237 (1964); **281**, 401 (1964); **285**, 542 (1965); **296**, 82 (1966); **296**, 100 (1966).
183. Hückel, E., *Z. Phys.* **70**, 204 (1931).
184. Huggins, R. A., in *Diffusion in Solids*, Nowick, A. S., Burton, J. J., editors, Academic (1975).
185. Hughbanks, T., Hoffmann, R., *J. Amer. Chem. Soc.* **105**, 3528 (1983).
186. Hulliger, F., *Structure and Bonding* **4**, 83 (1968).
187. Hulliger, F., in *Structure and Bonding in Crystals*, Vol. 2, O'Keeffe, M., Navrotsky, A., editors, Academic (1981).
188. Hulliger, F., Mooser, E., *Prog. Solid State Chem.* **24**, 330 (1965).
189. Hume-Rothery, W., *The Metallic State*, Oxford (1931).
190. Hyde, B. G., Andersson, S., *Inorganic Crystal Structures*, Wiley (1989).
191. Ibach, H., Lüth, H., *Solid State Physics*, Springer-Verlag (1990).
192. Inkson, J., *Many-Body Theory of Solids*, Plenum (1984).
193. Jafri, J. A., Newton, M. D., *J. Amer. Chem. Soc.* **100**, 5012 (1978).
194. Jahn, H. A., Teller, E., *Proc. Roy. Soc. A* **161**, 220 (1937).
195. James, A. C. W. P., Zahurak, S. M., Murphy, D. W., *Nature* **338**, 240 (1989).
196. Jamieson, J. C., *Science* **139**, 1291 (1963).
197. Jerome, D., Schultz, H. J., in *Extended Linear Chain Compounds*, Miller, J. S., editor, Vol. 2, Plenum (1982).
198. Johnson, D. A., *Some Thermodynamic Aspects of Inorganic Chemistry*, Cambridge (1968).
199. Jones, H., *The Theory of Brillouin Zones and Electronic States in Crystals*, 2nd edition, North-Holland (1975).
200. Jorgensen, J. D., Beno, M. A., Hinks, D. G., Soderholm, J., Volin, K. J., Hitterman, R. L., Grace, J. D., Schuller, I. K., Segre, C. U., Zhang, K., Kleefisch, M. S., *Phys. Rev. B* **36**, 3608 (1987).
201. Jung, D., Whangbo, M.-H., Herron, N., Torardi, C. C., *Physica C* **160**, 381 (1989).
202. Kamimura, H., editor, *Theoretical Aspects of Band Structures and Electronic Properties of Pseudo-One-Dimensional Solids*, Reidel (1985).
203. Kamimura, H., Oshiyama, A., in *Theoretical Aspects of Band Structures and Electronic Properties of Pseudo-One-Dimensional Solids*, Kamimura, H., editor, Reidel (1985).
204. Kertész, M., *Adv. Quantum Chem.* **15**, 161 (1982).
205. Kertész, M., *Int. Rev. Phys. Chem.* **4**, 125 (1985).

206. Kertész, M., Hoffmann, R., *Solid State Comm.* **47**, 97 (1983).
207. Kertész, M., Hoffmann, R., *J. Amer. Chem. Soc.* **106**, 3453 (1984).
208. Ketelaar, J. A. A., *Chemical Constitution; An Introduction to the Theory of the Chemical Bond*, Second Edition, Elsevier (1958).
209. Kettle, S. F. A., *Symmetry and Structure*, Wiley (1985).
210. Kittel, C., *Introduction to Solid State Physics*, 6th edition, Wiley (1986).
211. Klemm, W., Busmann, E. Z., *Anorg. Allgem. Chem.* **319**, 297 (1963).
212. Kobayashi, H., Kobayashi, A., in *Extended Linear Chain Compounds*, Miller, J. S., editor, Vol. 2, Plenum (1982).
213. Koster, G. F., *Solid State Physics* **5**, 173 (1957).
214. Kutzelnigg, W., *Angew. Chem. Int. Ed.* **23**, 272 (1984).
215. Lacorre, P., Torrance, J. B., Pannetier, J., Nazzal, A., Wang, P. W., Huang, T. C., *J. Solid State Chem.* **91**, 225 (1991).
216. Ladd, M. F. C., *Structure and Bonding in the Solid State*, Wiley (1979).
217. Lannoo, M., Friedel, P., *Atomic and Electronic Structure of Surfaces*, Springer-Verlag (1991).
218. Laves, F., *Theory of Alloy Phases*, American Society for Metals (1956).
219. Lax, M., *Symmetry Principles in Solid State and Molecular Physics*, Wiley (1974).
220. Lee, S., *Accts. Chem. Res.* **24**, 249 (1991).
221. Lee, S., *J. Amer. Chem. Soc.* **113**, 101 (1991).
222. Lee, S., *J. Amer. Chem. Soc.* **113**, 8611 (1991).
223. Lee, S., Rousseau, R., Wells, C., *Phys. Rev. B* **46**, 12121 (1992).
224. Leman, G., Friedel, J., *J. Appl. Phys.* **33**, 281 (1962).
225. Lennard-Jones, J. E., *Proc. Roy. Soc. A* **198**, 1, 14 (1949).
226. Lennard-Jones, J. E., Pople, J. A., *Proc. Roy. Soc. A* **202**, 166 (1950).
227. Lennard-Jones, J. E., Pople, J. A., *Proc. Roy. Soc. A* **210**, 190 (1951).
228. Levin, A. A., *Solid State Quantum Chemistry*, McGraw-Hill (1977).
229. Lomer, W. M., *Proc. Phys. Soc.* **80**, 489 (1962).
230. Longuet-Higgins, H. C., Salem, L., *Proc. Roy. Soc. A* **251**, 172 (1959).
231. Machin, E. S., in *Theory of Alloy Phase Formation*, Bennett, L. H., editor, AIME (1980).
232. Mao, H.-K., Hemley, R. J., *Science* **244**, 1462 (1989).
233. Marks, T. J., Kalina, D. W., in *Extended Linear Chain Compounds*, Miller, J. S., editor, Vol. 1, Plenum (1982).
234. Martynov, A. J., Batsanov, S. S., *Russ. J. Inorg. Chem.* **25**, 1737 (1980).
235. McCarley, R. E., Li, K.-H., Edwards, P. A., Brough, L. F., *J. Solid State Chem.* **57**, 17 (1985).
236. McWeeney, R., *Coulson's Valence*, Oxford (1979).
237. Meschel, S. V., Kleppa, O. J., *Met. Trans. A* **24**, 947 (1993).
238. Meyer, T. J., *Prog. Inorg. Chem.* **30**, 389 (1983).
239. Miedema, A. R., *Phillips Tech. Rep.* **33**, 149, 196 (1973).
240. Miedema, A. R., de Boer, F. R., Boom, R., *J. Less Common Metals* **41**, 283 (1975).
241. Miedema, A. R., de Boer, F. R., Boom, R., *J. Less Common Metals* **46**, 67 (1976).
242. Miedema, A. R., de Boer, F. R., De Châtel, P. F., *J. Phys. F* **3**, 1558 (1973).
243. Miller, J. S., editor, *Extended Linear Chain Compounds*, Vols. 1–3, Plenum (1982).
244. Miller, J. S., Epstein, A. J., *Prog. Inorg. Chem.* **20**, 1 (1976).
245. Mingos, D. M. P., *Nature (Phys. Sci.)* **236**, 99 (1972).
246. Moore, P. B., Araki, T., *Neues Jahrbuch für Mineralogie Abhandlungen* **132**, 91 (1978).
247. Mooser, E., Pearson, W. B., *Phys. Rev.* **101**, 1608 (1956).
248. Mooser, E., Pearson, W. B., *Acta Cryst.* **12**, 1015 (1959).
249. Moret, R., Pouget, J. P., in *Crystal Chemistry and Properties of Materials with Quasi-One-Dimensional Structures*, Rouxel, J., editor, Reidel (1986).

250. Morin, F. J., *Bell Syst. Tech. J.* **37**, 1047 (1958).
251. Moruzzi, V. L., Janak, J. F., Williams, A. R., *Calculated Electronic Properties of Metals*. Pergamon (1978).
252. Mott, N. F., *Metal–Insulator Transitions*, Taylor and Francis (1974).
253. Mott, N. F., Davis, E. A., *Electronic Processes in Non-Crystalline Materials*, 2nd edition, Oxford (1979).
254. Mott, N. F., Jones, H., *The Theory of the Properties of Metals and Alloys*, Oxford (1936).
255. Mueller, F. M., *Phys. Rev.* **53**, 1080 (1984).
256. Müller, A., *Inorganic Structural Chemistry*, Wiley (1993).
257. Müller, K. A., Bednorz, J. G., *Science* **237**, 1133 (1987).
258. Muller, O., Roy, R., *The Major Ternary Structural Families*, Springer-Verlag (1974).
259. Mulliken, R. S., *Chem. Rev.* **9**, 347 (1931).
260. Mulliken, R. S., *J. Chem. Phys.* **23**, 1833, 2343 (1955).
261. Nelson, D. L., George, T. F., editors, *Chemistry of High-Temperature Superconductors*, ACS Symposium volume, No. 377 (1988).
262. Nelson, D. L., Whittingham, M. S., George, T. F., editors, *Chemistry of High-Temperature Superconductors*, ACS Symposium volume, No. 351 (1987).
263. Okamato, H., Mitani, T., Toriumi, K., Yamashita, M., *Phys. Rev. Lett.* **69**, 2248 (1992).
264. O'Keeffe, M., *Acta Cryst. A* **33**, 924 (1977).
265. O'Keeffe, M., in *Structure and Bonding in Crystals*, Vol. 1, O'Keeffe, M., Navrotsky, A., editors, Academic (1981).
266. O'Keeffe, M., *J. Solid State Chem.* **85**, 109 (1990).
267. O'Keeffe, M., Hyde, B. G., in *Structure and Bonding in Crystals*, Vol. 1, O'Keeffe, M., Navrotsky, A., editors, Academic (1981).
268. O'Keeffe, M., Hyde, B. G., *J. Solid State Chem.* **44**, 24 (1982).
269. O'Keeffe, M., Valigi, M., *J. Phys. Chem. Solids* **31**, 310 (1970).
270. Orgel, L. E., *J. Chem. Soc.*, 4186 (1958).
271. Orgel, L. E., *Introduction to Transition Metal Chemistry*, 2nd edition, Methuen (London) (1966).
272. Orgel, L. E., Dunitz, J. D., *Nature* **179**, 462 (1957).
273. Parr, R. G., Pearson, R. G., *J. Amer. Chem. Soc.* **105**, 7512 (1983).
274. Parthé, E., *Crystal Chemistry of Tetrahedral Structures*, Gordon and Breach (1972).
275. Pauling, L., *J. Amer. Chem. Soc.* **51**, 1010 (1929).
276. Pauling, L., *The Nature of the Chemical Bond*, third edition, Cornell (1960).
277. Pearson, R. G., *J. Amer. Chem. Soc.* **91**, 1252, 4947 (1970).
278. Pearson, R. G., *Proc. Natl. Acad. Sci. (USA)* **83**, 8440 (1986).
279. Pearson, R. G., *J. Chem. Ed.* **64**, 561 (1987).
280. Pearson, W. B., *The Crystal Chemistry and Physics of Metals and Alloys*, Wiley (1972).
281. Pearson, W. B., in *Treatise in Solid State Chemistry*, Hannay, N. B., editor, Vol. 1, Plenum (1975).
282. Peierls, R. E., *Quantum Theory of Solids*, Oxford (1955).
283. Pettifor, D. G., *J. Phys. C* **3**, 367 (1970).
284. Pettifor, D. G., *Calphad* **1**, 305 (1977).
285. Pettifor, D. G., *Solid State Comm.* **28**, 621 (1978).
286. Pettifor, D. G., *Phys. Rev. Lett.* **42**, 846 (1979).
287. Pettifor, D. G., *J. Magn. Magn. Mat.* **15–18**, 847 (1980).
288. Pettifor, D. G., in *Physical Metallurgy*, Cahn, R. W., Haasen, P., editors, Elsevier (1983).
289. Pettifor, D. G., *Phys. Rev. Lett.* **53**, 1080 (1984).
290. Pettifor, D. G., *J. Phys. C* **19**, 315 (1986).
291. Pettifor, D. G., *Solid State Phys.* **40**, 43 (1987).

292. Pettifor, D. G., in *Electron Theory in Alloy Design*, Pettifor, D. G., Cottrell, A., editors, Institute of Materials (1992).
293. Pettifor, D. G., in *Materials Science and Technology Encyclopedia*, VCH (1992).
294. Pettifor, D. G., *Materials Science and Technology Encyclopedia*, Chapter 2, Vol. 1, VCH (1992).
295. Pettifor, D. G., Aoiki, M., *Structural and Phase Stability of Alloys*, Morán-López, J. L., editor, Plenum (1992).
296. Pettifor, D. G., Podloucky, R., *J. Phys. C* **19**, 285 (1986).
297. Pettifor, D. G., Weaire, D. L., editors, *The Recursion Method and its Applications*, Springer-Verlag (1985).
298. Phillips, J. C., *Rev. Mod. Phys.* **42**, 317 (1970).
299. Phillips, J. C., *Bonds and Bands in Semiconductors*, Wiley (1973).
300. Phillips, J. C., in *Treatise in Solid State Chemistry*, Hannay, N. B., editor, Vol. 1, Plenum (1975).
301. Phillips, J. C., in *Structure and Bonding in Crystals*, Vol. 1, O'Keeffe, M., Navrotsky, A., editors, Academic (1981).
302. Phillips, J. C., Kleinman, L., *Phys. Rev.* **116**, 287 (1959).
303. Phillips, J. C., Van Vechten, J. A., *Phys. Rev. B* **2**, 2147 (1970).
304. Pireau, J. J., Basilier, E., Gelius, U., Svensson, S., Malmqvist, P.-Å., Caudano, R., Siegbahn, K., *Phys Rev. A* **14**, 2133 (1976).
305. Poulsen, H. F., Andersen, N. A., Andersen, J. V., Bohr, H., Mouritsen, O. G., *Nature* **349**, 594 (1991).
306. Price, G. D., Price, S. L., Burdett, J. K., *Phys. Chem. Min.* **8**, 69 (1982).
307. Py, M. A., Haering, R. R., *Can. J. Phys.* **61**, 76 (1983).
308. Rabe, K., Kortan, A. R., Phillips, J. C., Villars, P., *Phys. Rev. B* **43**, 6280 (1991).
309. Rao, C. N. R., *Ann. Rev. Phys. Chem.* **40**, 291 (1989).
310. Rao, C. N. R., Pisharody, K. P. R., *Prog. Solid State Chem.* **10**, 207 (1977).
311. Rao, C. N. R., Raveay, B., *Accts. Chem. Res.* **22**, 106 (1989).
312. Ravy, S., Canadell, E., Pouget, J. P., in *The Physics and Chemistry of Organic Superconductors*, Saito, G., Kagoshima, S., editors, Springer-Verlag (1990).
313. Reichlin, R., Ross, M., Martin, S., Goettel, K. A., *Phys Rev. Lett.* **56**, 2858 (1986).
314. Reitz, J. R., *Solid State Physics* **1**, 2 (1955).
315. Renault, A., Burdett, J. K., Pouget, J-P., *J. Solid State Chem.* **71**, 587 (1987).
316. Robin, M. B., Day, P., *Adv. Inorg. Chem. and Radiochem.* **10**, 247 (1967).
317. Rouxel, J., editor, *Crystal Chemistry and Properties of Materials with Quasi-One-Dimensional Structures*, Reidel (1986).
318. Saito, G., Kagoshima, S., editors, *The Physics and Chemistry of Organic Superconductors*, Springer-Verlag (1990).
319. Schäfer, H., Eisenmann, B., Müller, W., *Angew. Chem. Int. Ed.* **12**, 694 (1973).
320. Schegolev, I. F., Yagubskii, E. B., in *Extended Linear Chain Compounds*, Miller, J. S., editor, Vol. 2, Plenum (1982).
321. Schmidt, P. C., Stahl, D., Eisenmann, B., Kniep, R., Eyert, V., Kübler, J., *J. Solid State Chem.* **97**, 93 (1992).
322. Schrieffer, J. A., *Theory of Superconductivity*, Benjamin (1964).
323. Seidle, A. R., in *Extended Linear Chain Compounds*, Miller, J. S., editor, Vol. 2, Plenum (1982).
324. Shannon, R. D., *Acta Cryst. A* **32**, 751 (1976).
325. Shannon, R. D., in *Structure and Bonding in Crystals*, Vol. 2, O'Keeffe, M., Navrotsky, A., editors, Academic (1981).
326. Sherwood, P., Hoffmann, R., *J. Amer. Chem. Soc.* **112**, 2881 (1990).
327. Shibaeva, R. P., in *Extended Linear Chain Compounds*, Miller, J. S., editor, Vol. 2, Plenum (1982).

328. Slater, J. C., *Phys. Rev.* **82**, 538 (1951).
329. Slater, J. C., Koster, G. F., *Phys. Rev.* **94**, 1498 (1954).
330. Sleight, A. W., *Science* **242**, 1519 (1988).
331. Smith, D. K., Klein, C. F., Austerman, S. B., *Acta Cryst.* **18**, 393 (1965).
332. Smith, J. V., *Geometrical and Structural Crystallography*, Wiley (1982).
333. Snow, E. C., Waber, J. T., *Acta Metall.* **17**, 623 (1969).
334. St. John, J., Bloch, A. N., *Phys. Rev. Lett.* **33**, 1095 (1974).
335. Streitweiser, A., *Molecular Orbital Theory for Organic Chemists*, Wiley (1961).
336. Stroud, D., in *Theory of Alloy Phase Formation*, Bennett, L. H., editor, AIME (1980).
337. Szabo, A., Ostlund, N. S., *Modern Quantum Chemistry*, Macmillan (1982).
338. Tian, Y., Hughbanks, T., *Inorg. Chem.* **32**, 400 (1993).
339. Tinkham, M., *Group Theory and Quantum Mechanics*, McGraw-Hill (1964).
340. Tokura, Y., Takagi, H., Uchida, S., *Nature* **337**, 345 (1989).
341. Tolbert, L. M., Ogle, M. E., *J. Amer. Chem. Soc.* **112**, 9519 (1990).
342. Tománek, D., Louie, S. G., Mamin, H. J., Abraham, D. W., Thomson, R. E., Ganz, E., Clarke, J., *Phys. Rev. B* **35**, 7790 (1987).
343. Toriumi, T., Wada, Y., Mitani, T., Bandow, S., *J. Amer. Chem. Soc.* **111**, 2341 (1989).
344. Torrance, J. B., Lacorro, P., Asavaroengchai, C., Metzger, R. M., *J. Solid State Chem.* **90**, 168 (1991).
345. Tosi, M. P., *Solid State Physics* **16**, 1 (1964).
346. Tossell, J. A., *Phys. Chem. Min.* **14**, 320 (1987).
347. Tossell, J. A., Gibbs, G. V., *Acta Cryst. A* **34**, 463 (1978).
348. Tremel, W., Hoffmann, R., *Inorg. Chem.* **26**, 118 (1987).
349. Tremel, W., Hoffmann, R., *J. Amer. Chem. Soc.* **109**, 124 (1987).
350. Tremel, W., Hoffmann, R., Silvestre, J., *J. Amer. Chem. Soc.* **108**, 5174 (1986).
351. Vaknin, D., Sinha, S. K., Moncton, D. E., Johnston, D. C., Newsam, J. M., Safinya, C. R., King, H. E., *Phys. Rev. Lett.* **58**, 2802 (1987).
352. van Arkel, A. E., *Molecules and Crystals in Inorganic Chemistry*, Interscience (1956).
353. Vanderah, T. A., editor, *Chemistry of High-Temperature Superconductors*, Noyes (1991).
354. van Houten, S., *J. Phys. Chem. Solids* **17**, 7 (1960).
355. Varma, C. M., *Solid State Comm.* **31**, 295 (1979).
356. Villars, P., *J. Less-Common Met.* **119**, 175 (1986).
357. Villars, P., Hulliger, F., *J. Less-Common Met.* **132**, 289 (1987).
358. Villars, P., Mathis, K., Hulliger, F., in *Structures of Binary Compounds*, de Boer, F., Pettifor, D., editors, Vol. 2, North-Holland (1989).
359. Wade, K., *Adv. Inorg. Chem. Radiochem.* **18**, 1 (1976).
360. Wang, E., Tarrascon, J-M., Greene, L. H., Hull, G. W., McKinnon, W. R., *Phys. Rev. B* **41**, 6582 (1990).
361. Watson, R. E., Bennett, L. H., *Phys. Rev. Lett.* **43**, 1130 (1979).
362. Watson, R. E., Ehrenreich, H., *Comments Solid State Phys.* **3**, 109 (1970).
363. Watson, R. E., Ehrenreich, H., Hodges, L., *Phys. Rev. Lett.* **24**, 829 (1970).
364. Weaire, D., in *Treatise in Solid State Chemistry*, Vol. 1, Hannay, N. B., editor, Plenum (1975).
365. Wells, A. F., *Models in Structural Inorganic Chemistry*, Oxford (1970).
366. Wells, A. F., *Structural Inorganic Chemistry*, 5th edition, Oxford (1984).
367. West, A. R., *Solid State Chemistry and its Applications*, Wiley (1984).
368. Whangbo, M.-H., *J. Chem. Phys.* **70**, 4963 (1979).
369. Whangbo, M.-H., *ibid.* **73**, 3854 (1980).
370. Whangbo, M.-H., in *Extended Linear Chain Compounds*, Miller, J. S., editor, Vol. 2, Plenum (1982).
371. Whangbo, M.-H., *J. Chem. Phys.* **75**, 4983 (1985).

372. Whangbo, M.-H., in *Crystal Chemistry and Properties of Materials with Quasi-One-Dimensional Structures*, Rouxel, J., editor, Reidel (1986).
373. Whangbo, M.-H., Canadell, E., *Inorg. Chem.* **26**, 842 (1987).
374. Whangbo, M.-H., Canadell, E., *J. Amer. Chem. Soc.* **114**, 9587 (1992).
375. Whangbo, M.-H., Foshee, M. J., *Inorg. Chem.* **20**, 113 (1981).
376. Whangbo, M.-H., Foshee, M. J., Hoffmann, R., *Inorg. Chem.* **19**, 1723 (1980).
377. Whangbo, M.-H., Hoffmann, R., *J. Amer. Chem. Soc.* **100**, 6093 (1978).
378. Whangbo, M.-H., Hoffmann, R., Woodward, R. B., *Proc. Roy. Soc. A* **366**, 23 (1979).
379. Whangbo, M.-H., Torardi, C. C., *Science* **249**, 1143 (1990).
380. Whangbo, M.-H., Torardi, C. C., *Accts. Chem. Res.* **24**, 127 (1991).
381. Wheeler, R. A., Whangbo, M.-H., Hughbanks, T., Hoffmann, R., Burdett, J. K., Albright, T. A., *J. Amer. Chem. Soc.* **108**, 2222 (1986).
382. Wherret, B. S., *Group Theory for Atoms and Molecules*, Prentice-Hall (1986).
383. Wigner, E., Huntingdon, H. B., *J. Chem. Phys.* **3**, 764 (1935).
384. Wijeyesekera, S. D., Hoffmann, R., *Organometallics* **3**, 949 (1984).
385. Williams, A. R., Gelatt, G. D., Janak, J. F., in *Theory of Alloy Phase Formation*, Bennett, L. H., editor, AIME (1980).
386. Williams, J. M., *Adv. Inorg. Chem.* **26**, 235 (1983).
387. Williams, J. M., *Prog. Inorg. Chem.* **33**, 183 (1985).
388. Williams, J. M., Beno, M. A., Carlson, K. D., Geiser, U., Kao, H. C. I., Kini, A. M., Porter, L. C., Schultz, A. J., Thorn, R. J., Wang, H. H., *Accts. Chem. Res.* **21**, 1 (1988).
389. Williams, J. M., Beno, M. A., Wang, H. H., Leung, P. C. W., Emge, T. J., Geiser, U., Carlson, K. D., *Accts. Chem. Res.* **18**, 261 (1985).
390. Williams, J. M., Schultz, A. J., Underhill, A. E., Carneiro, K., in *Extended Linear Chain Compounds*, Miller, J. S., editor, Vol. 1, Plenum (1982).
391. Wilson, J. A., *Structure and Bonding* **32**, 57 (1977).
392. Wilson, J. A., DiSalvo, F. J., Mahajan, S., *Adv. Phys.* **24**, 117 (1975).
393. Yates, K., *Hückel Molecular Orbital Theory*, Academic (1978).
394. Yee, K. A. Y., Hughbanks, T., *Inorg. Chem.* **30**, 2321 (1991).
395. Yee, K. A. Y., Hughbanks, T., *Inorg. Chem.* **31**, 1620 (1992).
396. Yin, M. T., Cohen, M. L., *Phys. Rev. B* **26**, 5668 (1982).
397. Zaanen, J., Sawatzky, G. A., Allen, J. W., *Phys. Rev. Lett.* **55**, 418 (1985).
398. Zunger, A., in *Structure and Bonding in Crystals*, Vol. 1, O'Keeffe, M., Navrotsky, A., editors, Academic (1981).

Index

Numbers in italics refer to pages with figures.